P9-BYA-101

GEOMORPHOLOGY FROM THE EARTH

Harper & Row Series in Geography,
D. W. Meinig, Advisor

GEOMORPHOLOGY FROM THE EARTH

Karl W. Butzer
THE UNIVERSITY OF CHICAGO

Harper & Row, Publishers
New York   Hagerstown   San Francisco   London

Cover Photograph of an area near the Cape of Good Hope by Karl W. Butzer

Sponsoring Editor: Ronald K. Taylor
Project Editor: Robert Ginsberg
Designer: Rita Naughton
Production Supervisor: Stefania J. Taflinska
Compositor: Progressive Typographers, Inc.
Printer and Binder: The Maple Press Company
Art Studio: J & R Technical Services Inc.

GEOMORPHOLOGY FROM THE EARTH

Copyright © 1976 by Karl W. Butzer

All rights reserved. Printed in the United States of America. No part of this book may be used or reproduced in any manner whatsoever without written permission except in the case of brief quotations embodied in critical articles and reviews. For information address Harper & Row, Publishers, Inc., 10 East 53rd Street, New York, N.Y. 10022.

Library of Congress Cataloging in Publication Data

Butzer, Karl W.
  Geomorphology from the Earth.

  (Harper & Row series in geography)
  Includes index.
  1. Geomorphology. I. Title.
GB401.5.B87   551.4   76-10863
ISBN 0-6-041097-3

To the memory of
Ferdinand von Richthofen
(1833–1905)

Contents

Preface

The face of this planet is unique.

A decade of space exploration has confirmed that Earth has a diversity of relief, of forms, and of processes that are unrivaled in our solar system. We now know that Mars, Venus, Mercury, and the moon, like Earth, have solid crusts that have been modified by tectonics and vulcanism. Mars also has thin polar ice caps and some apparent erosional valleys; future satellite missions will probably show a topography modeled by wind action. But the dominant features of our closest neighbors in space are meteorite impact craters, which are very rare on Earth.

When seen from the moon, Earth appears as a blue planet in a black sky. It is a good earth, not forbidding and inimical like the other planets. Earth's thick atmosphere and abundant water have not only permitted and sustained life, they have sculptured the surface in distinctive ways. Water sets off the continents from the ocean deeps. In its many forms— in the air, as ice, and on and under the land surface—water has helped to mold distinctive scenery in mountains, on plains, and at the shore. With the assistance of a very mobile crust, it has eliminated the impact scars of most meteorites. The resulting landscapes owe their origins to the internal dynamics of the planet and to its unique atmospheric agents.

Earth landforms and the unique processes that create them are not only fascinating to scientists—they also excite the natural curiosity of people everywhere. Impressions of scenery have influenced the esthetic and symbolic values of most cultures in the course of human history. Floods and earthquakes, or constellations of mountains and plains, have had a persistent and practical impact on economic and social life. But the opposite is equally true. Prehistoric farmers began the destruction of soils, and industrial societies have deprived millions of clean air and clear water by covering the land with concrete surfaces and erecting metallic skylines between polluted waterways and hillsides scarred by machinery.

The study of landforms should not be the privilege of a few. People of all ages and backgrounds should gain a new appreciation for the magnificent planet we live on. This conviction influenced the approach I chose and the materials I selected for writing *Geomorphology from the Earth*. My own classes include a high proportion of undergraduates that have no prior training in physical geography or geology, as well as graduate students in fields as diverse as archeology and biology. The lack of a suitable text has been acute, since most books in geomorphology are written for future geomorphologists. I believe instead that an understanding of the forms and processes that shape rock and soil on our planet is a necessary ingredient in any liberal education. To this end, a first course in geomorphology should be concerned with the essentials, the critical interactions, and the overall perspective.

Geomorphology has become a vibrantly dynamic area of research, and fundamental revisions of many prevailing ideas are repeatedly necessary in the light of new field observations, laboratory simulation of processes, or mathematical approaches. I have attempted to incorporate these into comprehensive but nonmathematical sections in the text dealing with hillslope evolution and stream valleys. Specialized works and examples of primary research are cited in the bibliographies to suggest avenues for more intensive study. The exciting research into plate tectonics has given new significance to oceanic forms and suggested the need for a fresh overview of structural landforms, accompanied by analytic maps of the continents. Geomorphic processes and forms are parts of the major global ecosystems, and I have outlined the essentials of what we now know about the key environments. Somewhat surprisingly, time and man continue to be neglected in many textbooks. Temporal perspectives are of vital importance in appreciating that some forms change slowly, almost imperceptibly, whereas others change rapidly, sometimes overnight. Last but not least, man is a central agent in shaping the face of the earth. I have stressed this point to show the significance of geomorphic processes in human ecosystems. It is important for students in ecology or the liberal arts to grasp the basic dynamics before they embark on matters of environmental policy. Hopefully, many will feel encouraged to take additional coursework in "applied" aspects such as watershed management or urban geomorphology.

Science today is international. Geomorphology is no exception, and many aspects of the subject cannot be adequately documented in the publications of any one language. While I do not expect that undergraduates will often find the time to cope with French or German articles, there is no reason to exclude these from the bibliographic selections. Such references give a fairer impression of the background research, and the harried writer of term papers can find the basic content of many foreign-language contributions made accessible through their

English summaries and by translating diagrams or photo captions with the aid of a dictionary. Units of measurement have been given in metric and nonmetric equivalents; since many of the values quoted are approximations, such conversions are also rounded off.

It remains for me to thank the many individuals who helped make this book a reality: those who contributed to the often tiring but always exciting field experiences, those who listened to me in class, and those who tolerated my being demanding or inaccessible in the process of writing. Michael Sabbagh provided stimulating discussions during the early stages of writing, and an interim draft profited greatly from a critical reading by M. Gordon Wolman. The next-to-final copy was improved with the assistance of a detailed commentary by Stanley Trimble, whose encouragement was very helpful. Last but not least, Frederick Simoons supervised the first introductory course I taught: I still detect his influence on some parts of this book. My appreciation to all.

Karl W. Butzer

Chicago and Flossmoor, Illinois
March 1976

GEOMORPHOLOGY FROM THE EARTH

Chapter 1
OF THE LAND SURFACE

Geomorphology is the study of the Earth surface. But this definition is so vague as to be almost meaningless. Instead it is more practical to see the land surface, and the soils from which landforms can rarely be separated, as part of the total environment. Within a dynamic system of interacting factors, geomorphology is best viewed from a balanced, total perspective and approached with an ecological bias.

Let the waters under heaven come together into a single mass, and let dry land appear.

GENESIS I, 1:9, *The Jerusalem Bible*

1-1. PERSPECTIVES

Somewhere in the murky depths of time this planet and its satellite came into existence. All theory of the origins of the solar system aside, it appears that Earth and moon were solid spheres by about 4.5 billion years ago. A crust, with low-density rocks, and providing environments potentially suitable for algal life, was present by 4 billion years ago, judging by isotopic dating of rocks found in northeastern Greenland. By 3.3 billion years ago chemical and organic evolution has proceeded to the point that planktonic photosynthesis was taking place in sea water, under an atmosphere having at least some oxygen. Some 400 million years ago the bare land surfaces began to be colonized by plants, and the atmosphere reached approximately its modern composition. Finally, around 100 million years ago the continents were mantled by vegetation to about the same degree they are now. What we see today by way of continental configuration, rocks, and land forms clearly did not happen overnight.

The atmosphere is a here-and-now phenomenon for most of us; at least, the weather today and tomorrow is of much greater interest than that of last year. Admittedly, many scientists are now concerned with the evolution of the atmosphere through time, and anyone from five year olds fascinated with dinosaurs to archeologists digging for Vikings or Early Man has thought about climatic change. Nor can any meteorologist deny that yesterday's or last year's weather affects that of tomorrow or next year. Nonetheless, the atmosphere is prone to such rapid change that everything seems to "turn over" completely in at most a few weeks. By contrast, the differences of general configuration, altitude, slope, and form that constitute Earth's "relief" are far more durable. The appearance of the crust in Virginia or Plymouth Bay has changed little since the early 1600s, except that trees have been cut down, towns and roads built. Zebulon Pike would still recognize the Colorado Rockies, despite the contrary efforts of strip miners, urban developers, and highway planners. Together with this permanence of land forms there is a deep-seated appreciation for the timelessness of the land. It makes those of us who study Earth's relief a little more retrospective than most.

Why do we care about the morphology of the land? Very few professionals in this field of study have an avid interest in rocks. Those who began with such an interest are probably still peering down microscopes at mineral structures. Few geomorphologists are "rockhounds," although most have learned to appreciate the implications of and clues provided by rocks. Instead, many of us became involved because of fascination with the dynamism of nature reflected in the array of earthquake, flood, and landslide catastrophes that cross newspaper headlines week after week. Others were turned on by the less practical but more esthetic appeal of splendid mountains or rivers catapulting across rapids into gorges beyond. A visit to art galleries in Chicago, Madrid, or Tokyo sould convince even a cynic that peoples of all cultures—and from both rural and urban backgrounds—enjoy scenic landscapes. It takes only a small step of awareness to transform the subconscious appeal of the land into keen perception. The surface of this planet is not only unique in the solar system, but it is fascinating.

To the initiated it provides the same acute sense of pleasure that is enjoyed by kenners of wine or music or anything else that can be shared by people.

The multitude of surface forms created over thousands or millions of years is, at first sight, confusing. Yet it is precisely the mystery of it all that has intrigued thoughtful observers from time immemorial. Much of this flavor is conveyed by one of the oldest stories we have, that of Gilgamesh's journey through the mountain. Gilgamesh was king of Uruk in Mesopotamia about 2700 b.c., and most of the poems about him were written down within years of his death. One tells us:

> He followed the sun's road to his rising, through the mountain. When he had gone one league the darkness became thick around him, and there was no light, he could see nothing ahead and nothing behind him. . . . After nine leagues he felt the north wind on his face, but the darkness was thick and there was no light. . . . After eleven leagues the dawn light appeared. At the end of twelve leagues the sun streamed out. There was the garden of the gods; all round him stood bushes bearing gems. Seeing it he went down at once, for there was fruit of carnelian with the vine hanging from it, beautiful to look at; lapis lazuli leaves hung thick with fruit, sweet to see. For thorns and thistles there were hematite and rare stones, agate, and pearls from out of the sea. . . .[1]

This would seem to be the metaphorical description of a limestone cave, adorned with an array of dripstone and flowstone formations. A different passage from Gilgamesh clearly alludes to an earthquake preceding a volcanic eruption: "We stood in a deep gorge of the mountain, and suddenly . . . the mountain fell, it struck me and caught my feet from under me. Then came an intolerable light blazing out. . . ."[2] There also is the description of the great Mesopotamian deluge, which reads like a description of the typhoon that brought the Bay of Bengal surging into the Ganges-Brahmaputra delta, Bangladesh, in February of 1971:

> For six days and six nights the winds blew, torrent and tempest and flood overwhelmed the world, tempest and flood raged together like warring hosts. When the seventh day dawned the storm from the south subsided, the sea grew calm, the flood was stilled; I looked at the face of the world and there was silence, all mankind was turned to clay. The surface of the sea stretched as flat as a roof-top. . . .[3]

The epic of Gilgamesh and its younger literary counterparts in the Bible were followed by countless descriptions of or theories for surface

---

[1] *The Epic of Gilgamesh*, translated with an introduction by N. K. Sandars, Harmondsworth, Penguin, 1964, pp. 96–97.
[2] Ibid., p. 77
[3] Ibid., p. 108

features given by scientists of the ancient world from Herodotus to Pliny and Ptolemy.

But it is more than description that we seek. Gilgamesh and his contemporaries were content to attribute what they saw to the gods, but the scholars of Greece and Rome rebelled against explanation based on cultural preconceptions. Instead they insisted on the "how" and the "why" that are the prerogative of every child before he learns to repress unwelcome questions. Without explanation we ourselves can never progress from landscape admiration to understanding. Yet with understanding it is possible to read a landscape much as one reads a book, except that the former is written in another script and another language. The basic alphabet must be learnt, to be able to decipher and recognize features. Then begins the longer process of familiarization with vocabulary and grammar, the matter of understanding phenomena and ultimately reconstructing cause and effect. In this a textbook should serve as alphabet key and dictionary, so that with sufficient practice in the real thing—from maps, air photos, from the ground, or from the air—the student can begin to read the landscape by himself. Once that language barrier has been overcome, the surface of the land takes on an entirely new meaning as a vital and exciting part of the total environment.

1-2. THE SUBSTANCE OF GEOMORPHOLOGY

Earth surface form is part of the environment—whether physical, biological, or total. Landscape artists never tamper with the integrity of the whole, and so it should be for all. Yet to get at the subject matter it is necessary, briefly, to abstract topographic relief and continental configuration from their functional context.

At first sight the relief and rocks of this planet are highly variable and infinitely complex. As gross phenomena, they range from the mountain chains and great plainlands of a continent such as North America to the submerged forms beneath the Indian Ocean. As features at a more tangible scale, they include the surfaces, slopes, and crests found in any smaller area, down to the very details etched on a local hillside. The distribution of mountains, valleys, and plains seems to be random, lacking any particular order. However, comparison shows that larger mountain chains occur in broad belts, separated by valleys or basins that run parallel to them. So, for example, the ranges of the South American Andes trend from north to south, with river valleys or basins between them. Similarly the vast plains of Africa are interrupted by belts of hill or low mountain country that follow distinctive patterns. This organized aspect to the earth's surface forms can also be recognized on a much smaller scale. Any two glaciated peaks in the Rocky Mountains somehow look alike, and two different hills developed in similar rock in a Brazilian landscape will probably also show similarities.

As objects of study, Earth surface forms can be usefully categorized according to scale: (1) At the most general level and largest scale, continental relief is characterized by broad form types such as mountain systems, hill or plateau country, and plains. These would include features such as the Appalachian Mountains or the Great Rift Valley, and they can be usefully labeled as *land form* (written as two words). (2) More detailed features of a smaller and more uniform area can be described as *landforms* (written as one word). Further distinctions can be proposed here, as a matter of convenience: *Macrolandforms* can be recognized from the air at 30,000 feet; *mesolandforms* can be distinguished from a distance of several hundred yards or a few miles on the ground; and *microlandforms* may only be visible at a few feet. Consequently, a hill might be described as a macrolandform, a hillside as a mesolandform, and the individual rills or striations on a hillside as microlandforms.

Although description and categorization are often useful as a prerequisite to proper analysis and interpretation, they seldom serve an end by themselves. Instead, the patterns of Earth surface forms can and should be studied and interpreted very much in the same way as a set of phenomena in physics or biology, not as an encyclopedia of unrelated facts, but as a rational system of cause and effect. A student of the Rocky Mountains is not concerned with each individual hill or stream, its location or name. Rather, he is interested in factors such as rock types, earth history, and climatic differences because these help explain the location of the mountain range, the alpine scenery of the high country, or the distinctive rolling hills and flat-topped mesas found among the foothills. In detail this student might be concerned with the characteristic shapes of mountain crests, hillslopes, or valley floors; the specific features accompanying a mountain glacier; the peculiar patterns of drainage lines, running between mountains or emerging from them through flat valleys and spectacular gorges. Again each of these features is, in the end, susceptible to explanation in terms of processes that operated in the past or that continue to operate today.

The forces that create and model land form and landforms include two types of processes: (1) the *internal* or *tectonic* agents that emanate from the Earth's interior, producing earthquakes, deformation of rocks, and volcanic activity; and (2) the *external* or *gradational* agents that result ultimately from climatic influences, leading to the breakdown of rocks, erosion and transport of loose surface material, and, finally, deposition of such rock products. All of these agents work comparatively slowly, and most of the Earth's surface forms are the results of gradual but cumulative changes over many thousands or millions of years. For this reason it is as important to understand the evolution of a landscape through time as it is to study the tectonic and gradational processes themselves that are active at the present day.

Landforms are modeled in crustal materials that are distinctive in composition and have their own historical background. Landforms evolve through time, at different rates, commonly in the wake of changing tectonic and gradational forces that also vary in relative importance. Consequently, an isolated hill in Venezuela may owe its existence to an injection of tough volcanic rock into the crust a billion years ago; as the adjacent, weaker rocks—themselves much older—rot and wear away, this hill is gradually exposed at the surface. The shape it ultimately takes depends on changing environmental conditions during 100 million years of exposure to gradational processes. During periods of humid climate, the adjacent rocks rot to great depths, and the volcanic plug also does not escape rounding and striation by chemical attack; during times of drier climate, the loose or softened decay products are periodically swept away by water. As a result, neither the origin nor the form of the hill can be adequately explained in a simple, modern context. At the other exterme is the example of an ephemeral sand ridge formed on a Mexican coast by the waves generated from a severe storm; built up almost overnight it may be destroyed by wave attrition a few months later. Even here, however, the availability of abundant sand was a prerequisite, and this may link the beach ridge to stream action bringing sand from distant mountains thousands of years ago. In other words, contemporary surface forms are just as unintelligible when divorced from a historical perspective as are biological phenomena without the background of organic evolution. This should not imply that landforms are organisms but that the responsible processes are multivariate along the temporal axis just as they are in a contemporary perspective.

The historical perspective is essential to interpretation of landforms. They are sculptured in and formed of earth materials. To this extent landforms are an integral part of the Earth's crust. But they are more than that. In most environments they merge with the soil and are mantled by plant life, whether native vegetation or cultivated crops. At the same time, the gradational agents, apart from gravity, are dictated by water, ice, and wind derived from or part of the atmosphere. These external forces are ubiquitous and effective, often outweighing the influences of rocks or tectonics. Last but not least, man's activity—as farmer, herder, builder, or miner—is apparent in most parts of the world. Consequently, landforms are an integral part of the ecosystem. The underlying processes are interrelated in a complex manner, creating a landscape that is both unquiet and dynamic.

The evidence of change is to be seen everywhere, as it was in Gilgamesh's day. Sand dunes move, beaches recede, rivers flood, and earthquakes or volcanoes wreak havoc. These processes are very real and contemporary, and their practical importance is significant for a broad spectrum of human society. At long last a major focus of research has now come to bear on contemporary processes, in part because understanding them is indispensable for historical interpretation, but perhaps even more importantly because of their practical implications. These implications are two-sided. In part man has viewed earth processes as a force of potential destruction. Yet in this time of renewed ecological awareness it is equally clear to many that man himself is the destroyer, as his methods of exploitation set in train physical processes that endanger his very basis for survival.

The study of the Earth's surface forms, and of the processes that shape them, constitutes the field of *geomorphology*. It is an *earth science*, and for this reason geologists and geographers have long collaborated to develop a field that is seldom confined to only one department at a university. By deliberate choice or *de facto* use geomorphology has at different times and in different countries had a different scope, and the contributions of geographers and geologists have complemented each other in varying patterns. Geologists have, on the whole, tended to emphasize the significance of rock types and earth history for landscape evolution, and many modern geologists are particularly interested in the mechanisms of tectonic or gradational processes. Geographers, on the other hand, have commonly emphasized the nature of surface forms and the relationships between forms and processes. And, in recent years, they have shown particular interest in assessing the role of man as a geomorphic agent.

Despite a certain amount of philosophical or methodological discussion, by both geomorphologists and partisan methodologists, the many streams of geomorphological study have never been properly rationalized, either within a historical or a contemporary perspective. Geomorphological studies of all kinds today range through a wide multidisciplinary plane, with geology, soil science, hydrology, geophysics, and geography as coordinates, and include empirical and mathematical-quantitative dimensions. As a result, it is not really possible to draw valid lines between the contributions that different groups of researchers have made to geomorphology. In fact, it is undesirable to set up or retain artificial boundaries at a time when science has profited most from interdisciplinary collaboration and when university compartmentalization is increasingly recognized as an obstacle to both effective teaching and research.

1-3 THE SEARCH FOR SYNTHESIS

More often than not, geomorphology has been concerned with empirical knowledge, and the most lasting contributions are those based on intensive or extensive field observation. In view of the complexity of forms everywhere, and the complicated interactions of many unlike processes, it is small wonder that synthesis has been difficult. Natural processes responsible for a single feature can run the full range of tectonic and gradational forces, plus biogenic agents such as vegetation and man. Unlike in chemistry or physics, in which an experiment can be simulated successfully by reducing the number of variables under closed laboratory conditions, there almost always are more variables than equations in earth processes. Even in the laboratory it is difficult or impossible to scale down materials and processes in size or to replicate the effect of time. As a result, the earth sciences are not "exact," and there seldom are neat or foolproof answers, despite the increased use of geophysical techniques in observation and of mathematical or statistical approaches in analysis. Just as in the case of environmental ecology, of which geomorphic processes are an integral part, cause and effect are often unpredictable.

There is then no acceptable superstructure of theory, approach, or synthesis in geomorphology. The fascination of the immediate object under study has precedence. In addition, most geomorphologists are staunch individualists who, like historians, tend to reject universal theories with vehemence. So much by way of apology. A glance at the development of geomorphologic ideas shows that despite an auspicious launching in the nineteenth century the discipline then went into intellectual doldrums from which it has only recently recovered.

The "heroic" age of the earth sciences extended from the mid-eighteenth to the late nineteenth century. The nature and the impact of the primary geomorphic agents were recognized during this period. Perhaps the most surprising discovery of all concerned the action of glaciers and the realization that much of the Northern Hemisphere had been sculptured by moving ice in comparatively recent times. Prior to about 1875 the origins of geomorphology are inextricably interwoven with the evolving field of geology. The external or gradational processes were studied with the primary purpose of understanding sedimentation phenomena apparent in the geological record. Geomorphology, still undefined, became a tool rather than a goal of geology. Then, in 1886, Ferdinand von Richthofen, an uncle of the Red Baron of World War I flying fame, published what might be called the first text on the principles of geomorphology under the title *Guidebook for Scientific Travellers.*[4] In 1894 Albrecht Penck followed this with a two-volume work, *Morphology of the Earth Surface,*[5] which, if it were today translated into English with injection of the appropriate jargon, would probably cause little surprise among most book reviewers or

---

[4] *Führer für Forschungsreisende*, Berlin, Oppenheim, 1886, 745 pp.
[5] *Morphologie der Erodoberflache*, Stuttgart, Engelhorn, 1894, 2 vol., 471 and 696 pp.

teachers. Somehow, once the basic encyclopedia of facts had been assembled, progress in conceptualization seems to have slowed down.

However, a study of the substantive writings in the journals shows
remarkable gains in the decades after 1900—in factual knowledge, in
revision of old premises, and in new techniques or approaches. This is
particularly true since about 1940. Consequently the fault would seem
to lie with those writers of geomorphologic treatises that continue to
ignore significant advances in the field often made several decades earlier. Perhaps even more inhibiting have been the effects of a chain of
general texts that since 1911 have subjected the discipline to a theoretical model bearing no resemblance to empirical reality. A dispas-
sionate evaluation must place the primary responsibility with William
Morris Davis, who taught at Harvard from 1876 to 1912 and influenced a number of zealous disciples in North America, Britain, and
France. Beginning in 1899 Davis gave geomorphology a new method of
synthesis that could be learned in three simple lessons: youth, maturity, and old age. Given a set of initial conditions, every landscape
should evolve through a sequence of "stages" characterized by certain
geometrical forms (Figure 1–1) to a predictable end product. This

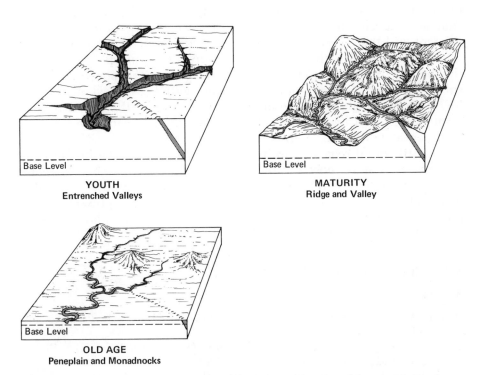

YOUTH
Entrenched Valleys

MATURITY
Ridge and Valley

OLD AGE
Peneplain and Monadnocks

Figure 1–1. Stages of the "Cycle of Erosion" as Postulated by W. M. Davis.

sequence was "normally" systematic, inevitable, and irreversible, although several complications were built in. Known as the "cycle of erosion," it was forced upon all landscape evolution in humid lands, with special adaptations for deserts, glaciated highlands, and coasts. Decades of semantic bickering followed, during which true age was confused with geometry, and historical evolution replaced by a priori, mechanistic sequences. None of the underlying theoretical points was tested empirically. In the end, an observational science was reduced to a parlor game of inductive reasoning that could be mastered by a freshman student.

Davisian geomorphology, as it was called, died a death of sterility several decades ago, despite the lingering stage-and-cycle approach of most textbooks. Productive researchers in geomorphology have mainly shied away from synthesis, while the majority of fieldworkers are pragmatists. This may be all to the good and has at least served to clear the air. In the meantime a surge of intensified field studies and experimentation in new techniques has been unimpeded by theoretical ballast. More refined mathematical and statistical concepts have been applied and found to be of increasing value. Synthesis itself has been analyzed, and new theoretical models have been devised and tested. All this has begun to change geomorphology, and modernization has indeed been long overdue. But as in any other field in rapid expansion and development, theoretical frameworks are practically nonexistent and there is little consensus as to direction or approach.

Consonant with the pluralistic nature of geomorphology today, this book will attempt to present several, balanced perspectives on the subject matter. Contemporary processes will be viewed in an ecological context, giving due emphasis to the role of the soil. Land-use implications will be stressed, no less than the historical evolution of continental landforms. Surely this is the only realistic approach, at least until sound reasons can be provided for a foolproof hierarchical system of what is most significant or most effective. The viewpoint subscribed to here is that neither description nor explanation, neither form nor process, neither rock nor climate, neither the present nor the past is *by itself* adequate. Each deserves its proper place; each is thought provoking in its own right.

1-4. SELECTED REFERENCES

A. General

Bloom, A. L. *The Surface of the Earth.* Englewood Cliffs, N.J., Prentice-Hall, 1969, 152 pp.

Brundsen, Denys, and J. C. Doornkamp, eds. *The Unquiet Landscape.* Bloomington, Ind., Indiana University Press, 1975, 176 pp.

Clark, Kenneth. *Landscape into Art.* Boston, Beacon, 1969, 148 pp.

Curran, H. A., P. S. Justus, E. L. Perdew, and M. P. Prothero. *Atlas of Landforms*, 2nd ed. New York, Wiley, 1974, 140 pp.
Dury, G. H. *The Face of the Earth*. Harmondsworth, Penguin, 1959, 223 pp.
Garner, H. F. *The Origin of Landscapes*. New York, Oxford University Press, 1974, 734 pp.
Hamblin, W. K., and J. D. Howard. *Exercises in Physical Geology*, 4th ed. Minneapolis, Minn., Burgess, 1975, 233 pp.
Hidore, J. J., and M. C. Roberts. *Physical Geography: A Laboratory Manual*. Minneapolis, Minn., Burgess, 1974, 203 pp.
Holz, R. K., ed. *The Surveillant Science: Remote Sensing of the Environment*. Boston, Houghton-Mifflin, 1973, 390 pp.
Monkhouse, F. J. *A Dictionary of Geography*. London, Arnold, 1965, 344 pp.
Pitty, A. F. *Introduction to Geomorphology*. New York, Barnes & Noble, 1971, 526 pp.
Snead, R. E. *Atlas of World Physical Features*. New York, Wiley, 1972, 158 pp.
Sparks, B. W. *Geomorphology*, 2nd ed. London, Longman, 1974, 530 pp.
Vitaliano, Dorothy, *Legends of the Earth: Their Geologic Origins*. Bloomington, Ind., Indiana University Press, 1974, 305 pp.
Wyckoff, Jerome. *Rock, Time, and Landforms*. New York, Harper & Row, 1966, 372 pp.
York, Derek, and R. M. Farquhar. *The Earth's Age and Geochronology*. Elmsford, N.Y., Pergamon, 1974, 178 pp.

B. Critical Surveys

Butzer, K. W. Pluralism in geomorphology. *Proceedings, Association of American Geographers*, Vol. 5, 1973, pp. 39–43.
Clayton, K. M. Geomorphology, a study which spans the geology-geography interface. *Journal of the Geological Society*, Vol. 127, 1971, pp. 471–476.
Dury, G. H. *Perspectives on Geomorphic Processes*. Association of American Geographers, Commission on College Geography, *Resource Paper*, No. 9, 1969, 56 pp.
Dury, G. H. Some views on the nature, location, needs and potential of geomorphology. *Professional Geographer*, Vol. 24, 1972, pp. 199–202.
Dury, G. H. Some current trends in geomorphology. *Earth-Science Reviews*, Vol. 2, 1972, pp. 45–72.
Zakrzewska, Barbara. Trends and methods in landform geography. *Annals, Association of American Geographers*, Vol. 57, 1967, pp. 128–165.

C. History of Geomorphology

Adams, F. D. *The Birth and Development of the Geological Sciences*. New York, Dover, 1954, 506 pp.
Albritton, C. C., ed. *Philosophy of Geohistory*. New York, Wiley, 1974, 400 pp.
Bunbury, E. H. *A History of Ancient Geography*. New York, Dover, 1959, Vol. 1, 666 pp.; Vol. 2, 743 pp.
Chorley, R. J. A reevaluation of the geomorphic system of W. M. Davis. In R. J. Chorley and P. Haggett, eds., *Frontiers in Geographical Teaching*. London, Methuen, 1965, pp. 21–38.

Chorley, R. J., A. J. Dunn, and R. P. Beckinsale. *The History of the Study of Landforms*. Vol. 1. *Geomorphology Before Davis*. London, Methuen, 1964, 678 pp.

Davis, W. M. *Geographical Essays*. New York, Dover, 1954, 777 pp. Part 2 includes the key papers published 1899–1906.

Hettner, Alfred. *The Surface Features of the Land: Problems and Methods of Geomorphology*. Translated from the German original of 1921 by P. Tilley. London, Methuen, 1972, 193 pp.

King, L. C. *Morphology of the Earth*, 2nd ed. Darien, Conn., Hafner, 1967, 726 pp. Chaps. 5 and 6 present alternative views to Davis' cycle of erosion, based primarily on the views of W. Penck.

Penck, Walther. *Morphological Analysis of Land Form*. Translated from the German original of 1924 by Hella Czech and K. C. Boswell. London, Macmillan, 1953, 429 pp.

D. Applied Geomorphology

Coates, D. R., ed. *Environmental Geomorphology*. Binghampton, N.Y., State University of New York, Publications in Geomorphology, 1971, 262 pp.

———. *Environmental Geomorphology and Landscape Conservation, III: Non-Urban Regions*. New York, Wiley, 1973, 496 pp.

Cooke, R. U., and J. C. Doornkamp, *Geomorphology in Environmental Management*. New York, Oxford University Press, 1974, 413 pp.

Detwyler, T. R., and M. G. Marcus, eds. *Urbanization and Environment: The Physical Geography of the City*. Belmont, Ca., Duxbury, 1972, 287 pp.

Hails, John, ed. *Applied Geomorphology*. New York, Wiley, 1976, in press.

Legget, R. F. *Cities and Geology*. New York, McGraw-Hill, 1973, 624 pp.

McKenzie, G. D., and R. O. Utgard. *Man and His Physical Environment: Readings in Environmental Geology*, 2nd ed. Minneapolis, Minn., Burgess, 1975, 388 pp.

E. Mathematical Approaches

Chorley, R. J., ed. *Spatial Analysis in Geomorphology*. New York, Harper & Row, 1972, 392 pp.

Chorley, R. J., and B. A. Kennedy. *Physical Geography: A Systems Approach*. Englewood Cliffs, N.J., Prentice-Hall, 1971, 370 pp.

Doornkamp, J. C., and C. A. M. King. *Numerical Analysis in Geomorphology*. London, Edward Arnold, 1971, 372 pp.

Scheidegger, A. E. *Theoretical Geomorphology*, 2nd ed., Berlin, Springer, 1970, 333 pp.

F. Basic Sources

Geographical Abstracts "A" (Geomorphology) and *"B"* (Soils, Biogeography and Climatology). Norwich, University of East Angelia. Series A (as "Geomorphological Abstracts") since 1960, B since 1966.

Zeitschrift für Geomorphologie. Berlin, Borntraeger. Since 1956. With a variety of thematic supplementary volumes. An international journal of geomorphology; the majority of the articles in many issues are in English.

Revue de Géomorphologie dynamique. Paris, Masson. Since 1950. The only other specialized journal, with some papers in English.

G. Some Journals That Periodically Include Significant Articles on Geomorphology

American Journal of Science
Annals, Association of American Geographers
Bulletin, Geological Society of America
Catena: An Interdisciplinary Journal of Geomorphology, Hydrology, Pedology
Eiszeitalter und Gegenwart
Geographical Journal
Geographical Review
Professional Papers, U.S. Geological Survey
Quaternary Research
Revue de Géographie physique et de Géologie dynamique
Transactions, Institute of British Geographers

Most periodicals dealing with soils are specialized, regionally oriented, or otherwise dominated by national organizations in terms of interest and methodology. Of most general interest to geomorphologists are the British *Journal of Soil Science* and the American *Soil Science.*

Chapter 2

THE LITHOSPHERE

The topmost shell of the planet consists of several layers of differing rocks that are all relatively rigid. These are grouped into the lithosphere, *which has a thickness of 40 to 60 miles. The rocks of the lithosphere are the medium of which landforms are made. These rocks can be classified according to an infinity of minerals, only a few of which are relatively common. They can also be classified by origin into volcanic or igneous, reworked or sedimentary, and altered or metamorphic rocks. Here again there are innumerable kinds and varieties, but the more important rocks are limited in number. It is useful to be familiar with the common minerals and rocks because they break down differently and provide distinct residual products, features that are important in the course of geomorphic change and soil development. The lithosphere is stretched across the* asthenosphere, *a plastic zone of the Earth's mantle that is 80 to 100 miles*

thick. As a result of this unstable foundation, the litho-
sphere is repeatedly deformed by earthquakes and vol-
canic activity. These processes are introduced in this
chapter, and their significance for creating landforms
is described in Chapters 14 and 15.

> And ever, as we go, there is some new
> pinnacle or tower, some crag or peak,
> some distant view of the upper plateau,
> some strange-shaped rock, or some deep,
> narrow side canyon.
>
> JOHN WESLEY POWELL, *The Exploration of*
> *the Colorado River.*

2-1. INTRODUCTION

The study of rocks and minerals is not a central theme of geomorphology. Neither is the composition of Earth's crust. However, the rocks
exposed at the surface provide the stage setting for geomorphic processes to act on, and landforms will inevitably be affected by the mineralogy and attitude of surface rocks. In fact, in order to understand
the effectiveness of geomorphic agents or to explain many landforms,
it is useful if not essential to have a basic working knowledge of
Earth's crustal materials.

The two fundamental properties of the rocks that constitute the
Earth's crust are (1) *lithology* and (2) *structure*. Lithology refers to features such as mineral composition; size and hardness of constituent
mineral grains; degree and type of cementation or attitude of rock
beds; degree and type of vertical jointing and horizontal bedding, if
any; and internal deformation of the rocks.

This chapter will describe the interior of the planet, the composition of the crust, and the nature of minerals and rocks. After this background data, attention will then be focused on the origin of rocks, on
volcanic activity, and on deformation of rock beds.

2-2. INTERNAL STRUCTURE OF EARTH

In 1864 the French science-fiction writer Jules Verne wrote his best-
seller, *Twenty Thousand Leagues Under the Sea.* It gives a fictional account of an expedition to the center of Earth, of the technical problems
faced and the observations made. Today, over a century later, the internal structure of the Earth is still poorly understood. Since even the

deepest oil rigs have only probed to a depth of a few miles, most knowledge about the interior of our planet has had to be inferred from indirect observations.

Some data concerning the crust have been obtained by measuring the minute differences of Earth gravity and magnetism that can be observed from place to place on the surface. The behavior of earthquake waves has also been closely studied, since earthquake waves are transmitted through different materials at different rates. This provides information about rock density and rigidity in the subsurface. Such data can be supplemented by fragments of rocks that are not normally found in the crust but that are brought up in the wake of volcanic activity. Finally, the mineral composition of meteorites, which constitute debris from the solar system, has also been used to infer the nature of the deep interior of the planet. All of these classes of information have been analyzed and interpreted by geophysicists to obtain the traditional model illustrated by Figure 2–1.

The deep interior, or *core*, is believed to consist largely of iron, mixed with sulfur or silicon, at temperatures of up to 10,000° F (5,500° C). The specific gravity is about 13, compared with an average of 2.8 for the rocks of Earth's crust. The *inner* core, with a radius of about 800 miles (1,250 km), is thought to be near the melting point or partly molten, while the *outer* core, forming a second concentric shell about 1,400 miles (2,250 km) in thickness, is liquid.

The third concentric zone, constituting the greater part of the Earth's bulk, is the *mantle*, with a thickness of 1,750 miles (2,800 km).

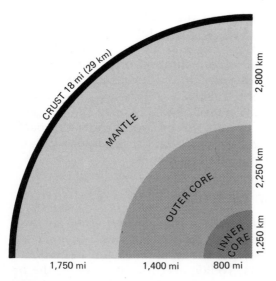

Figure 2–1. The Interior of the Earth.

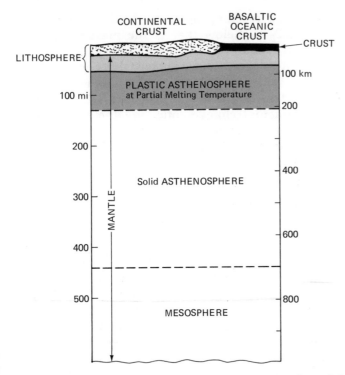

Figure 2–2. Schematic Section of the Upper Mantle and Crust.

Specific gravity of the outer mantle rock averages about 3.0 to 3.5, but this value increases to 4.5 and more with depth. The mantle is mainly solid and probably consists of very heavy minerals rich in magnesium and iron. Temperatures range from 1,300 to 5,000° F (675–2,750° C). Significantly, however, part of the upper mantle is capable of plastic deformation (Figure 2–2), yielding and adjusting to external pressures by very slow flowage. In some ways, this is analogous to ice, which, although hard, will flow slowly if unequal pressures are applied.

The surface layer or *crust* forms a veneer averaging only 15 to 20 miles (24–32 km) thick, and varying from 3 to 6 miles (5–10 km) in thickness under the oceans, 18 to 40 miles (30–65 km) under the continents. The contact between the crust and the mantle is marked by a discontinuity in chemical composition or crystal structure. The crust itself is often subdivided into two layers: (1) The discontinuous upper zone, corresponding to the major continental outlines, has an average specific gravity of 2.65. (2) The lower zone is continuous, underlying the continents and exposed on the ocean floors; specific gravity averages 3.0.

Together the crust and the uppermost mantle form the *lithosphere,* composed of very brittle rock reaching down to depths of 40 to 60 miles (60–100 km) (Figure 2–2). Below that is the plastic *asthenosphere*, which has a thickness of 80 to 100 miles (130–160 km). Since the lithosphere is stretched thinly over an asthenosphere that can flow and so relieve internal and external stress, it is not surprising that the lithosphere is easily deformed. This characteristic probably holds the clue to the repeated and fundamental changes affecting the face of the Earth during the course of earth history: rock deformation, continental uplift or submergence, mountain building, and volcanic eruptions (see Chapters 14 and 15 for further discussion).

2-3. COMPOSITION OF EARTH'S CRUST

It may seem unexpected that oxygen is the most common chemical *element* of the rocks at the surface of the earth. As a basic constituent to most minerals, it accounts for 47 percent of the exposed rocks. Next in importance is silicon (28 percent), while aluminum, iron, calcium, sodium, potassium, and magnesium together account for another 24 percent. All other elements add up to 1.3 percent only.

Chemical elements are seldom found in isolation but, instead, combine to form more stable compounds. Certain compounds are crystallized, varying in color, crystal structure, hardness, and solubility. These are the *minerals* from which rocks are made.

The principal minerals include the following groups:

1. The *feldspars* are subdivided into the *potash* (mainly potassium aluminum silicate) and *plagioclase* feldspars (mainly sodium and calcium linked with aluminum silicates). Both types of feldspar are physically hard but are liable to chemical alteration, particularly in the case of the plagioclase subgroup. Feldspars are usually white, gray, or pink in color and account for 39 percent of the rock-forming minerals of the continental surfaces.

2. *Quartz* is a tough, inert mineral, also known as silica (silicon dioxide), of white or varied color. As the principal constituent of most sands, quartz accounts for 28 percent of the surface rock minerals.

3. *Clay minerals* and *micas* comprise a complex group with a common mineral base of aluminum silicate, variously linked with potassium, magnesium, iron, or hydrogen. The micas are flaky, elastic, and transparent and are subdivided into clear *muscovite*, fairly resistant to chemical change, and dark *biotite*, moderately resistant to chemical breakdown. The clay minerals are very fine particles that can only be recognized under powerful magnification. When found in concentration, uncompacted clay minerals are earthy and soft. Clays result from a long succession of chemical alterations of other minerals and correspondingly change slowly in response to further chemical at-

tack. Together the micas and clay minerals constitute 18 percent of the rock-forming minerals at the surface.

4. *Calcite, dolomite,* and *aragonite* consist of calcium carbonate, including some magnesium in the case of the dolomite. Physically hard, carbonates are generally prone to solution, and calcite is readily dissolved in natural waters. They are commonly white. These minerals account for 9 percent of the surface rock minerals.

5. *Iron* minerals include widespread earthy varieties such as yellowish-brown *limonite* and reddish-brown *hematite,* as well as black, metallic forms (magnetite and some hematites). Hardness varies, while chemically the oxides are moderately stable (limonite) or almost inert (hematite). They constitute 4 percent of the surface minerals.

6. *Amphiboles, pyroxenes,* and *olivine* constitute the most abundant of what are conveniently called the *ferromagnesian* minerals, complex iron, magnesium, and calcium silicates. Physically hard, these dark-colored minerals are readily altered chemically. Although the ferromagnesians account for less than 2 percent of the surface rock minerals, they are major minerals in the lower layer of the crust and in the mantle. Their importance as auxiliary rock-forming minerals at the surface is also considerable.

The six major groups of minerals just mentioned constitute over 90 percent of the surface rocks of the continents. The remaining types, including the economic minerals, are of limited significance for land-form development, with exception of the more abundant salts. The latter include *halite* or common salt (sodium chloride), *gypsum* (a hydrated form of calcium sulfate), *anhydrite* (calcium sulfate), as well as other water-soluble chlorides, bicarbonates, and sulfates.

2-4. ROCKS AND MINERALS

The rocks of Earth's crust are built up of particular combinations of minerals. The individual minerals range from microscopic in size to large crystals several inches long. These mineral grains may be closely imbricated with one another, or they may be held together by a cement that partially fills the available spaces between adjacent mineral particles. The most common *rock* types of the continental surface are only five in number, and their major mineral composition is as follows:

1. *Shale,* composed mainly of clay minerals, commonly mixed with some fine quartz grains (52 percent)
2. *Sandstone,* composed of quartz sand, occasionally mixed with feldspars (15 percent)
3. *Granitic rocks* (granodiorite and granite), composed of quartz

and potash feldspars with some amphibole or biotite mica (15 percent)
4. *Limestone* and *dolomite,* mainly calcite or dolomite (7 percent)
5. *Basaltic rocks,* composed of plagioclase feldspars, biotite mica, pyroxenes, and olivine (3 percent)

Together these major rock types account for 92 percent of the surface rocks.

Rocks are defined according to their chemical and physical properties: mineral content; size of mineral grains; type of cement, if any; and rock origin. Even the first three of these properties are, to some degree or other, a reflection of rock origin. Consequently, rocks are commonly classified according to their origin. Three major classes are recognized: ✳

1. *Igneous rocks* are formed from magmas, that is, molten mixtures of minerals, often rich in gases, found deep within the lithosphere. Such rocks fall into *acidic* and *basic* types. The acidic variety is rich in quartz and potash feldspars (e.g., granite) while the basic variety is rich in ferromagnesian minerals (e.g., basalt). Igneous rocks can also be subdivided into *extrusive* and *intrusive* groups. Extrusive igneous rocks include volcanic lavas and ash that have been ejected onto the Earth surface. Such magmas cool rapidly and lose their gases, so that crystal structures are fine—commonly submicroscopic and occasionally glassy. Intrusive igneous rocks include magmas that were once injected into the upper part of the crust, where they have cooled and solidified gradually. Gases were lost slowly. Crystal structures are large and clearly visible to the eye, so that intrusives are *crystalline* in appearance.

2. *Sedimentary* rocks are composed of waste products from older rocks. The materials are deposited mechanically or chemically in seas or lakes, at the shore, in streams or under ice, and elsewhere on the land surface. Sedimentaries are subdivided into *clastic* rocks, formed by mechanical aggregation of minerals (e.g., shale or sandstone), and *organic* or *chemical* rocks, formed through precipitation of solubles (e.g., limestone).

3. *Metamorphic* rocks have undergone change, bringing about a new crystal structure and formation of new minerals. Such metamorphism can affect all kinds of rocks—igneous, sedimentary, and other metamorphics. The changes may result from heat and pressure, which increase with depth in the lithosphere. During the course of earth history, surface rocks have repeatedly been displaced downward through crustal deformation, exposing them to intense pressures and great heat. Metamorphosis can also result from permeation of a rock by gases or fluids from adjacent magmas or by mineral-bearing groundwater.

2-5. LAVAS AND PYROCLASTIC ROCKS

During the winter of 1064 to 1065 A.D., the Sinagua people living north of Flagstaff, Arizona, were alarmed by earthquakes occurring many times a day. On one day a column of steam and gas, looking like smoke, appeared from a crack in one Indian's field. Soon the quantity of this smoke increased, and explosions began to throw up rock fragments. Then a steep mound was built up, with violent bursts hurling out fiery projectiles every few seconds. A cloud of fine rock dust and gases built up over the area, resembling a small-scale nuclear explosion. Tons of rock pellets, shot out from the new volcanic cone, rained down nearby, piling up several tens of feet a day. Finer dust was carried away by south and southwest winds, spreading out as much as 20 miles (32 km) before it, too, settled down on the surface.

By the time the cone was a few hundred feet high, lavas broke out from fissures near the base of the crater. This molten rock was a basalt that flowed slowly down inclined surfaces, dragging along older rocks with it. The liquid cooled as it flowed so that the top crust often tore open and rolled over, solidifying into tumbled masses of clinker with a sharp, jagged surface.

The Sinagua people evacuated the area in the face of this awesome spectacle, leaving their homes to be buried by ash, cinders, and lava. Archeologists and geologists, working together, have been able to reconstruct these events leading to the formation of what is now Sunset Crater National Park. Within less than a year the steep-sided cinder cone had built up to a height of 1,000 feet (300 m). Around the crater itself alternating lava flows and massive explosions, spewing out masses of cinder and ash, accumulated desolate fields of black rock. For a few years thereafter the earth continued to rumble while steam, cinders, and lava again and again burst out of new cracks in the surface. Small spatter cones, lava squeeze-ups, and mounds of cinders built up in this way, forming low hills and hummocks around the main crater.

Some time later the volcanic activity died away completely and lichens, mosses, and grasses took root on the bare stretches of lava, cinders, and ash. The ash was very fertile, finely divided, and it absorbed water well. As a result, the disaster proved to be a blessing in disguise. The new volcanic soils became the center of a veritable land rush by 1100 A.D., with Indians from all parts of Arizona moving into the areas north and northwest of Sunset Crater. A century after the eruption about 8,000 people lived there in a number of multistoried pueblo villages.

Many of the inferences concerning the life of Sunset Crater are derived from empirical observations made of a new volcano, Paricutín, which first made its appearance in central Mexico on February 20, 1943. The activity of Paricutín was filmed over several years, providing a unique record of volcanic activity. In its heyday the single vent of the volcano discharged 16,000 tons of steam and 100,000 tons of lava per day, together with an unknown amount of cinders and ash. By the end of two years a cone 1,500 feet (450 m) high, some 10 miles (16 km) wide at its base, and with a total mass of 150,000 million tons had been built up. Significant activity ceased after nine years, and the volcano is dormant today.

Sunset Crater and Paricutín have many parallels (Figure 2–3), some less fortunate, such as the destruction of Pompeii and Herculaneum in south Italy by the violent eruption of Mount Vesuvius in 79 A.D. Vulcanism of this kind is most common in mountainous areas prone to frequent earthquakes. Fractures in the lithosphere tap magma reservoirs at depth, and molten rock periodically finds its way up to the surface. If the magma happens to be very basic in character, it contains fewer gases and commonly flows out more quietly. The eruptions of Mauna Loa, on the main island of Hawaii, are of this type.

The more explosive variety of volcano ejects lavas that are more acidic and much richer in gases. As the molten rock reaches the surface, the steam and other gases burst out so that the liquid frequently explodes into particles of white-hot and red-hot material. These particles usually cool and harden before they reach the ground. Larger

Figure 2–3. Cinder Cones and Fresh Lava Flows Reflecting Modern Vulcanism in the Kenya Rift Valley, South of Lake Rudolf. (Karl W. Butzer)

fragments, such as volcanic "bombs" and cinder, fall down immediately, while finer ash may be transported over long distances by wind. The most spectacular eruption of this kind known in historical times was that of Mount Tamboro, an island crater off Indonesia. The whole crater blew itself up in 1815, spewing 36 cubic miles (95 cu km) of rock into the atmosphere and destroying the top 4,100 feet (1,250 m) of the volcano. Analogous to some nuclear explosions, ash fallout was still noticeable around the globe years later. Even more destructive was the glowing ash cloud that killed 28,000 people on the island of Martinique in 1902. The total of volcanic debris that is hurled out of craters constitutes the class of *pyroclastic* rocks, of which compacted volcanic ash, known as *tuff*, is the most important. The rough, jagged lavas associated with violent eruptions are less fluid and solidify at higher temperatures than basic lavas.

Lavas dominated by plagioclase feldspars and ferromagnesian minerals constitute *basalt.* The primary minerals are all susceptible to chemical alteration and gradually decompose into clay minerals, oxides, and some soluble products. Basalts and pyroclastics of similar mineralogy consequently provide rather fertile soils. When potash feldspars are prominent or dominant, the lava is acidic and belongs to the group of *felsites.* The quartz grains present remain inert, while the potash feldspars are converted into residual products at a slow rate. As a result, felsite and allied pyroclastic rocks produce sandier soils of more limited fertility. Of the fine-grained, extrusive vulcanics, basalts are far more common than felsites, and there are extensive areas in the Americas, in Africa, and Asia where basalt flows and associated ash falls accumulated to a thickness of several 1,000 feet (1 km or more) during the geological past.

Since lavas and volcanic ash are commonly laid down in successive layers, they may show bedding, or *stratification.* When lavas are extruded in great masses, however, they show little or no horizontal stratification. Instead, in the case of basalts, cooling from the top down results in a peculiar vertical arrangement known as columnar jointing.

2-6. CRYSTALLINE ROCKS

Magmas that are injected into the crust but that fail to reach the crust before cooling, are part of an unsensational, hidden vulcanism. But if they are unearthed, millions of years later, their comparative resistance may lead to preservation of bizarre forms such as Shiprock, New Mexico, or a mighty mountain range such as the Sierra Nevada, whose peaks separate the states of California and Nevada.

These intrusive vulcanics are classified according to their size and original location (Figure 2–4). Small crack fillings in older rocks con-

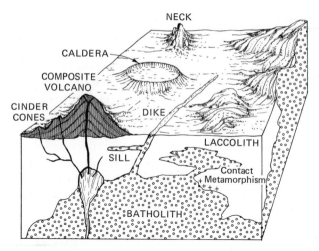

Figure 2–4. Landforms Associated with Intrusive and Extrusive Vulcanism.

stitute *veins,* whereas larger masses of magma spreading upward and along a vertical joint or fracture are called *dikes.* Molten rock once forced in between horizontal beds of older strata becomes a *sill.* The pipes of magma leading up to a former crater may be preserved long after the crater itself has been eroded away. Such pipes can then be exposed as a stump of resistant rock, known as a volcanic *neck.*

Whereas dikes, sills, and necks seldom include more than several thousand cubic yards of magma, other intrusive masses occur with much greater dimensions. Magma may be forced up into a part of the crust, expanding as a mushroom-shaped wedge between rock strata. This is called a *laccolith.* The blister so created by the uparching of surface rocks may have an area of several hundred square miles. Larger still is a *batholith,* a mass of magma with the dimensions of a mountain range.

The minerals of the intrusive volcanics are coarse grained, and the various rock types are classified according to their mineral composition. The intrusive counterpart of basic basalts is a *gabbro,* while the more acidic variety, with some quartz and fewer feldspars, is a *diorite. Granites* are the crystalline rocks corresponding to the felsite group. As in the case of the extrusives, gabbro and diorite break down into fine-grained residual products of high fertility. Granite, on the other hand, produces considerable quantities of coarse quartz sand as it weathers, while the potash feldspars present break down slowly. Granitic soils, therefore, tend to be sandy and inert.

As a result of their origin in the molten state, intrusive rocks are massive, that is, they lack stratification. Instead of the horizontal bedding planes common to sedimentary rocks, intrusive rocks are prone to develop vertical *jointing systems* when they become exposed at the surface.

2-7. CLASTIC SEDIMENTARY ROCKS

Some 200 million years ago a small dinosaur plodded across some muddy sands in what is now the Connecticut Valley. In 1802 a farmer found a block of red sandstone in his fields, showing its footprints. It was the first discovery of the lost age of the reptiles. The dinosaurs had lived near river flats, and when they walked across the muddy surface, their feet sank in, leaving deep imprints. Later the mud dried out and baked in the sun. Finally, the flats were again covered with water, and muddy sand filled the tracks, preserving their impressions in the sediment.

The clastic sedimentary rocks consist of particles ranging in size from less than 1/10,000 of an inch (0.002 mm) to boulders weighing a ton or more. What they all have in common is that they consist of material carried together from older rocks by running water, gravity, ice, wind, or waves. The size of individual particles is determined by the medium of transport as well as by distance from the source. Fine-grained particles accumulate beneath standing or slowly moving waters or through the action of wind, while sand, pebbles, or cobbles may be moved through a swift stream or by waves at the shore. Water and wind tend to sort out particles by weight so that mineral grains of a particular size are concentrated. Gravity movements usually produce little sorting, while ice picks up loose materials of all sizes, including boulders up to several tons in weight. Whatever the agency of transport, after the particles come to rest, they are gradually compacted as free space between particles is reduced. Consolidation may follow as grains are cemented together. In this way a sand becomes a sandstone, a clay a shale, and so on.

The basic names of clastic sediments and rocks are defined according to dominant particle size:

Size of Particles	Unconsolidated Particles	Consolidated Rock
Finer than 1/10,000 inch (under 0.002 mm)	Clay	Claystone
1/10,000 to 1/1,000 inch (0.02 to 2.0 mm)	Silt	Siltstone
1/1,000 to 1/10 inch (0.02 to 2.0 mm)	Sand	Sandstone
Coarser than 1/10 inch (over 2.0 mm)	Gravel, i.e., rounded pebbles	Conglomerate
	Angular rock fragments	Breccia

Mineral composition, the second variable of the clastic sedimenta-ries, is highly variable. Clay-sized particles may include clay minerals, oxides, or quartz; silt-sized particles: quartz, feldspars, calcite, dolo-mite, micas, ferromagnesians; gravel: quartz or fragments of any other rock type. Clay and silts are generally of mixed mineralogy, but sands may be dominated by a particular mineral, for example, quartz sand, arkosic sand (rich in feldspars), lime sand (consisting largely of calcite or dolomite grains). Gravel is often labeled according to the dominant rock type, for instance, basalt gravel or limestone gravel.

The third distinguishing factor is the cement that may be present. Cement may form within the rock through recrystallization between adjacent mineral grains. A silica cement commonly accumulates between quartz grains in this way, a lime cement between calcite grains. The binding material may also impregnate the rock from out-side through the agency of mineral-bearing groundwater. Calcite or ferruginous (iron-oxide) cements form analogously, consolidating a variety of shales, sandstones, or conglomerates.

The resistance of a clastic sedimentary to physical breakdown and chemical decomposition depends on both its mineral composition as well as the nature of the cementing substance. The contribution of such rocks to an agricultural soil depends largely, however, on the minerals present.

2-8. CHEMICAL AND ORGANIC SEDIMENTARY ROCKS

Rock materials dissolved in water can be precipitated in a number of ways. The most common rocks of this kind are *limestones* deposited at the bottoms of lakes and seas, through inorganic precipitation as well as through accumulation of organic carbonates from coral, mollusks, algae, and the like. Another form of limestone is the *travertine* depos-ited in caves, including stalactites, stalagmites, and flowstone (see Chapter 15). Limestones generally consist of calcite, with a number of impurities such as quartz, clays, and oxides. Limestone frequently changes into *dolomite* during the course of time.

As limestones at the surface break down, calcite is ultimately dis-solved and largely carried away in solution. The clayey residue of former impurities is then available to form a soil of moderately high fertility. Dolomites weather more slowly, since the calcium-magnesium carbonate dissolves less readily. Although less frequent than limestone, *salts* are precipitated chemically in coastal lagoons or shallow desert lakes. Salt pans of this origin may be frequent in arid environments.

A last category of sedimentary rocks includes the accumulations of partially decayed plant matter, called *peat,* which ultimately produces *coal.* Although coal itself has practically no significance for landforms, the earliest stages of development include bogs in which peat forms under waterlogged conditions. Peat bogs are common in low-lying, cool-humid lands.

The gradual, layer-by-layer accumulation of clay, silt, sand, gravel, or limey oozes that build up sedimentary rocks account for their characteristic stratification. The different layers or *strata* are commonly separated by breaks or *bedding planes* that result from interruptions of deposition or from differences in particle size. Most bedding planes were more or less horizontal at the time the rock materials were deposited. In addition to horizontal bedding there are inclined beds deposited onto or down an inclined surface, such as may be found in river beds, along the shore, and in sand dunes. Horizontal and inclined bedding frequently alternate within a single rock. As a result of subsequent crustal deformation, horizontal sedimentaries may be warped or tilted at any angle. In addition to bedding planes, sedimentaries tend to develop vertical joint systems near the surface. At ground level such joint systems normally show geometric patterns of rectangular or trapezoidal form.

2-9. METAMORPHIC ROCKS

The broad class of metamorphic rocks is particularly complex, since metamorphosis varies in its impact, depending on original mineralogy, and the intensity and duration of the period of change.

Some metamorphics represent sedimentary rocks that have been further compacted and subjected to recrystallization. So, for example, shale is changed into a hard rock, *slate*, that is suitable for roof shingles and floor stone. Similarly, limestone is recrystallized into marble. Sandstones and conglomerates may be permeated with silica to form remarkably tough *quartzites*.

More intensive change will produce rocks that are often difficult to relate to their original source. A *schist* is composed largely of clay minerals and micas set in a characteristic, very fine, wafery structure known as foliation. *Gneiss* resembles granite in its overall composition, but micas are more prominent, and the minerals are all elongated in parallel lines, giving the rock a lineated structure.

The resistance of metamorphic rocks varies according to mineralogy, compaction, or cementation. Slate and quartzite are rather durable and produce stony soils of limited fertility. Marble, on the other hand, is as soluble as limestone. Schist and gneiss are prone to chemical breakdown, with gneiss behaving much as a granite. Many schists also weather like shales.

The recrystallization of minerals that takes place during metamorphosis generally destroys bedding planes and joint systems, so that marble or quartzite may show no conspicuous stratification. In some rocks the old bedding planes and joints are filled out with secondary minerals, such as quartz. Rocks with a well-developed foliated structure, such as slate or schist, or with a lineated structure, such as gneiss, are not truly stratified. Nonetheless, as they subsequently break up on the surface, they can behave very much as a stratified rock.

2-10. EARTH MOVEMENTS

The preceding sections outlined the basic attributes of the rocks of the lithosphere. This lithosphere is unstable and mobile and frequently subject to a variety of tectonic forces. These internal forces are simply referred to as *Earth movements* and can be subdivided into two categories: (1) crustal deformation or *diastrophism*, through folding, fracturing, uplift, or subsidence; and (2) *vulcanism*, with modification of terrain by newly extruded magmas or the appearance of older intrusive rocks at the surface. The overall effect of tectonic activity is to create or accentuate the irregularity and roughness of the Earth's surface—to increase relief by mountain building, vulcanism, or uplift in one area, by downwarping or subsidence in another.

The major types and forms of vulcanism have been discussed. The two primary mechanisms involved in crustal deformation are bending under pressure, and fracturing under tension or compression.

Intensive crustal bending or crumpling, under lateral pressure from one or two sides, gives rise to *folds*. The upfolds (*anticlines*) and downfolds (*synclines*) commonly run parallel to each other (Figure 2–5). Warping is a gentler form of folding that affects larger areas, leading to uplift or upwarping on the one hand, to subsidence or downwarping on the other.

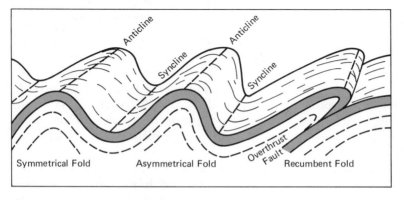

Figure 2–5. Fold Structures.

When stresses are too sudden and intensive in relation to the rigidity of rock surface, the crust can fracture, with adjacent strata permanently dislocated. This process leads to *normal faults,* due to tensional stresses, or *reversed* faults, due to compression of strata (Figure 2–6). Displacement or *throw* along a fault may be vertical or horizontal or both. If horizontal movement is also involved, it is called a tear or lateral, *strike-slip* fault.

Folding and faulting often occur together, particularly where crustal deformation is intensive. In extreme cases, special overturned or *recumbent* folds with *overthrust* faults can develop (Figure 2–5). Large-scale faulting frequently occurs independently of crustal folding, and the crust may be divided into individual blocks as a result of *block faulting.* Raised fault blocks are called *horsts,* depressed fault blocks *grabens* (Figure 2–6).

Crustal deformation that involves folding and faulting, or block faulting, is known as *orogeny,* or *mountain building.* The Appalachians, the Rocky Mountains, and the Alps are examples of orogenic deformation. Uplift or subsidence on a continental scale, with or without crustal warping or tilting, is known as *epeirogeny.* The repeated emergence from or submergence beneath the sea of vast land surfaces during the geological past has largely been a result of such epeirogenic movements.

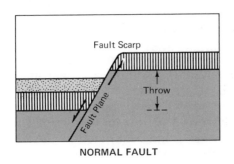

NORMAL FAULT

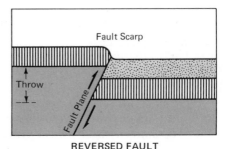

REVERSED FAULT

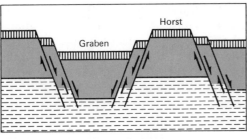

BLOCK FAULTING

Figure 2–6. Fault Structures.

2-11. CRUSTAL DEFORMATION TODAY AND YESTERYEAR

On May 22, 1960, an earthquake shattered southern Chile, destroying whole towns and damaging 450,000 homes. The violent shock created a gigantic wave, or *tsunami*, in the Pacific Ocean that, flooding in on the Chilean coast, drowned out many low-lying villages. Some 4,000 people lost their lives. Earthquakes of this severity serve as a reminder that crustal deformation is active in many parts of the world. Many Earth tremors are an indication of crustal displacement along fault lines. In the case of the catastrophic Alaskan earthquake of March 27, 1964—with damage estimated at $750 million—the original shock wave emanated from a point 12 miles (19 km) beneath the surface. But the vertical throw recorded at the surface was a bare 6 feet (less than 2 m).

The fault displacements of some earlier earthquakes are more impressive. During the notorious 1906 quake at San Francisco, roads and fences were displaced horizontally by up to 21 feet (6 m), while an 1899 earthquake in Alaska uplifted one stretch of shoreline by 47 feet (14 m). Severe tremors recur frequently in the vicinity of major fracture lines, so, for example, the San Andreas Fault in California.

Less spectacular but equally effective is contemporary folding. Pipes and roads in an oilfield near Bakersfield, California, bent and buckled up at a rate of 1 inch (2.5 cm) per year over a 17-year period until they were finally broken. On a much larger scale, epeirogenic uplift has been observed in Scandinavia, in some areas amounting to over 650 feet (200 m) in about 10,000 years, or about 1 foot (30 cm) every 14 years.

The evidence of present-day crustal deformation provides a concrete idea of dimensions and rates for faulting, folding, and warping. Mobility is concentrated in certain Earth regions, and such changes are either gradual or spasmodic, their effects being cumulative over long periods of time. Earthquake activity and vulcanism today (Figure 2–7) are concentrated in the mountain belts that originated in the recent geological past, roughly during the past 65 million years. The very fact that severe earthquakes and vulcanism persist in these mountain systems demonstrates that this particular period of mountain building is still going on. Geological history has been marked by many active, orogenic phases separated by longer periods of quiescence. Different parts of the globe were affected in each case. So, for example, the Appalachians were originally folded during an earlier period of mountain building, 250 to 300 million years ago.

These successive spasms of orogeny are closely related to mechanisms within the interior of the planet. Breakdown of radioactive elements such as uranium and thorium in the core and mantle produces prodigious heat, and only the enormous pressure of overlying rock materials keeps the mantle from becoming totally liquid. Some geo-

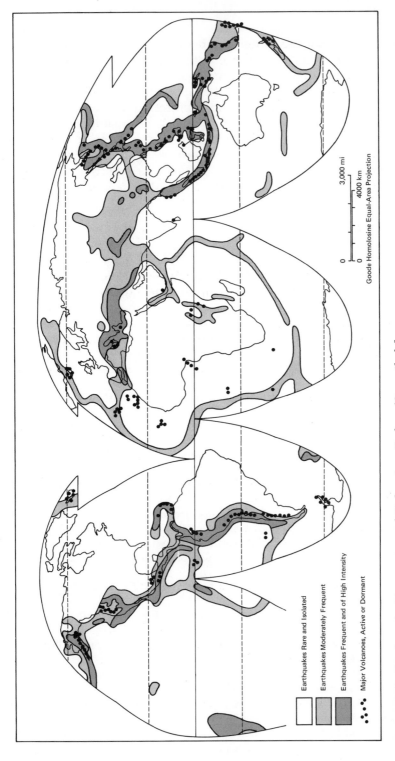

Figure 2–7. Earthquake Frequency and Vulcanism Today. (Compiled from various sources, including Bederke and Wunderlich, 1968. Goode base map copyright by The University of Chicago Department of Geography.)

Earthquakes Rare and Isolated

Earthquakes Moderately Frequent

Earthquakes Frequent and of High Intensity

Major Volcanoes, Active or Dormant

3,000 mi

0

4000 km

0

Goode Homolosine Equal-Area Projection

physicists believe that these high temperatures and pressures produce abrupt molecular changes in the mantle rock, leading to changes of volume and accompanying deformation of the crust above. Others believe that the plastic material of the upper mantle circulates at an infinitesimally slow rate in titanic convection currents that drag against the crust, distorting and rupturing it with sudden releases of energy. Vast reservoirs of boiling magma are then tapped, filling new cavities within the crust or pouring out along faults, fissures, and joints. The mechanisms of crustal deformation are discussed further in Chapters 13 and 14, in light of the theory of plate tectonics.

2-12. SELECTED REFERENCES

Anderson, D. L., C. Sammis, and T. Jordan. Composition and evolution of the mantle and core. *Science*, Vol. 171, 1971, pp. 1103–1112.

Bederke, Erich, and H. G. Wunderlich. *Atlas zur Geologie.* Mannheim, Bibliographisches Institut, 1968, 75 pp. With key in five languages.

Blatt, H., G. Middleton, and R. Murray. *Origin of Sedimentary Rocks.* Englewood Cliffs, N.J., Prentice-Hall, 1972, 634 pp.

Bullard, F. M. *Volcanoes in History, in Theory, in Eruption.* Austin, Tex., University of Texas Press, 1962, 441 pp.

Dott, R. H., and R. L. Batten. *Evolution of the Earth.* New York, McGraw-Hill, 1971, 649 pp.

Ernst, W. G. *Earth Materials.* Englewood Cliffs, N.J., Prentice-Hall, 1969, 150 pp.

Fielder, G., and L. Wilson, eds. *Volcanoes of the Earth, Moon and Mars.* London, Elek, 1974, 128 pp.

Folk, R. L. *Petrology of Sedimentary Rocks.* Austin, Texas, Hemphill, 1974, 182 pp.

Garrels, R. M., and F. T. MacKenzie. *Evolution of Sedimentary Rocks.* New York, Norton, 1971, 397 pp.

Gilluly, J., A. C. Waters, and A. O. Woodford, *Principles of Geology*, 3rd ed. San Francisco, Freeman, 1968, 687 pp.

Green, J., and N. M. Short, eds. *Volcanic Landforms and Surface Features: A Photographic Atlas and Glossary.* New York and Berlin, Springer, 1971, 522 pp.

McAlester, A. L. *The History of Life.* Englewood Cliffs, N.J., Prentice-Hall, 1968, 152 pp.

Mason, B. H. *Principles of Geochemistry*, 3rd ed. New York, Wiley, 1966, 329 pp.

Ollier, Cliff. *Volcanoes.* Canberra, Australian National University Press, and Cambridge, Mass., M.I.T. Press, 1969, 177 pp.

Rhodes, F. H. T., H. S. Zim, and P. R. Shaffer. *Fossils: A Guide to Prehistoric Life.* New York, Golden Press, 1962, 160 pp.

Robertson, E. C., J. F. Hays, and L. Knopff, eds. *The Nature of the Solid Earth* New York, McGraw-Hill, 1972, 678 pp.

Seyfert, C. K., and L. A. Sirkin. *Earth History and Tectonics: An Introduction to Historical Geology.* New York, Harper & Row, 1973, 504 pp.

Smiley, T. L. The Geology and Dating of Sunset Crater, Flagstaff, Arizona. Albuquerque, *New Mexico Geological Society Guidebook*, Ninth Field Conference, 1958, pp. 186–190.

Sparks, B. W. *Rocks and Relief.* London, Longmans, 1971, 404 pp.

Twidale, C. R. *Structural Landforms.* Canberra, Australian National University Press, and Cambridge, Mass., M.I.T. Press, 1971, 247 pp.

Chapter 3

WEATHERING

The rocks of Earth's crust appear to be very durable, and, of course, they are. But wherever exposed, they are subject to slow but effective alteration by weathering. Rocks are broken down into smaller particles, or they rot away or even dissolve in their entirety. Mechanical disintegration works through the freezing of water in pore spaces or cracks; through salt crystallization; through dilation after release of bounding rock pressures; or even through root wedging. The particles produced consist mainly of sands, small splices, and large slabs that provide many surfaces for solution by dilute acids or for chemical alteration into soft residual products. Mantles of regolith produced by weathering of all types cover about 90 percent of the continents. Most geomorphic processes operate exclusively on and in such weathering mantles, or in materials derived

*from them. However, as Chapter 4 will show, biota
take over the upper part of the regolith in most environments, producing soils that are also part of the
biosphere.*

> **The atmosphere's work on rock is slow,
> but ubiquitous and ultimately trium-
> phant.**
>
> JEROME WYCKOFF, *Rock, Time, and Land-
> forms*

3-1. WEATHERING AND THE REGOLITH

One of the most fundamental of geomorphic processes is the disintegration and decomposition of rocks exposed at Earth's surface. This
process is known as *weathering*, since it is commonly the direct result
of atmospheric agents attacking exposed rock or the consequence of
climate working directly through the plant cover and a variety of
microorganisms. Weathering may be a mechanical or physical process,
leading to disintegration of rock into smaller components or constituent mineral grains, without major chemical change. Weathering may
also proceed through chemical alteration, leading to solution or alteration of rock into weaker minerals or soluble products, favoring gradual decomposition.

As a result of weathering, the top layer of the Earth's crust is
broken down in size and altered chemically, forming a mantle of loose
and partially decomposed rock known as *regolith* (Figure 3–1). The existence of regolith is a prerequisite for many of the gradational agents,
providing loose materials of a suitable size for running water, wind, or
waves to pick up and remove. Weathered rock products are constantly
forming over all land surfaces, although the rate varies considerably
from place to place, depending on the nature of the surface rock, the
type of vegetation, and the local climate. Consequently, the depth of
the regolith will vary, in part according to the rate and type of weathering, in part depending on the rate of removal by gradational
agents. Although intact *bedrock* is frequently exposed on hillsides, cliff
faces, and in desert or arctic regions, about 90 percent of the Earth's
surface is mantled by regolith or other unconsolidated materials. This
means that the gradational agents generally operate on a mantle of
loose rock or weathered rock products.

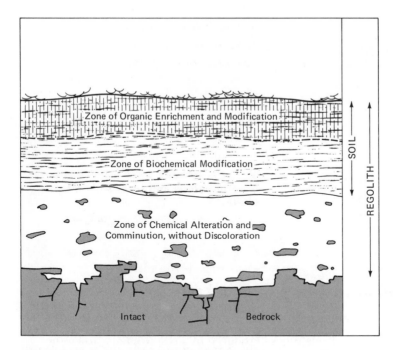

Figure 3–1. Soil, Regolith, and Weathering.

3-2. MECHANICAL WEATHERING BY CRYSTAL GROWTH

Frost Weathering. Frost is the most important agent of mechanical weathering. As water freezes, it expands, with an increase in volume of 9 percent and, if confined, pressures of up to 2,000 pounds per square inch (125 kg/sq cm) can develop. There is some water in the joints, cavities, or pore spaces of most rocks, and a hard freeze will lead to the crystallization of ice in cracks and between individual mineral grains. The resulting pressures gradually pry rock particles apart or weaken mineral bonds. Ultimately, large fragments or small particles are loosened and detached.

Such frost weathering is most effective in climates with abundant moisture and a high frequency of strong frosts that alternate with intervals of thaw. The water ensures ice expansion deep within rock fissures, so that the rock is wedged apart, breaking up into slabs or boulders (Figure 3–2). Frequent, hard frosts accelerate the process. Light frosts, on the other hand, only affect the rock surface and so limit the impact to individual mineral grains or small rock splices at the surface. When rock moisture is scarce, freezing is confined to water in pore spaces, restricting disintegration to surface grains. Since the products of frost weathering vary considerably in size, it is useful to

Figure 3–2. Masses of Talus Blocks Produced by Frost Wedging when the Basin Below Harbored a Glacier. Sierra de Urbión, central Spain. (Karl W. Butzer)

distinguish frost wedging, leading to cracking and splitting—with detachment of blocks, slabs, or splices—from grain-by-grain *disintegration*→whereby freeze-thaw changes crumble rock surfaces into their constituent mineral particles.

The effectiveness of frost weathering is not only determined by climatic variables. Rocks with well-developed joint systems or bedding planes are particularly susceptible to frost shattering, white porous rocks favor ice crystal growth in pore spaces, resulting in grain-by-grain disintegration. Shales are also susceptible to frost weathering, since, when moist, the clay minerals are surrounded by a microscopic film of water. This moisture can freeze, and fine laminae or splices may be peeled off the surface. Most resistant to frost weathering are massive rocks of low porosity, with few joints or bedding planes. Massive limestones and dolomites, many igneous rocks, as well as quartzite, are particularly resistant to this type of mechanical weathering.

Frost weathering is not confined to rock but can also operate within loose regolith. As a wet, clayey soil freezes, the increase in volume may result in expansion at the expense of unfrozen, soft regolith below. Mixing and churning may result, particularly in subarctic climates. When the soil subsequently thaws out, it becomes soft and highly erodible, due to reduced cohesion. Ice crystals may grow into small ice needles within the regolith, particularly under surface rocks, when moisture freezes to the bottom of rocks that cool rapidly. Rock and soil particles, as well as asphalt and concrete roads, can be raised or buckled by *needle ice* in this way. Since the rock or soil grains never quite fall back into their former positions, stones gradually move upward within the regolith, collecting on top of the ground. This explains the old farm saying that "stones grow" and explains the presence of many rocks on cultivated fields in cooler climates.

Salt Weathering. Salts are abundant in the rocks of dry lands, since rainfall does not remove all solubles. Aerosolic atmospheric salts, derived from evaporation of wave spray in distant oceans, is washed out during occasional rains and accumulates in the ground or on rocks. At the same time, evaporation of rock moisture, carrying salts in solution, takes place on or just beneath the surface of the rock. The salts crystallize here, and the growing crystals in pores and cracks create local tension of a force in excess of rock tensile strengths. Evans (1970) has assembled a great array of evidence to show that this seemingly insignificant agent is indeed effective in arid and coastal settings and other environments where salt is relatively abundant in rocks. Crystallization in the pores breaks small flakes off the rock surface or leads to granular disintegration (see Figure 3–3), similar to that produced by frost weathering. Salt weathering is commonly associated with cavernous rotting or honeycombing of rocks (*tafoni*) particularly along shaded cliff bases and on the undersides of boulders where moisture is conserved longest in desert climates. Additional stresses are created as changes in humidity or temperature favor alternating expansion and contraction of salts that encrust mineral grains.

3-3. OTHER FORMS OF MECHANICAL WEATHERING

Pressure Release. One of the most potent agents of mechanical weathering, next in importance only to frost wedging, is built right into the rock itself. Intrusive igneous rocks that solidify beneath the surface or sedimentaries that are consolidated or metamorphosed at depth do so under very considerable pressures. The rock is fully adjusted to this pressure, so that if and when it is subsequently exposed at the surface, it dilates. This *expansion as a result of unloading* produces vertical joints as well as cracks between concentric layers parallel to the rock surface. This dilation begins long before the rock is totally exposed and so prepares joining systems suitable for attack by other types of weathering. The visible results are large slabs that spall off as surfaces are exposed (Figure 3–4) or as cliffs are undermined by erosion. The effects of pressure release can also be observed in relatively short periods of

Figure 3–3. Granite Disintegrating Grain by Grain to Produce a Coarse Gritty Sand. Intensive chemical weathering in southeastern Australia has destroyed rock structure well below the surface, leaving intact "core-boulders" and intrusive quartz veins (finger). Salt weathering may produce similar residues but mainly affects rock surfaces, with little alteration of minerals. (Courtesy of C. D. Ollier.)

Figure 3–4. Giant Slabs of Rock Spalled off by Pressure Unloading at the Foot of an Inselberg on the Serengeti Plains of Tanzania. (Karl W. Butzer)

time in quarries, where "quarry rupture" or "rock burst" detaches masses of rubble from freshly cut rock faces.

Thermal Tension. Another form of mechanical weathering was, in the past, generally attributed to alternate heating and cooling of rock. Since heat conductivity of rock is low, rock surfaces can heat up rapidly in the sun, while the interiors remain cool. Temperature differences of 30 to 100° F (17 to 55° C) may develop, with a slight expansion of the surface rock and resulting tension between the external shell and the interior. At night radiational cooling establishes a reversed temperature gradient. It is to be expected that, over long periods of time, the resulting rock fatigue should lead to spalling. Additionally, if the rock is of variable mineralogy, each mineral type will heat up in the sun at a different rate, creating another form of differential expansion and stress. Granular disintegration should result. Both varieties of differential *thermal expansion and contraction* would be accelerated through sudden heating by fires or cooling by rain, creating small fractures perpendicular to the rock surface.

Cyclical temperature changes and insolation of sufficient intensity to promote thermal tension can only be expected in climates with considerable radiation and with little or no vegetation. Consequently, such forces should be most effective in desert areas, and, indeed, spalling and granular disintegration can be observed in deserts having little or no frost. It is significant, however, that laboratory experiments have failed to reproduce this mechanism, and it is now widely thought that in warm environments pressure release by unloading is primarily responsible for spalling, salt weathering for granular disintegration.

Biological Agents. Another form of mechanical weathering is due to biological agents. Tree roots, for example, can pry apart rock joints or fractures, very much as frost wedging can. The fine root hairs of many plants also exert pressures on individual mineral grains in the process of extracting water and nutrients, thereby aiding in granular disintegration.

3-4. PRODUCTS OF MECHANICAL WEATHERING

The different agents of mechanical weathering produce a variety of products. Depending on whether granular disintegration, microsplicing, spalling, or block wedging is involved, the size of the comminuted products varies from single mineral grains in the sand grade, over small laminae or splices, to slabs, blocks, or boulders. Such sand, grit, or rubble generally is angular in shape and in the cemented state would form a breccia. Finer particles in the clay and fine silt grade, less than about 1/2,000th of an inch (0.01 mm) in diameter, are seldom produced by mechanical weathering.

The major role of mechanical weathering in most areas is to complement and aid chemical weathering, by greatly increasing the area of rock surface exposed to chemical attack. In some environments, such as in the subarctic or in mountain lands, mechanical weathering assumes primary importance. Similarly, wherever chemical weathering is impeded by insufficient moisture, in dry regions, mechanical agents are more prominent (Figure 3–5).

3-5. AGENTS OF CHEMICAL WEATHERING

Chemical weathering always requires water and warmth and, in some cases, dissolved acids or oxygen. This means that at least a certain amount of moisture must be present and that temperatures must be above the freezing point. For most chemical reactions high temperatures are desirable.

There are two basic types of chemical weathering:

1. Soluble materials, such as various salts calcite, or dolomite, are corroded and ultimately dissolved. Salts, limestone, and dolomite are attacked in this way, together with several soluble alteration products that result from other forms of chemical breakdown.

2. Other minerals, such as feldspars, micas, and ferromagnesians, are recombined chemically, leading to the formation of soft residual materials—such as clay minerals and oxides—or of soluble products that can subsequently be dissolved.

Solution. Some minerals may be dissolved in pure water, while others require water containing dilute acids. Simple solution by water will only affect the chlorides and bicarbonates and some sulfates. This means that salt and gypsum can be easily washed out of the soil or regolith.

Solution by dilute acids is considerably more potent. Several kinds of acids are available in soil water. Most important of these is carbonic acid, derived from fermenting plant materials, from respiration by plant roots, as well as from atmospheric carbon dioxide acquired by rainwater.[1] Humic acids, a group of poorly understood organic acids derived from alteration of plant remains by organisms, play a role as well. Rotting of certain kinds of animal matter, or oxida-

---

[1] $H_2O + CO_2 \rightarrow H_2CO_3$

or

water + carbon dioxide $\rightarrow$ carbonic acid.

Figure 3–5. The Effects of Differential Resistance on Weathering Along the Walls of the Grand Canyon. Shaley beds are relatively soft and produce the gentler gradients, while resistant sandstones form the steep faces. (Karl W. Butzer)

tion of ferrous sulfides, can produce sulfuric acid, while several other, minor acids of inorganic or organic origin can also be present.

The overall effects of solution by dilute acids are dissolution of calcareous cements and decomposition of limestone or dolomite (Figure 3–6). Calcium carbonate combines with carbonic acid to produce calcium bicarbonate, a salt that is readily soluble in water.[2] Disassociated mineral grains or "inherited" residues such as clays, oxides, and quartz may then be all that remains of limestone.

Solution in the presence of carbonic acid also accelerates breakdown of feldspars in igneous rocks and attacks soluble products derived from partial decomposition of feldspars or ferromagnesian minerals.

Decomposition. Chemical recombination and decomposition of feldspars, micas, and ferromagnesians proceed mainly through *hydration, hydrolysis,* and *oxidation,* roughly in that sequence. In addition there are further biochemical processes that will be discussed in Chapter 4.

Figure 3–6. Limestone Cliff Face Fluted and Grooved by Solution, Mallorca. Evergreen-oak woodland in foreground. (Karl W. Butzer)

[2] $CaCO_3 + H_2CO_3 \rightarrow Ca(HCO_3)_2$

or

     calcium carbonate + carbonic acid $\rightarrow$ calcium bicarbonate.

Hydration is a matter of water molecules attaching themselves electrically to mineral molecules, thus weakening the mineral bonds. This prepares the way for further attack. Hydration also affects some minerals by swelling, with considerable loss of hardness. The resulting products create stresses through expansion and are often soluble. Salts, as well as some clays, swell rapidly when water is absorbed; this is another superficial form of hydration involving little or no permanent change. Interestingly, laboratory experiments simulating thermal expansion and contraction have succeeded when water was added, suggesting that superficial hydration may be a prerequisite to that form of mechanical weathering.

Hydrolysis is an effective form of deep rot. Molecular recombinations take place between minerals and the soil or rock water.[3] The products are weak and continue to break down into soft and very fine residuals or into solubles that are added to the soil or rock water and made available to plant roots.

Oxidation is a logical follow-up in the wake of hydrolysis. Silicates or carbonates of iron or manganese combine further with water and oxygen to produce iron or manganese oxides (Fe_3O_4, FeO, FeS_2)[4] that may also be attacked and altered.

3-6. PRODUCTS OF CHEMICAL WEATHERING

In terms of rock disintegration, chemical weathering is not only more intensive and thorough than any mechanical process but, on the average, penetrates to greater depths (Figures 3–3 and 3–7). It can attack bare rock surfaces, working simultaneously with frost weathering or pressure release. But it is most effective in a weathering mantle where moisture remains long after rains and where organic acids are available to accelerate chemical reactions. Carbonic and humic acids as well as water develop within the root zone and infiltrate well down into regolith and bedrock, attacking rock at depth. As a result chemi-

---

[3] With increasing temperatures water molecules disassociate into their respective ions: $H_2O \leftrightarrows H^+ + OH$. Ionized water can perform as a reactant rather than as a solvent. Here is an example of how a feldspar mineral can be altered:

$KAlSi_3O_8 + HOH \rightarrow HAlSi_3O_8 + KOH$

potash feldspar + ionized water → aluminum silicate + potassium hydroxide.

The aluminum silicate so produced is unstable and will normally break down further. One possibility is as follows:

$2HAlSi_3O_8 + 5HOH \rightarrow Al_2Si_2O_5(OH)_4 + 4H_2SiO_3$

aluminum silicate + ionized water → kaolinite + colloidal silica (bound with water)

The kaolinite is a stable clay mineral, but the KOH produced by the earlier reaction is a base, soluble in carbonic acid:

cal weathering is not necessarily impeded by the existence of a regolith, whereas mechanical agents become fairly inactive once a substantial regolith has formed.

Climate and lithology obviously influence chemical weathering.
The rates of most chemical reactions double or treble with every 18° F
(10° C) increase of temperature. It is to be expected, therefore, that decomposition of rocks containing ferromagnesians, micas, and plagioclase feldspars will be optimal in warm, humid climates. In middle
latitudes such alteration is restricted to the warmer part of the year,
whereas in the tropics it will be effective throughout the year, as long
as sufficient moisture is available. Carbon dioxide is most soluble in
cold water, near the freezing point, so that carbonate solution is quite
effective in cool waters. The solubility of common salt is hardly affected by temperature. In every case, however, freezing of water puts
an effective halt to chemical weathering. As a result, liquid water and
soluble minerals are the prime prerequisites for solution.

The effects of chemical weathering are rather distinct from those
of mechanical weathering. At first, corrosion or alteration will produce
rounded edges or pitted surfaces on crude rock already disintegrated
through mechanical agents. Ultimately, fine-grained products accumulate within the weathering mantle (Figure 3–7). In the case of limestone or dolomite, the bulk of the rock is removed by solution, leaving
only a residue of inherited clays or quartz. Most of the other minerals,
namely, feldspars, micas, and ferromagnesians, are converted to new
and different alteration products:

1. Fine-grained *clay minerals*, most of which are fairly stable end
products, constitute a vital soil component. Most, although not all,

---

$$2KOH + H_2CO_3 \rightarrow K_2CO_3 + 2H_2O$$

potassium hydroxide + carbonic acid → potassium carbonate + water

The potassium carbonate is quite soluble. The process of hydrolysis is, then, a matter of
H^+ (from ionized water) replacing other ions, such as K^+, Na^+, Ca^+, or Mg^+, that form
part of mineral structures. Through a number of chain reactions, relatively stable residual products are created. Dilute acids aid in speeding up the rates of hydrolysis and attacking any soluble alteration products.

[4] An example of oxidation can be given for the case of a ferromagnesian mineral. Hydrolysis first creates a ferrous form of iron:

$$MgFeSiO_4 + 2HOH \rightarrow Mg(OH)_2 + H_2SiO_3 + FeO$$

olivine + ionized water → magnesium hydroxide + colloidal silica + ferrous oxide

Oxidation then goes to work on the FeO:

$$4FeO + 3H_2O + O_2 \rightarrow 2Fe_2O_3 \cdot 3H_2O$$

ferrous oxide + water + oxygen → ferric oxide (bound with water).

This form of ferric oxide is also an example of hydration. The mineral in question is
commonly represented by rust, a relatively soft yellow or brown form of iron.

Figure 3–7. Chemical Alteration Penetrates Deeply into Igneous Rocks in Humid, Tropical Environments. The clean, smooth granite face is thoroughly rotted, but "core-boulders" of intact rock stand out in this road-cut, Hong Kong. (Courtesy of C. K. Leung.)

clay-sized particles are clay minerals. Chemically, these are very complex, and they have different properties: Clays of the kaolinite group do not expand after absorbing moisture, while clays of montmorillonite type swell appreciably when wet, contracting when dry.

2. Iron and aluminum oxides (Fe_2O_3 and Al_2O_3), known as *sesquioxides*. These are another important regolith component, responsible for much of the discoloration in soil horizons. Most of the sesquioxides are present as components of clay mineral structures.

3. *Silica* in the mobile, jellylike form, known as amorphous or colloidal silica. This soluble silica is mainly attached to complex clay-mineral molecules.

4. *Mineral nutrients* released into the soil water as part of the solubles and thus made available as a source of plant nourishment. These include calcium, potassium, magnesium, and ammonia.[5]

5. Inert *residual materials*, mainly resistant quartz sand.

3-7. A SYNOPSIS OF INORGANIC WEATHERING

The many, interacting forces of weathering can be summarized in terms of each process, how it operates, and the resulting product. Essentially, all of the changes outlined in Table 3–1 and in the preceding discussion involve weathering by inorganic agents or processes, although the mechanical effects of plant roots are also included. A deliberate omission is the organic role of the soil mantle in changing the rate and direction of chemical weathering, a topic to be discussed in Chapter 4.

3-8. AN EXAMPLE OF WEATHERING: MESA VERDE

The cliff dwellings of Mesa Verde, in southwestern Colorado, were built shortly after 1200 A.D. in great cavelike hollows. The caves had formed in the vertical walls of a system of canyons, cut 1,000 feet (300 m) deep by intermittent tributaries of the Mancos River. Accessible only by precarious ladders or rock-cut steps from the bluffs above, these shelters provided a safe refuge for the peaceful Pueblo Indians from marauding invaders then sweeping across the Colorado Plateau. Multistoried buildings were eventually constructed in the gigantic caverns, some of which housed entire villages (Figure 3–8).

When I first visited Mesa Verde in 1963, I was stunned by the size of these caverns and equally puzzled by their origin. Two years later I collected rock samples and searched for telltale evidence of different weathering types among the caves. After discussion with geologists working in the area, I visited other canyons in Arizona, where caves

---

[5] A nitrogen compound with the chemical formula NH_4^+.

Table 3–1. A SYNOPSIS OF WEATHERING

PROCESS	HOW IT OPERATES	RESULT
I. MECHANICAL WEATHERING		
Crystal Growth (creates pressures)	1. As water freezes and ice forms (*frost weathering*) a) Ice crystals grow and wedge apart rock fissures and joints	Blocks wedged apart along vertical joints; slabs or splices spall off along cracks or bedding planes parallel to rock surface
	b) Freezing of capillary water (adhering to mineral grains) and gravitational water (in pore spaces) exerts pressures on intermineral structures	Grain-by-grain disintegration of mineral aggregates at surface
	2. As salts crystallize (*salt weathering*) Salt crystals grow as moisture evaporates, creating pressures in pore spaces and cracks	a) Grain-by-grain disintegration of surface particles b) Rock shell undermined; splices or spalls break off parallel to rock surface
Thermal Tension (creates differential stress)	1. As rock surfaces are heated by sun (or fire) 2. As rock surfaces are cooled by reradiation (or rain) Rapid expansion or contraction of surface rock weakens mineral bonds between shell and rock interior, creating fissures vertical and parallel to surface (aided by hydration)	a) Segments of rock shell spall off as slabs or splices b) Grain-by-grain disintegration of surface particles
	3. As different minerals heat at different rates Intermineral stress as result of differential expansion or contraction	Grain-by-grain disintegration of surface particles
Pressure Release (by unloading)	1. Release of vertical pressure as surface rocks are eroded Dilation of upper rock masses, favoring development of vertical jointing and weakening of surface shell	a) Joint systems enlarge b) Splices or slabs spall off
	2. Release of lateral pressure as valleys are cut and slopes under- Dilation of exposed rock faces with growing weakness along joints and bedding planes	Splices, slabs, or blocks spall off

PROCESS	HOW IT OPERATES	RESULT
Root Tension		
1. As large roots grow	Joints and bedding planes wedged apart	Joint systems enlarged
2. As small roots absorb water and nutrients	Mineral bonds weakened	Grain-by-grain disintegration
II. CHEMICAL WEATHERING		
Solution		
1. In pure water	Chlorides, bicarbonates, and some sulfates dissolved	Salts removed in solution
2. By dissolved acids	a) Limestone and dolomite altered and dissolved except for impurities b) Bases produced during hydrolysis are altered and dissolved	Carbonates and bases removed, leaving earthy residual of insoluble clays, oxides, or quartz sand
Hydration		
Water molecules electrically attached to mineral grains	a) Swelling, leading to mechanical stresses within mineral matrix b) Formation of weaker or soluble, hydrated minerals	a) Prerequisite or auxiliary process to hydrolysis b) Creation of certain clay minerals
PROCESS	HOW IT OPERATES	RESULT
Hydrolysis		
Hydrogen ions from ionized water replace potassium, calcium, magnesium, or sodium ions in mineral combinations	Feldspars, ferromagnesian minerals, and micas decompose	Fine residuals (including clay minerals, hydroxides, carbonates, and colloidal silica) as well as inert quartz sand and other unweathered minerals (released by rock deterioration)
Oxidation		
Minerals with iron or manganese recombine with oxygen and water	Silicates or carbonates of iron or manganese, as well as ferrous forms of iron, altered to oxides	Various oxides that form part of an earthy residual product

Figure 3–8. Mesa Verde Cliff Face. Note bedding of rocks, the cave houses, and a rock-carved water chute. (Karl W. Butzer)

and rock arches were also well developed. Eventually, a hypothetical picture emerged of how such caves can form. The key to the story is the nature of the bedrock.

At Mesa Verde the dominant rock type of the cliffs is a medium-grained sandstone, generally quite porous except for occasional shaley horizons richer in silt or clay. Cementation is weak, partly calcareous. The rock strata are gently inclined, the faces almost vertical.

Several clues were apparent:

1. Small caves or hollows in the rock face coincided with bedding planes of the rock. Many of these hollows arched up above shaley horizons: It appeared that water seeping down through the porous and permeable sandstone was caught just above these less permeable strata. Local concentrations of seepage seemed to be responsible for the occurrence of caves or hollows.
2. The floors or ledges of the small caves were covered with disintegrated sand that, in shaded locations, was often wet with moisture-holding salts. Some salts were also precipitated on rock faces. The salt itself was probably introduced in part by wind action, and the available rainfall in this semiarid climate was inadequate to remove it all. Salt weathering was evidently favorable to grain-by-grain disintegration in damp hollows and caves, particularly those that received seepage moisture.
3. Within the body of the rock itself, seepage was also dissolving the calcareous cement. This could be deduced from hard coatings of white lime ($CaCO_3$) on the ceilings of caves or overhangs.
4. Occasionally, defunct potholes were present on ledges just below caves. This suggested that at times, in the past, much water poured over the rim of the bluffs down the cliffs. Such localized water chutes could form after protracted rains and might help erode hollows at appropriate points on the cliff face.
5. All of the canyon walls showed evidence of repeated spalling of rock on a large scale. The effect was to round off the edges of the caves (Figure 3–9) and to give all detached blocks of rock a roundish shape. This spalling was probably initiated by the effects of pressure release after those deep canyons had been cut. However, the trigger action seems to have been provided by frost weathering. Park rangers have in fact noted that most rock falls occur during the early spring, after freeze-thaw changes.

Each of these factors or processes was involved in the origin of the Mesa Verde caverns, against a backdrop of weak sandstones and a semiarid climate. As the streams cut down into the rock, pressures were released on the sheer walls of rock being exposed. Unloading pre-

Figure 3–9. Evidence of Fresh Spalling Just Above Cliff Cavern at Mesa Verde. (Karl W. Butzer)

pared these faces for frost weathering by widening the rock joints. Seepage of water through the rock, from the surface above, dissolved the cement of the sandstone, creating damp areas where rock water concentrated above shaley lenses. Here salt weathering and solution led to granular disintegration, with the formation of deep hollows at selected spots. Water chutes poured down into or through some of these after rains, speeding up their growth. Finally, some of the caverns attained the unusual dimensions suitable for the cliff dwellings that were built in the thirteenth century.

Rates of change have probably been very slow. There have been few damaging rockfalls since the cliff houses were abandoned about 1280 A.D., and the toeholds and footholds cut into the rock by the Pueblo Indians are still remarkably well preserved.

This example, which is complemented by a paper by Schumm and Chorley (1966), helps to illustrate how intimately weathering is related to landforms. At the same time, it shows how the different agents of weathering work in association and are often difficult to separate.

3-9. SELECTED REFERENCES

Birkeland, P. W. *Pedology, Weathering, and Geomorphological Research.* New York and London, Oxford University Press, 1974, 285 pp.

Carroll, D. *Rock Weathering.* New York, Plenum, 1970, 203 pp.

Chorley, R. J., ed. *Water, Earth and Man.* London, Methuen, and New York, Barnes & Noble, 1969. Especially Chap. 3. II (pp. 135–155).

Evans, I. S. Salt crystallization and rock weathering: a review. *Revue de Géomorphologie dynamique,* Vol. 19, 1970, pp. 153–177.

Jackson, M. L. Chemical composition of soils. In F. E. Bear, ed., *Chemistry of the Soil.* New York, Reinhold, pp. 71–141.

Keller, W. D. *The Principles of Chemical Weathering,* 2nd ed. Columbia, Mo., Lucas Bros., 1962, 111 pp.

———. Processes of origin and alteration of clay minerals. In C. I. Rich and G. W. Kunze, eds., *Soil Clay Mineralogy.* Chapel Hill, N.C., University of North Carolina Press, 1964, pp. 3–76.

Reiche, Parry. *A Survey of Weathering Processes and Products. University of New Mexico Publications in Geology,* No. 3, 1962, pp. 1–95.

Schattner, Isaac. Weathering phenomena in the crystalline of the Sinai in the light of current notions. *Bulletin, Research Council of Israel,* Vol. 10G, 1961, pp. 247–266.

Schumm, S. A., and R. J. Chorley. Talus weathering and scarp recession in the Colorado Plateaus. *Zeitschrift für Geomorphologie,* Vol. 10, 1966, pp. 11–36.

Strakhov, N. M. *Principles of Lithogenesis.* Translated from the Russian by J. P. Fitzsimmons, S. I. Tomkieff, J. E. Hemingway. Edinburgh, Oliver and Boyd, 1967, 245 pp.

Chapter 4

THE SOIL MANTLE

The soil is a veneer of mixed mineral and organic matter that differs in appearance, composition, and biology from the underlying regolith. Weathering within the soil mantle is modified from that affecting bare rock: Temperatures fluctuate little in the soil, but moisture is retained long after rain. Even so the chemical weathering to be expected is profoundly modified by the dynamism of the biological components. As part of the nutrient cycle, plants extract nutrients that are later returned, decomposed, and recombined as humus through the action of the teeming microorganisms in the soil. The resulting clay-humus molecules allow the soil to aggregate and thereby to provide access for air and moisture; they also increase water retention and the capacity of the soil to hold nutrients, thus increas-

*ing fertility and retarding chemical weathering in alka-
line soil environments. However, in an acidic soil the
nutrient bases have been largely lost and the resulting
organic acids accelerate chemical weathering. In other
words, biochemical weathering differs from simple inorganic processes: Alteration is retarded and buffered
in a neutral or alkaline soil (where bases are abundant
or climate is on the dry side) but accelerated in acidic
soils (on rocks poor in nutrients, under acidic vegetation, or in permeable substrata in cool, wet climates).
Soil processes favor the development of distinctive sets
of vertical horizons in different macroenvironments as
well as in special settings that reflect local factors,
such as impeded drainage or the presence of salt or excessive lime. This environmental approach is adopted
for classification of soil types. The concluding section
outlines related aspects of soil fertility.*

4-1. WHAT MAKES THE SOIL?

For most people soil is synonymous with "dirt." As Bridges puts it:
"Farmers or horticulturists till it, engineers move it about in
Juggernaut-like machines, little boys dig in it, and mothers abhor it
for being dirty."[1] But the soil is much more than the top of the regolith, or the inorganic product of chemical weathering. It is both a part
of the lithosphere and the biosphere. There would be no soil without
the hordes of microorganisms that range in size from bacteria to ants
and earthworms. Equally essential are the dead and living plants and
animals in and on the soil. Plants extract minerals from the soil and
gradually return them as organic matter that is decomposed with the
aid of the microorganisms. In effect, biological and biochemical processes are primarily responsible for the conspicuous differences
between the soil and the mineral regolith, with its inorganic weathering processes.

By way of definition, the soil is the shallow zone of intermixed
mineral and organic matter, exhibiting one or more horizons that
differ from the underlying regolith in morphology, particle size, chemical composition, and biological characteristics. In humid mid-
latitudes the depth of such horizons can be measured in inches or feet
(several cm to several m). In subarctic or desert environments soil

[1] E. M. Bridges, *World Soils*, London and New York, Cambridge University Press, 1970,
p. 5.

horizons may be almost entirely absent, while they may be several tens of feet (10 m) deep in the humid tropics. The soil is one of the most important elements of the physical environment. Soils, in the restricted sense of the word, cover about 75 percent of Earth's land surface, providing a natural resource of inestimable value. And, without the soil mantle, the course of geomorphic activity would be radically different in many environments.

One major concern in studying the soil mantle centers on the interacting factors that produce soils and thereby account for the broad distributions of similar soils in areas of comparable terrain, climate, plant cover, or rocks. Again, there are important interrelationships between soils and geomorphic forces: Soils modify the gradational processes, and, in turn, the external agents of erosion and deposition affect the soil mantle and its development. And last but not least, the agricultural value of soils is significant, because soils respond closely to manipulation and management, good or bad. They may be carefully tended and preserved or even improved from one generation to another, or they may be destroyed by erosion, through a lack of understanding or even wanton carelessness.

4-2. FACTORS OF SOIL DEVELOPMENT

The soil zone represents the product of mechanical, chemical, and biological changes. Accordingly, there are many factors that determine the type and intensity of weathering as well as the nature of the organic components. They include:

Parent Material. Rock type, particularly mineral composition and cementation, influence the form and rate of weathering. So, for example, in a humid environment a granodiorite with some plagioclase feldspars is far more susceptible to alteration than a granite outcrop with potash feldspars exclusively.

Jointing, bedding planes, porosity, and permeability are also important from several points of view. The amount of soil and rock moisture depends to a great degree on *porosity*, the ratio of pore volume to total volume. This is high in regolith and unconsolidated sediments, or in shales and sandstones, but quite low in igneous and metamorphic rocks. The *permeability* refers to the passage of moisture, together with jointing, bedding planes, and foliation, if any. In general, unconsolidated sediments and regolith (particularly sands and gravels) are highly permeable, while igneous and metamorphic rocks as well as shales are not. Highly permeable materials may not provide much opportunity for chemical attack, since water will percolate too rapidly as, for example, in a sand or gravel. Extensive jointing or bedding planes usually favor chemical weathering as well as frost wedging in sedimentary rocks.

Soil Temperature and Moisture. Depending on radiation, air temperature, and precipitation, the agents of weathering vary from one environment to another. As a result, most limestones will be resistant in desert environments, whereas they are readily weathered in a humid, mid-latitude setting. However, the local climate, as officially described from climatic statistics obtained from an artificially sheltered box at 5 feet (1.5 m) elevation, is not representative of the soil itself. The peculiarities of the soil microclimate deserve enumeration, since they will profoundly affect the processes of weathering.

Temperatures in the soil or rock medium are seldom identical with those of the air. Daily temperature changes are rarely noticeable to depths of more than 5 or 10 inches (12 or 25 cm), and alternate freezing and thawing at the surface may not be effective to depths of more than a few inches. On the other hand, there may be twice as many night frosts at the soil surface during spring and autumn than are registered by the regulation thermometer at 5 feet (1.5 m). Similarly, on a sunny day, ground temperatures commonly are 20 to 50° F (11 to 27° C) or more higher than shade air temperatures at regulation height. Although little is known about actual soil and rock temperatures, the immediate surface is subject to considerable and rapid extremes of temperature, whereas subsoil temperatures are fairly constant, exhibiting few day-to-day changes.

Depending on the availability of atmospheric water and the porosity and permeability of rock or regolith, water may be held in the soil by several forces. *Hygroscopic* water is attached to the outside of mineral grains, in microscopic dimensions only. It is often loosely linked with the soil and rock molecules. Even soil that appears to be completely dry contains a percentage of hygroscopic water by weight. *Capillary* water is wholly absorbed within fine soil or rock materials or attached to individual particles by surface tension. Although it resists gravitational attraction, capillary moisture can be moved upward or sideways by evaporation at the surface or through pressures exerted by plant roots. This is the moisture present in the soil many days after rain, often long after the soil surface appears to be dry. *Gravitational* water fills the pore spaces more or less completely and migrates downward after the infiltration of rainwater. In humid lands it serves to replenish the groundwater.

Although water is vital for any form of chemical weathering, the exact amount of capillary or gravitational water necessary for moderately effective weathering is not known. But soil moisture, since it is conditioned by porosity and permeability, is not directly proportional to precipitation received. Even dew can form an important source of moisture, particularly in dry areas.

Terrain and Drainage. Topography has an obvious twofold significance for soil development (see Figure 4–11 later in the chapter):

(1) Whether drainage of the soil zone and regolith is free or impeded will affect all types of weathering. So, for example, flooding or water-logging will inhibit mechanical weathering and oxidation, while favoring hydrolysis. (2) Ideally, soil development and erosion are in equilibrium. However, erosion may match soil development on steep slopes, possibly preventing the maturation of a deep soil. On the other hand, soils at the base of a steep slope may be repeatedly buried by soil or crude regolith eroded further upslope. As a result, development of characteristic soil zones may be impeded in areas with steep slopes or in rough terrain with considerable relief. Soil features that reflect slope and drainage will soon be discussed further.

Plant Cover and Soil Microorganisms. The biota, including plant cover, organic matter, and animal life in the soil, modify the type and rate of weathering. Further, leaf litter and rooting limit erosion while increasing the percentage of rainfall that infiltrates the soil. The role of the biota will be outlined in more detail.

Man. Through his manipulation of soils man also is an important agent in soil development. In the positive sense, his attempts to meliorate the soil by addition of fertilizers compensate for or increase the organic matter and mineral plant nutrients, in part removed in the course of cultivation. In this way the aeration and capacity for moisture retention remain high or improve. In the negative sense, soil use without compensation tends to reduce the organic content and lead to soil compaction, with reduced moisture capacity. In the case of sandy soils, planting of needle-leafed trees favors increased acidity with leaching out of mineral nutrients and accelerated chemical weathering. Last but not least man is a potential agent of physical soil destruction, through erosion.

Time. This factor plays its part in soil development by ultimately allowing most of the work to produce tangible results, although at variable and differing rates.

4-3. ORGANIC MATTER AND THE NUTRIENT CYCLE

A part of the soil consists of growing roots and lower plants (molds, algae), as well as of living animals (mainly insects). Equally significant are the decaying plant and animal products: leaves, stems, bark, tree trunks, animal droppings, and the like. As such materials decay, they break down into water, carbon dioxide, nitrogen compounds (ammonia), other mineral nutrients, and new organic derivatives collectively called humus.

The amount and the kind of organic matter in the soil depend on the available nutrients and on the nature of the plant community. Trees and especially grasses that require many mineral nutrients will eventually return these to the soil as they die and decompose. Most of these nutrients are bases, and their return to and partial retention by the soil favors a basic or alkaline soil reaction. Plants that require few base nutrients return few to the soil. As a result, their decay products tend to be acidic, favoring removal of bases in the soil by percolating rainwaters. The cycle of organic matter (Figure 4–1) returned to the soil, and gradually incorporated within it, eventually will produce three organic horizons in the topsoil:

1. A surface horizon of forest litter, consisting of undecomposed organic matter
2. An intermediate fermentation horizon, consisting of partly decomposed leafy mold

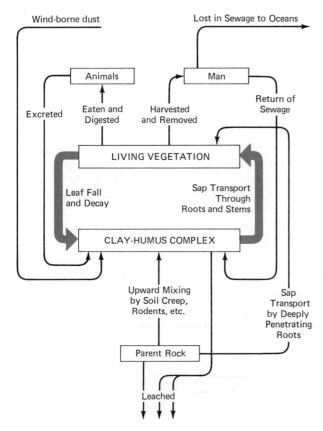

Figure 4–1. The Nutrient Cycle. (From S. R. Eyre, *Vegetation and Soils*, 1963, p. 43, with permission of Edward Arnold Publishers, London.)

3. A basal, humic mineral horizon, consisting of fully decomposed organic matter, in close association with mineral grains.

The transformation of decaying plant and animal products into humus is the role of the teeming microorganisms of the soil. To illustrate their quantity and diversity, the following example can be cited for 1 cubic foot (0.026 cu m) of organic, woodland topsoil:[2]

Microfauna

one-celled protozoa	45 million
subinsects, including eelworms and rotifers	4 million
other insects (millipedes, woodlice, larvae, ants, beetles, etc.)	60 thousand
earthworms	150
	Total weight up to 4 ounces (120 gm)

Microflora

bacteria	1.3 billion per ounce
fungi (molds) and algae	up to 10 ounces (300 gm)

Roots of Higher Plants	up to 30 ounces (900 gm)

The number and the variety of microorganisms decrease sharply in very acid or waterlogged or very dry soils. In such microenvironments organic matter is composed slowly and tends to build up correspondingly. In neutral or moderately alkaline soils there are many microorganisms, so that plant debris is rapidly incorporated into the soil.

The nutrient cycle and soil microorganisms are generally unable to maintain the organic content of agricultural soils where much of the vegetation cover is regularly removed. As a result farmers find it necessary to constantly reincorporate organic matter into the soil. Man therefore plays a major role in the nutrient cycle.

4-4. CLAY-HUMUS AND SOIL REACTION

The integrated organic matter or *humus* of a soil is a dark-brown, gelatinous substance that coats the mineral grains, entering into complex chemical relationships with the clay minerals. The resulting clay-humus molecules are a vital factor within any soil. So, for example, these colloids absorb moisture and account for much or most of the

[2] See Herbert Franz, *Feldbodenkunde*, Vienna, Fromme, 1960, pp. 134–136.

water-retentive properties of the soil. At the same time, humus reduces the stickiness of wet and the hardness of dry soil. Since the humus swells when wet and shrinks when it dries, it permits shrinking and cracking of the soil mass into aggregates during dehydration (see the discussion on soil structure, section 4-7). In this way channels of access are provided for air and moisture, permitting ventilation of the carbon dioxide produced by decaying organic materials and root respiration. These aggregation qualities are vital for the functioning of the soil microorganisms and for the biological renewal and maintenance of agricultural soils.

Equally significant is the fact that the clay-humus molecules have spare valences, electrical links onto which positively charged molecular ions can be attached. In a purely mineral soil the soluble *base nutrients* produced by hydrolysis, such as potassium, magnesium, and ammonia, would be rapidly washed out. When clay-humus molecules are present, many of these solubles link up and exchange molecules as shown in Figure 4–2. Thus a part of the mineral bases released by weathering or decay of organic matter is retained, at least temporarily, within the subsoil. Rainwater percolation slowly but surely washes these base nutrients out, but if the mineral soil and the organic matter have abundant nutrients, a balance is maintained. If this is not so, however, the bases attached to the clay-humus molecules will gradually be replaced by hydrogen ions.[3] When the mineral bases are displaced in this way, *leaching* takes place, and the soil reaction becomes increasingly acidic. Leached bases and other solubles may be precipitated in the subsoil or washed out entirely.

When the soil reaction or pH is 7 or more, the clay-humus molecules are saturated with base nutrients (*base saturation*), and the soil is

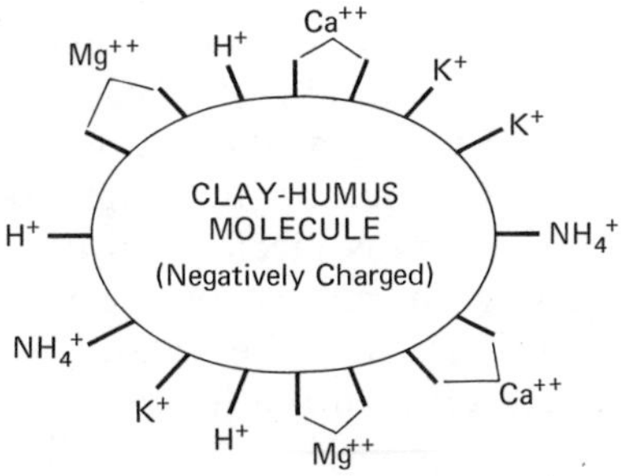

Figure 4–2. A Clay-Humus Molecule with Adsorbed Ions of Hydrogen, Calcium, and Plant Nutrients. (After S. R. Eyre, from Bridges, *World Soils*, 1970, p. 12, with permission of Cambridge University Press.)

[3] The H^+ is derived from ionized water ($H_2O \leftrightharpoons H^+ + OH$) or from acids in the soil.

said to be alkaline. If the pH lies between 4 and 7 (moderately acidic to neutral), mineral bases as well as hydrogen ions are attached to the clay-humus molecules. If the pH is below 4, the soil is highly acidic and all spare valences are occupied by hydrogen ions.[4] The actual quantity of base nutrients present is not indicated by the pH value alone but also depends on the number of available valences (*exchange capacity*) of the clay-humus molecules. This varies in part according to the nature of the clay minerals—clays such as montmorillonite or il-lite have a high capacity (kaolinite has a low capacity) for ion ex-change. However, the organic component greatly increases exchange capacity and accounts for 30 to 90 percent of the exchange capacity of humic mineral soils.[5]

The significance of the clay-humus for soil fertility is readily apparent. It is also important in other ways: Chemical weathering is effectively retarded or conditioned in neutral or alkaline soils, whereas it will often be accelerated in acidic soils. This, in effect, distinguishes *biochemical* weathering in the soil zone from simple chemical weath-ering of inorganic regolith or bare rock.

4-5. PROCESS AND MACROENVIRONMENT

There are several sets of processes that move energy and materials through the soil body and produce certain soil characteristics in dif-ferent environments. For moderately to well-drained soils there are three major kinds of soil development or pedogenesis: podsolization, calcification, and latozation.

On bedrock that is very poor in nutrients, under certain kinds of acidic vegetation (e.g., pine, spruce, heather) or with a cool-wet cli-mate and highly permeable soil, *podsolization* is the common type of soil development (Figure 4–3). Base nutrients are few to begin with or are otherwise leached out as fast as they form. Many microorganisms will not tolerate the resulting acid environment, and clay-humus pro-duction is low. Instead, organic acids are provided from the abundant forest litter and fermenting organic matter. The acidity and the ex-treme chemical reactions resulting from this vicious circle lead ulti-mately to dissociation of the clay minerals, with release of the ses-quioxides. These are washed out of the topsoil and into the subsoil. In the same fashion, clay-humus may also be *eluviated* into a lower-lying *illuvial* horizon.

Podsolization is, then, characterized by eluviation of sesquioxides

---

[4] With a pH of 7, base saturation is 100 percent; pH 6.0, 75 percent; pH 5.5, 50 percent; pH 5.0, 25 percent; pH 4.0, 0 percent.
[5] Exchange capacity (in units of milliequivalents per 100 gm) commonly ranges from 3 to 15 for kaolinite, 15 to 40 for illite and chlorite, 80 to 100 for montmorillonite, and up to 200 for humus.

HEAVY RAINS

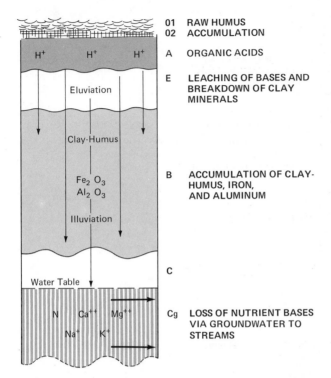

01	RAW HUMUS
02	ACCUMULATION
A	ORGANIC ACIDS
E	LEACHING OF BASES AND BREAKDOWN OF CLAY MINERALS
B	ACCUMULATION OF CLAY-HUMUS, IRON, AND ALUMINUM
C	
Cg	LOSS OF NUTRIENT BASES VIA GROUNDWATER TO STREAMS

Figure 4–3. The Podsolization Process. Horizon designations at left are explained in section 4-6. (Based in part on Bridges, 1970.)

and clay-humus from the upper horizons, illuviation into lower horizons. It is preceded by leaching of base nutrients and eventually produces a distinctive soil type.

When bases are naturally abundant and when the climate is on the dry side, there may be a net accumulation of bases, particularly of calcium carbonate and, in some areas, salt. This counterpart of leaching is commonly associated with calcification (Figure 4–4) (see section 19-2). It is the basis for distinguishing two major soil groups: *pedocals*, with lime enrichment, and *pedalfers*, leached of carbonates and with possible eluviation.

In areas with very intensive hydrolysis and permanent soil moisture and warmth, another type of soil development may be apparent. Clay mineral dissociation can occur, with selective shedding of the colloidal silica and concentration of the sesquioxides in the upper soil. The mobile silica is eluviated or otherwise crystallized out into quartz, destroying the plasticity of the soil clays. At the same time, inert ses-

LIMITED RAINFALL

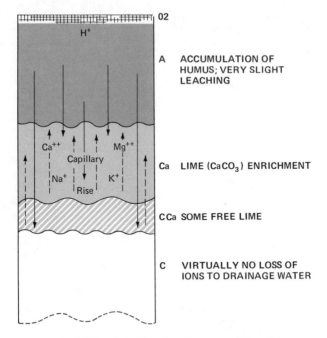

Figure 4–4. The Calcification Process. (Based in part on Bridges, 1970.)

quioxides build up in the soil. This process is known as *latozation* and may ultimately produce hardpans known as laterite (Figure 4–5).

4-6. SOIL PROFILES AND HORIZONS

Most soils developed under intermediate conditions of moisture and temperature exhibit subdivisions or horizons when seen in a vertical section, as, for example, along a fresh highway cut. These horizons can usually be distinguished on the basis of color, but the differences are more fundamental and involve size and type of mineral materials, as well as amount and type of organic matter. The basic horizons recognized by the U.S. Department of Agriculture are as follows (Figure 4–6):

O1 *Litter horizon:* Loose leaves and organic debris, largely undecomposed. The original form of vegetative matter is visible to the naked eye.

O2 *Fermentation horizon:* Organic debris partially decomposed or matted. The original form of most plant or animal matter cannot be recognized with the naked eye.

HEAVY RAINFALL

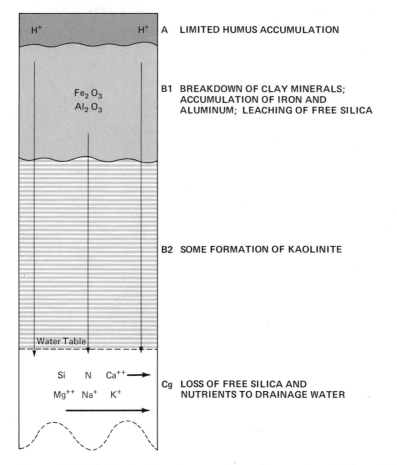

A   LIMITED HUMUS ACCUMULATION

B1 BREAKDOWN OF CLAY MINERALS;
     ACCUMULATION OF IRON AND
     ALUMINUM; LEACHING OF FREE SILICA

B2 SOME FORMATION OF KAOLINITE

Cg LOSS OF FREE SILICA AND
     NUTRIENTS TO DRAINAGE WATER

Figure 4–5. The Latozation Process. (Based in part on Bridges, 1970.)

A   *Humus horizon:* A dark-colored horizon with a high content
        of organic matter, intimately associated with the mineral
        fraction.

E   *Eluvial horizon:* A light-colored leached layer above the B
        horizon, with loss of clay, iron, or aluminum and concen-
        tration of quartz and other resistant minerals.

B   *Illuvial* or *color horizon:* Subsoil that is more yellow, brown,
        or red in color than the horizons above and below it, as a
        consequence of several possible changes: (1) illuvial con-
        centration of clay, iron, aluminum, or humus; (2) weath-
        ering of the parent material in place so as to produce clay
        minerals, or lead to superficial discoloration or visible
        change in structure.

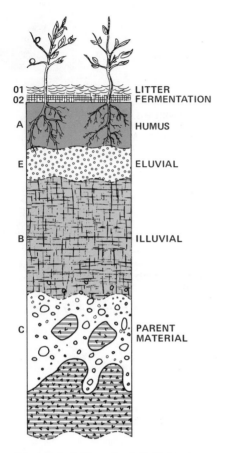

01 LITTER
02 FERMENTATION
A HUMUS
E ELUVIAL
B ILLUVIAL
C PARENT MATERIAL

Figure 4–6. Standard Soil Horizons and Profile.

C   *Parent material:* Unconsolidated regolith similar to that from which the overlying A and B horizons formed, but lacking the properties diagnostic of A and B.

There are a variety of transitional horizons between the A, B, and C horizons, as well as some special types or subtypes including: Ca or calcium *carbonate horizons,* Sa or *salt horizons,* and g or *waterlogged horizons.* These additional letters are used either alone, to designate distinct horizons, or, in the case of intermediate properties, as a suffix to other standard notations, as, for example, BCa or Cg.

The soil *profile* refers to the pattern of horizons present, for example, O1-O2-E-B-C or O2-A-Ca-C. Since the absence or presence of certain horizons as well as their individual development reflect environmental conditions, soil profiles can convey information on climate, vegetation, parent material, and drainage. This applies to both the macroenvironmental contrasts outlined in the previous section as well as to the mesoenvironmental influences utilized to distinguish additional soil types further below. Soil profile changes with respect to their virgin condition may also provide clues about erosional processes or human interference. So, for example, clayey B horizons exposed at the surface in the eastern United States generally indicate destructive erosion as a result of European farming techniques. In short, profiles provide a valuable criterion for the classification of soils.

4-7. OTHER SOIL PROPERTIES

Soil Texture. *Soil texture* refers to the particle sizes of the individual soil grains composing the soil. The proportion that each size grade contributes to the total soil mass can be determined mechanically in the laboratory or by approximation in the field. The soil texture is then defined by the relative proportions of sand, silt, and clay. Sandy soils are said to be coarse textured; clayey soils, fine textured. Medium-textured soils are called loamy. A few of the more common textural classes may be described in approximate terms (see Figure 4–7):

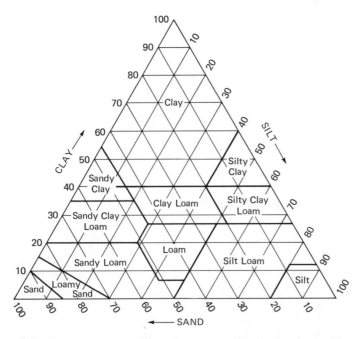

Figure 4–7. Standard Diagram for Determining Soil Texture. Clay values (in percent) run horizontally up the left, silt values diagonally down from the right, sand values diagonally along the base. Reading along any two axes automatically gives the third variable and indicates the overall textural class.

Sand	85 percent or more sand
Loamy sand	70–85 percent sand
Sandy loam	50–70 percent sand
Loam	less than 50 percent sand, 25–50 percent silt, remainder clay
Silt loam	50–80 percent silt and less than 25 percent clay
Clay loam	approximately equal proportions of sand, silt, and clay
Silty clay	80–100 percent silt and clay, in about equal proportions
Clay	40 percent or more clay, less than 45 percent sand, remainder silt

In fact, four more intermediate classes exist in the textural range between loams and clays.

Soil texture is significant for permeability and cohesion. A coarse-textured soil allows water to percolate rapidly, and little capillary moisture is retained for subsequent use by plants. At the same time, soil cohesion is minimal. On the other hand, very fine-textured soil is almost impermeable and prone to waterlogging, while soil aeration is poor. A medium-textured soil has intermediate characteristics that are more suitable for plant growth, root penetration, and soil productivity.

Soil Humus Types. There are major differences in the quality and state of soil humus. Some organic material (*raw humus*) has been broken down physically but not chemically, remaining unintegrated with the mineral soil. It is commonly acid and the product of an acidic environment with limited microorganisms. Other, intermediate grades of humus have been ingested by insects and loosely intermixed with the mineral soil fraction. Plant remains can still be recognized microscopically. Optimal is that humus that has been completely reduced to amorphous colloids by repeated earthworm ingestion. Most agricultural soils are of this *mild humus* type, which is indicative of a good nutrient cycle.

Soil Structure. *Soil structure* refers to the aggregation of the mineral grains and humus into larger units. These are the soil clods turned up by the farmer's plow, and the natural structure can be easily seen by dropping a piece of dry soil and then picking up a handful of the natural aggregates (Figure 4–8). Structure is vital for soil permeability, aeration, cohesion, and fertility. It consequently affects the water balance of the soil, the infiltration capacity of rainwater, as well as erodibility. Structure may be determined by the amount of humus and clay minerals, by drainage conditions, or by climatic factors.

Several structural types mark dense or compact subsoils with limited aeration and drainage. Some are tightly fitted, cubelike aggregates (*blocky*) that become dense and almost impermeable when

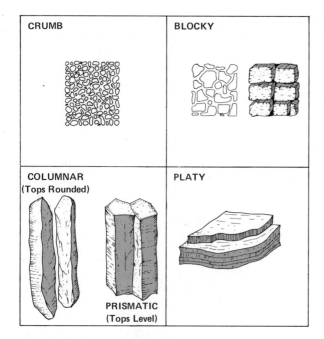

Figure 4–8. Soil Structure Refers to Characteristic Shapes of Dry Aggregates.

wet. Others are arranged as large vertical columns or slabs (*columnar* or *prismatic*), moderately favorable for infiltration, drainage, and aeration. Still others form horizontal platelets of soil (*platy* structure) that are fairly impermeable. At the other extreme, lacking sufficient clay and silt, soils are loose and noncohesive.

The optimal structure for agricultural soils then consists of small, spheroidal aggregates of irregular shape, produced by earthworm excretion. This *crumb* structure maintains aeration and permeability of the soil, even when moderately wet. Furthermore, soil aggregation can be changed through cultivation. In particular, loss of crumb structure may result from excessive use with the depletion of organic matter and nutrients. As a consequence, fertility is drastically reduced, the soil is compacted, and water acquisition and transmission are reduced, favoring accelerated erosion. Conversely, man may enhance soil structure by proper treatment.

Soil Color. Soil color is never an infallible guide to soil characteristics, but it can be quite useful for identifying distinct soil horizons. Color may be accurately determined in the moist or dry state by color charts. Several of the more indicative colors—diverging from those of the parent material—are as follows:

Grayish or blackish shades in the topsoil generally indicate the presence of humus or compounds incorporating base nutrients.

Brownish, yellowish, and reddish shades in the subsoil usually indicate the presence of sesquioxides, as a result of illuviation or
weathering in place.

Brownish surface horizons may reflect both humification and
weathering.

White subsoil zones frequently indicate concentrations of lime,
gypsum, or other salts when compact, or an advanced state of
eluviation when uncohesive.

Grayish, greenish, or bluish subsoil horizons commonly reflect
ferrous iron compounds and permanent waterlogging, while rust
spots and stains due to the presence of limonite indicate seasonal waterlogging.

4-8. TOWARD A CLASSIFICATION OF SOILS

Soil classification poses a problem, since soils vary according to several major factors, some of which, like topography, vary from place to
place. Three basic approaches can be taken to this question of soil classification:

1. An empirical approach, dictated by agricultural considerations,
must necessarily classify soils on a more minute scale, emphasizing
soil productivity, fertility, and management. The Wisconsin Soil
Survey, for example, recognizes 22 major soil categories within the
state of Wisconsin, each of which occupies anywhere from 300,000 to
over 4 million acres. Yet for practical use, primarily in agriculture,
these categories must be subdivided into 157 subtypes.

2. A systematic approach is based on categories of observed soil
properties ordered in a hierarchical system. The prototype is the *Seventh Approximation*, a comprehensive classification introduced in 1960
and later adopted by the U.S. and Belgian soil surveys. At the highest
level of generalization, 10 soil orders are now defined, primarily on the
presence or absence of specific diagnostic horizons and the degree of
development of these horizons. The lowest levels of classification include the whole suite of agricultural categories utilized by regional
organizations.

3. A genetic approach attempts to recognize regional similarities
of soil development and soil profile, emphasizing broad environmental
factors (vegetation and climate), as well as local factors such as
drainage, topography, and parent material. The three primary categories of classification reflect different degrees of dependence on
local peculiarities.

Although these three approaches are fully reconcilable, it is necessary to adopt one for the sake of convenience. The Seventh has many
theoretical advantages, particularly to pedologists. But it is also ab-

stract, rigid, and cluttered up with cumbersome terms. Many of the soil orders and all of the suborders can only be identified after elaborate laboratory analyses. Furthermore, the soil classes are based entirely on morphological criteria, to the almost total exclusion of parent material, environmental conditions, or historical context. This sets practical limitations to the use of the Seventh by geomorphologists or soils geographers. For purposes of comparison, the basics of the Seventh are given in Appendix A, together with their approximate equivalents.

The genetic classification followed here distinguishes three orders of zonal, intrazonal, and azonal soils. These are defined on the basis of environmental parameters although there are corresponding differences in soil profile and morphology.

Zonal soils form under "normal" conditions of slope, drainage, and parent material and exhibit broad regional similarities that reflect the distribution of vegetation and climate. Excluded would be waterlogged bottomlands, rough uplands, as well as parent materials deficient in plant nutrients (e.g., sandstones, quartzites), highly permeable (e.g., loose sand and gravel), or with highly specialized mineral nutrients (e.g., salts, limestone). The suborders of zonal soils are defined by very general environmental criteria:

1. Shallow, humic soils of polar and alpine regions
2. Shallow, calcified soils of arid regions
3. Humic, calcified or leached soils of semiarid and subhumid grasslands
4. Podsolized soils of boreal and temperate forests
5. Intermediate soils of subtropical woodlands and tropical savannas
6. Latosolic soils of tropical forests.

Some of the great soil groups that make up these suborders will be outlined and discussed in Chapters 17 to 20.

Intrazonal soils develop under special conditions that produce salt or waterlogged horizons, or they have superabundant concentrations of carbonates. They include:

1. *Saline soils* of poorly drained arid and coastal regions (Figure 4–9)
2. Poorly drained, *hydromorphic* soils of humid lands:
   Gley soils are soils with peculiar subsoil horizons of bright, mottled aspect (oxidation) or of pale gray, blue, or greenish color (reduction), reflecting seasonal or permanent waterlogging (Figure 4–10)
   Bog soils are highly organic soils resulting from accumulation of muck or peat in acidic, swampy environments
   Hardpan soils can form as a result of podsolization in cool,

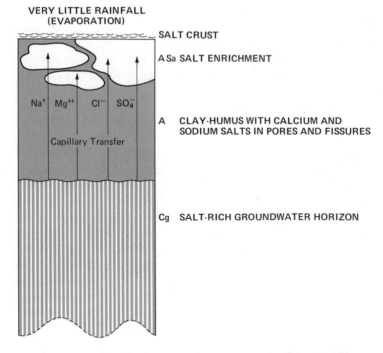

**VERY LITTLE RAINFALL
(EVAPORATION)**

SALT CRUST

A Sa SALT ENRICHMENT

Na⁺ Mg⁺⁺ Cl⁻ SO₄⁻

Capillary Transfer

A   CLAY-HUMUS WITH CALCIUM AND
     SODIUM SALTS IN PORES AND FISSURES

Cg   SALT-RICH GROUNDWATER HORIZON

Figure 4–9. Salinization. (Based in part on Bridges, 1970.)

humid lands or of latozation in tropical humid environments, mainly as a result of sesquioxide accumulation in the subsoil, related to groundwater conditions

3. Highly calcareous, *calcimorphic* soils with lime accumulations, developing on permeable or soft limestones as well as in semi-arid environments on lime-rich parent materials.

In contrast to the zonal and intrazonal orders, *azonal soils* lack a real profile due to some inhibitive factor. They include:

1. *Lithosols:* weak soils on stony terrain, usually on steep slopes, where erosion keeps pace with or exceeds the rate of soil development
2. *Regosols:* minimal soil veneers on dry sands or gravel, usually in desert environments
3. *Alluvial soils:* indistinct profiles in lowlands prone to flooding, where soil development cannot keep pace with deposition.

A last concept of value in soil classification is the topographic sequence or soil *catena*. Given identical parent material along an arbitrary cross-section, topography may vary from a swampy valley floor to a rolling upland. In response, soils commonly show a gradual tran-

HEAVY RAINFALL

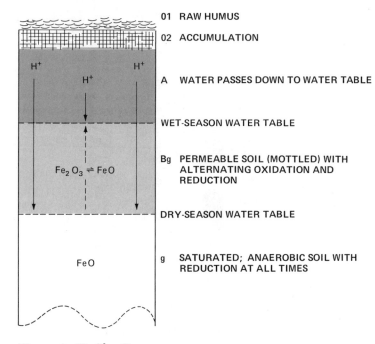

01 RAW HUMUS

02 ACCUMULATION

H^+ H^+

H^+

A   WATER PASSES DOWN TO WATER TABLE

WET-SEASON WATER TABLE

$Fe_2O_3 \rightleftharpoons FeO$

Bg PERMEABLE SOIL (MOTTLED) WITH
    ALTERNATING OXIDATION AND
    REDUCTION

DRY-SEASON WATER TABLE

FeO

g   SATURATED; ANAEROBIC SOIL WITH
    REDUCTION AT ALL TIMES

Figure 4–10. Gley Processes.

sition of horizon number and depth, topsoil and subsoil texture, as well as drainage conditions (Figure 4–11).

4-9. SOIL FERTILITY

The soil is a natural resource, particularly for the cultivation of crops to feed the world's billions. Maintenance of soil productivity has always posed a challenge, even to the earliest farmers in the Near East of some 10,000 years ago. Soil fertility is not a simple matter, however. Depending on the level of technology employed and the crops planted, a particular soil will behave differently. Most farmers know that some crops, including cotton and corn, rapidly exhaust a soil, while others, such as alfalfa, peas, and beans, revitalize it. Yet depending on soil type and chemistry, these critical crops are not necessarily identical everywhere.

Perhaps more important still are farming practices. Early prehistoric farmers used hoes or digging sticks to prepare their fields. Since only a shallow soil zone was utilized, fertility was rapidly reduced, as

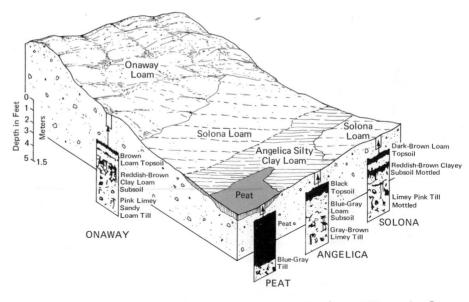

Figure 4–11. A Soil Catena on Glacial Deposits in Northeast Wisconsin. On-away soils (A-E-B-C) are on higher ground and well drained, while Solona soils (A-E-Bg-Cg) are found on gently sloping, lower ground with periodic waterlogging. Angelica soils (O-A-g) grade into peats (O-g) on the bottomlands. (After Beatty et al., 1964, p. 169. With permission of the Wisconsin Soil Survey.)

we know from the primitive agriculture practiced in some underdeveloped areas today. After one or two crop seasons the soil is exhausted and then allowed to revert back to natural vegetation for a period of 20 to 30 years. The invention of the plow, about 5,000 years ago, revolutionized agriculture in some parts of the Old World. Whether of wood or iron, the plow turns up a deep furrow and effectively mixes the topsoil. Crops can therefore utilize a larger segment of the soil, and successful planting can be continued for longer periods of time. Simultaneously, other means were discovered to maintain soil fertility:

1. Addition of organic fertilizers, at least for select plots of high-value produce
2. Plowing in of field stubble, after the harvest, to replenish the organic content and nitrogen level
3. Crop rotation, whereby fields are successively planted with different crops, including one season in three of alfalfa or legumes (pod plants such as beans, peas, and lentils) which favor nitrogen-producing bacteria
4. Lying fallow, that is, having a field in hay or pasture for one or two years out of three.

Large-scale application of fertilizers was limited to wealthy farmers or commercial enterprises until quite recently. Mineral fertilizers such as nitrogen had to be imported at considerable expense and generally were not available to subsistence farmers. Only the industrial development of chemical fertilizers could change this picture, as it has during the past generation or two. Yet even today commercial fertilizers are beyond the means of most small farmers in many parts of the world. Unfortunately, these are also the areas of greatest population pressure.

Soil fertility involves a wide variety of factors: The first factor is the major mineral nutrients. The most basic requisite for soil productivity is an adequate level of four minerals: nitrogen, phosphorous, potassium, and calcium.

The soil microorganisms of a healthy soil produce nitrogen in the form of ammonia, but excessive cultivation or natural leaching can reduce the supply precariously. Plowing in of stubble and manure or planting of nitrogen-replenishing crops is a natural way to supplement this crucial mineral. Nitrate fertilizers may still have to be added, however. These are obtained from bird guano, sodium nitrate deposits, or industrial ammonia.

Phosphorous is present in all organic matter as well as in the weathering residue of many igneous and some sedimentary rocks. But it often requires artificial replenishment in agricultural soils, since it is difficult to break down and assimilate. Sedimentary phosphate rocks (calcium phosphate) are mined for this purpose. Potassium, in the form of potash (potassium oxide, or K_2O), is released slowly in the soil from wood ash and during the hydrolysis of feldspars. Soils poor or lacking in potassium can be fertilized by mineral potash. Calcium or lime is frequently added to very acidic or to highly alkaline soils; ground-up limestone of local origin is commonly used for this purpose.

The second aspect of soil fertility relates to the minor mineral nutrients. Further bases or metals required in small quantities in agricultural soils include magnesium, sulfur, iron, manganese, and zinc, together with trace elements such as boron, copper, chlorine, molybdenum, iodine, and cobalt. Many of these are vital as microscopic elements in soil fertility or as catalysts in the organic chemical reactions of plant growth. Although present in sufficient abundance in most soils, some or many of these elements may be deficient with serious consequences. Extremely podsolized or latosized soils are the most common offenders.

The third factor is humus and microorganisms. The humus type, preferably mild, and the amount of clay-humus, are critical aspects of agricultural soils. They have important implications for soil reaction and base nutrient content, while affecting permeability, water retention, aeration, maintenance of structure and nutrient cycle. The different categories of soil microorganisms, each with a different role, are similarly involved.

Fourth is soil reaction, base-saturation, and exchange capacity. The acidity or alkalinity of the soil may be sufficiently extreme to impede plant growth seriously. This is very rare, however. Instead, the soil reaction gives a measure of the degree of base saturation, a partial index of soil fertility. The exchange capacity expresses another part of this relationship.

Fifth, soil texture, structure, and drainage are each significant with respect to air and water circulation, and they condition water acquisition, storage, and transmission. Soil moisture—in excess, in proper amounts, in deficiency—is obviously critical for plant growth. Aeration is also vital, since most plant roots require a supply of free oxygen. Most plants do not root in permanently waterlogged subsoil as a result. Excessive carbon-dioxide concentration in the soil air is also undesirable. Last but not least, beneficial microorganisms invariably require good aeration.

Sixth, water, either atmospheric precipitation or irrigation water, has obvious significance for soil fertility.

Altogether, soil fertility and productivity are far more than a matter of mineral supply. The organic content and the dynamism of the soil are fundamental. Depletion of organic matter commonly goes hand in hand with a reduction of the soil microorganisms. As a result, the role of the vital clay-humus molecules is affected, favoring increased leaching of mineral nutrients with a possible increase of soil acidity. Excessive cultivation then leads to a deterioration of soil structure, with compaction, reduced aeration, and a less favorable moisture balance. This further reduces the number and variety of microorganisms and the soil dynamism in general. Finally, these phenomena serve to make the soil more erodible. Evidently, all of these factors are interrelated, and soil exhaustion and deterioration are part of a vicious circle of events. Apart from the addition of manure and mineral or chemical fertilizers, careful soil management is required. The details of such management vary from one soil type to another and from one environment to another.

4-10. SELECTED REFERENCES

Arkley, R. J. Climates of some great soil groups of the western United States. *Soil Science*, Vol. 103, 1967, pp. 389–400.

Beatty, M. T., et al. The soils of Wisconsin. In *1964 Wisconsin Blue Book*. Madison, Wis., University of Wisconsin, 1964, pp. 149–170.

Bidwell, O. W., and F. D. Hole. Man as a factor of soil formation. *Soil Science*, Vol. 99, 1965, pp. 65–72.

Birkeland, P. W. *Pedology, Weathering, and Geomorphological Research.* Oxford and New York, Oxford University Press, 1974, 285 pp.

Black, C. A. *Soil-Plant Relationships*. New York, Wiley, 1957, 332 pp.

———. *Methods of Soil Analysis*. Madison, Wis., American Society of Agronomy, Series in Agronomy, No. 9, 1965, pp. 1–1572.

Brewer, R. *Fabric and Mineral Analysis of Soils*. New York, Wiley, 1964, 407 pp.

Bridges, E. M. *World Soils*. London and New York, Cambridge University Press, 1970, 89 pp.

Buckman, H. O., and N. C. Brady. *The Nature and Properties of Soils*, 6th ed. New York and London, Macmillan, 1960, 567 pp.

Bunting, B. T. *The Geography of Soil*. Chicago, Aldine, and London, Hutchinson, 1967, 213 pp.

Cornwall, I. W. *Soils for the Archaeologist*. London, Phoenix House, 1958, 230 pp.

Cruickshank, J. G. *Soil Geography*. Newton Abbot (Devon), David and Charles, 1972, 256 pp.

Deevey, E. S. Mineral cycles. *Scientific American Offprint*, No. 1195 (September 1970), 12 pp.

Delwiche, C. C. The nitrogen cycle. *Scientific American Offprint*, No. 1194 (September 1970), 12 pp.

Drew, J. V., ed. *Selected Papers in Soil Formation and Classification*. Soil Society of America, Special Publication, No. 1, 1967, 428 pp.

Dudal, R. *Definitions of Soil Units for the Soil Map of the World*. Rome, FAO, World Soil Resources Report 33, 1968, pp. 1–72.

Eyre, S. R. *Vegetation and Soils: A World Picture*. London, Edward Arnold, and Chicago, Aldine, 1963, 324 pp.

Farnham, R. S., and H. R. Finney. Classification and properties of organic soils. *Advances in Agronomy*, Vol. 17, 1965, pp. 115–162.

Geiger, Rudolf. *The Climate near the Ground*, 2nd ed. Cambridge, Mass., Harvard University Press, 1961, 611 pp.

Hunt, C. B. *Geology of Soils: Their Evolution, Classes, and Uses*. San Francisco, Freeman, 1972, 344 pp.

Jenny, Hans. *Factors of Soil Formation*. New York, McGraw-Hill, 1941, 281 pp.

Kubiena, W. L. The Classification of Soils. *Journal of Soil Science*, Vol. 9, 1958, pp. 9–19.

U.S. Department of Agriculture. *Soils and Men*. Department of Agriculture Yearbook, 1938. Washington, D.C., GPO, 1938, 1232 pp.

U.S. Department of Agriculture, Soil Survey Staff. *Soil Classification: A Comprehensive System (7th Approximation)*. Washington, D.C., GPO, 1960, 244 pp.

Chapter 5

HILLSLOPE SCULPTURE (I): BY GRAVITY

Only a small part of the Earth's land surface is directly affected by streams, glaciers, or shorelines. The major part consists of an infinite variety of slopes in rock, regolith, or soil. The sculpture of these "hillslopes" is not just a matter of river systems, wind, or moving ice. We tend to overlook the fact that gravity operates everywhere, inexorably, in dry or wet masses of soil or rock. Occasional landslides or other mass movements overwhelm distant mountain villages in Peru or suburban developments in California. But the less appreciated and more gradual forms of gravity erosion daily transfer millions of tons of rock down mountainsides and remove equal quantities of soil and regolith from

*cultivated lands into the drainage sluices provided by
the Earth's rivers. Next to the surface denudation effected by mass movements we also underestimate the
importance of running waters before they enter
streams. Chapter 6 turns to this question of rainfall,
infiltration, and surface runoff, and then considers
complex slope development.*

5-1. WHAT ARE HILLSLOPES?

World landforms have what could be described as a lowest common
denominator: slope. Every surface, whatever its origin, is composed of
one or more slopes of variable inclination, orientation, length, and
shape. But of the many possible slope types, the geomorphologist is
most commonly interested in these slopes that extend from a valley
floor up onto the interfluves of a watershed. These would include (1)
the upland interfluves, (2) the valley sides, whether smoothly sloping
or cliffed, and (3) the sloping valley floor. All of these surfaces together
can be described as *hillslopes,* even though dimensions may range from
the miniature slopes of a gully system or highway trench to an entire
mountainside.

On a particular hillside, slope inclinations vary almost continuously from bottom to top or from one point to another along a single
contour. However, just what the land surface looks like can be conveyed by translating the visual impression and actual measurements
of select components into categories based on a few, simple identifying
characteristics. For example, Savigear (1965) has suggested that:

A *flat* is a horizontal surface or one inclined at an angle of less
than 2 degrees.
A *slope* is a surface area included between 2 degrees and 40 degrees.
A *cliff* is a near-vertical face or one inclined at more than 40 degrees.

These definitions can be integrated into a hillslope model (Figure 5–1),
adding several further components (Savigear, 1965) as follows:

The *crest-slope* is the upper, convex part of a hillslope, possibly
grading onto upland flats.
The *mid-slope* is the central, gently curving or straight (rectilinear)
part of a hillslope, possibly interrupted or replaced by cliffs.
The *foot-slope* is the lower, concave part of hillslope, commonly
grading into valley-bottom flats.

The change of slope inclination at the crest or foot-slope may be

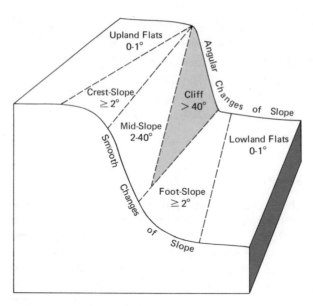

Figure 5–1. Slope Elements.

abrupt and *angular,* or it may be *smooth* (Figure 5–1). If both the crest-and foot-slopes are angular, the slope will be straight sided or recti-linear (Figure 5–2). If the crest-slope is smooth and the foot-slope

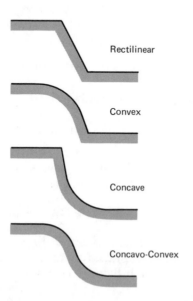

Figure 5–2. Basic Categories of Slope Forms.

angular, the slope will be *convex* in shape. On the other hand, if the crest-slope is angular and the foot-slope smooth, the slope will be *concave* in appearance. Finally, if both the crest- and foot-slopes are smooth, the slope will be *concavo-convex*. In detail, any general slope may be interrupted by minor, steplike *facets* (angular in shape) or *segments* (smoothly curved). A classification for the resulting *complex* slopes has been suggested by Ahnert (1970).

In the broadest sense, hillslopes account for a large part of the Earth's land surface. Excluded are only the extensive upland or lowland flats characteristic for some parts of the continents. The predilection of the geomorphologist for focusing much of his attention on hillslopes is, of course, a matter of perspective. He could, instead, devote his attention to stream patterns, to microlandforms formed in a particular rock type, or to large-scale erosional surfaces. But it is at the intermediate scale of hillslopes that processes can perhaps be best observed and successfully related to forms. Cause and effect are most apparent here, and the differences dictated by environment, relief, or rock type can be most readily seen.

This and the following chapter deal with the development of hillslopes. The principal gradational agents responsible include (1) mass movements, involving transport of material on slopes under the influence of gravity and (2) running water, both as diffuse surface flow and as concentrated runoff in rills and streams. In some environments wind or moving ice also contributes to slope evolution. But since neither wind nor ice is a universal agent of hillslope development, these chapters concentrate on the mechanisms and impact of mass movements and of surface runoff. After outlining these processes and their function in a simple setting, we will examine the interplay of the several forces in the development of hillslopes.

5-2. DENUDATION VERSUS DISSECTION

The materials of the soil mantle and regolith represent the products of mechanical, chemical, and biochemical weathering. As time goes on, a part of this weathered material is removed through erosion and transported variable distances by one or other of the gradational agents. If this were not so, hillslopes would eventually be choked and buried in soil or regolith. Instead, on all but steep slopes, the depth of the weathering mantle is generally quite similar on a particular rock type within a given environment. This suggests that there is a crude equilibrium between rates of weathering and erosion on most surfaces. Ultimately, of course, materials that are eroded will be deposited in the form of sediment—downhill, downstream, or downwind.

From this basic argument we can evolve three axiomatic principles that are fundamental in geomorphology:

1. Rates of weathering may or may not balance rates of erosion. If weathering proceeds faster, soil and regolith will accumulate on hillslopes until a new balance obtains. If weathering does not match erosion, soil and regolith are gradually removed, exposing bare hillslopes (Figure 5–3).

2. Erosion is counterbalanced by deposition. On a local scale and on a short-term basis, eroded materials are repeatedly transported short distances and deposited again. On a long-term basis erosion will dominate in one area, deposition in another. Consequently, there are zones of net deposition corresponding to each zone of net erosion (Figure 5–4).

3. Erosion and deposition may affect an entire surface, or they may be concentrated or restricted. Three-dimensional surface *denudations* should therefore be distinguished from linear or channeled erosion, of which *dissection* by running water or moving ice is the most important (Figure 5–5).

In the following chapters it will be useful to keep the idea of weathering-erosion ratios in mind and to distinguish between short- and long-term erosion, as well as between surface denudation and linear erosion.

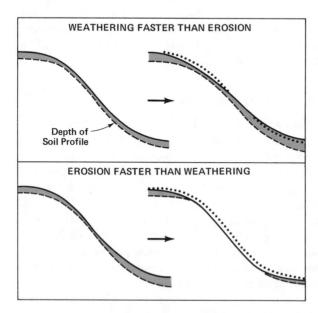

Figure 5–3. Slope Development Related to Rates of Weathering and Erosion.

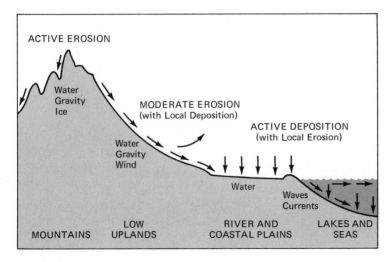

Figure 5–4. Erosion and Deposition in a River System.

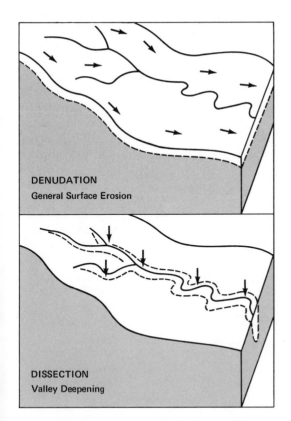

Figure 5–5. Distinct Types of Erosion Represented by Denudation and Dissection.

5-3. MASS MOVEMENTS ON STEEP SLOPES
OF HIGH RELIEF

Slope denudation by mass movements is effected by a great variety of transport mechanisms that range from a simple rockfall from a cliff face to slow creep of soil and regolith down a gentle slope (see Table 5–1). Methods of transfer include individual slides or falls, restricted to part of a slope, as well as slow or rapid flows of material that can affect an entire hillside. Some of these movements are rapid, spectacular, or even catastrophic; others are barely or not at all perceptible. The material transported may consist of fine soil, of regolith or unconsolidated sediment, or of compact bedrock. It may be dry, or it may be lubricated by water or perhaps affected by ice.

Whatever the mechanism of movement and the type of material involved, mass movements result from ruptures of slope stability. The instability so generated may be temporary or long term, and it may be restricted in occurrence or widespread. The factors that control slope stability include: (1) slope angle, (2) slope height, (3) the nature of soil, regolith, and bedrock, and (4) the nature and abundance of soil or rock moisture. These four major variables make it difficult to formulate general rules for slope stability or instability. Thus, for example, the laws that govern rotational slumping or planar block gliding on a hillside are quite different from those effecting slow, plastic deformation of lubricated soil or regolith. For each type of mass movement there is but one basic prerequisite: sufficient energy to overcome both mass inertia and friction.

In outlining the more common forms of mass movement it is practical to distinguish two broad categories, those involving compact

Table 5–1. A CLASSIFICATION OF MASS MOVEMENTS

TYPE OF MOVEMENT	TYPE OF MATERIAL	
	BEDROCK	SOIL OR REGOLITH
Falls	Rockfall	Soilfall
Slides { few units	Block slump (rotational) Block glide (planar)	Soil slump
Slides { many units	Rockslide	Debris slide
Slow flows	Talus creep Block streams	Soil creep Solifluction
Rapid flows	Dirty snow avalanche	Rapid earthflow Mudflow
Complex	Combinations of materials or type of movement	

(Based on Sharpe, 1960, and Highway Research Board, 1958)

rocks and those involving unconsolidated materials (see Table 5–1). In a very general way, mass movements in compact rocks are most characteristic of steep slopes with moderate to high relief and an incomplete mat of vegetation. Such conditions apply best in mountain lands, areas of highly irregular terrain, and in many desert and subpolar environments. Gravity deformations of soil or regolith are more typical of gentle to moderate slopes, preferably with a cover of vegetation. This would apply to most landscapes with limited or moderate topographic expression.

Mass movements in compact rock leave a number of distinct traces or landforms, some a result of slow changes through time, others a result of sudden catastrophic events. While conspicuous and often spectacular, these types of mass movement are of limited importance outside of mountain environments. The remainder of this section outlines these processes and the features they create.

Rockfalls, Talus, and Block Streams. Sheer cliffs or steep rocky hillsides are prone to falls of isolated stones or larger masses of rock, possibly involving many tons of material. Loosening and detaching of rock can result from mechanical or chemical weathering along fissures, bedding planes, or jointing systems. Apart from the simple rockfall below an overhang, mass inertia is overcome by frost burst, differential settling, earthquakes, and by water or wind. Whatever the initial trigger action, the resulting aggregates fall freely or tumble downslope. The impact of multiple collisions may dislodge other rock particles farther down or impart new energy to the stranded debris of older rockfalls. In the process rock fragments, splices, slabs, or blocks are reduced in size and ultimately come to rest at the foot of the cliff or along the lower mid-slope.

With time, successive rockfalls and a constant dribble of loose stones accumulate slope deposits collectively called *talus*. Consisting of angular rock waste, with little or no fine matrix, talus may form discontinuous veneers as well as extensive sheets or aprons along lower hillsides (see Figure 3–2). It may also form cones at selected points of concentration, where rock waste is funneled together. Talus surface slopes commonly lie between 26 and 36 degrees, with gradients varying according to the distance of free fall, the angularity and size of the rock particles, as well as the thickness of the talus accumulation.

The slow movement within a talus sheet or cone is known as *talus creep*. Individual rock particles will settle a little downslope with the energy provided by fresh rock impacts. A more general mechanism, however, is provided by simple diurnal temperature changes: The rock waste expands when heated and contracts as it cools again. On a slope this means that many of the rock particles shift a little downslope as they return to their original size and position. The effect is cumulative over a period of years, leading to a net migration of rock waste. When ice forms between the rock fragments, as water freezes, there is even greater pressure due to the expansion of water as ice forms, with contraction as the thaw sets in. Consequently, freeze-thaw changes provide another mechanism for talus creep. Downslope, surface gradients decrease to 25 degrees or less and the talus may be fixed by vegetation and soil mantle (Figure 5–6). At that point, talus creep is commonly replaced by other types of mass movement such as soil creep or solifluction.

Rock burst and other forms of rockfall are accelerated in subpolar or high mountain climates subject to intensive frost weathering. Other factors of bedrock and topography being equal, talus creep is accelerated in cold environments with frequent freeze-thaw alternations, particularly when there is abundant interstitial water or ice. Under these circumstances movement can be accelerated to a threshold where flows of rock fragments form tongues that extend downvalley. Typical rock diameter is about 6 inches (15 cm), but much larger blocks may be common or even dominant. These rock streams, rock glaciers, or block streams, as they are variously called, have an average surface

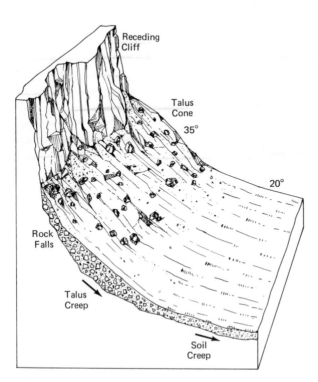

Figure 5–6. Rockfalls, Talus, and Creep. (Modified after Sharpe, 1960.)

slope of 9 to 18 degrees. Longitudinal or concentric ridges are often evident, parallel to the margins of these accumulations (see Figure 5–10 later in the chapter). Block streams move along existing valleys and may also form beneath outcrops of resistant bedrock. Occasionally, large spreads of similar blocks will mantle a hillside, forming a kind of block field.

Block Slumping and Gliding. On moderately steep slopes compact rock can be dislodged and transported by sliding mechanisms. The mass itself may consist of a single block or of several contiguous units. *Block slumping* (Figure 5–7) involves backward rotation and is distinguished from simple *block gliding* (Figure 5–7), with planar motion. In general, block movements of these two types proceed slowly and intermittently over long periods of time, although some may be completed in a single, rapid slip.

Block slumping and gliding represent a subsidiary process to talus accumulation and can provide a substitute mechanism on intermediate slopes unsuitable for rockfalls. They may also be preconditioned by slope undermining through stream or wave cutting. Sliding of all kinds is commonly facilitated by lubricated clays or regolith. It can be most frequently observed in cold or humid environments.

Rockslides and Debris Slides. The most destructive and rapid mass movements involve slumping or gliding of large masses of rock or other debris (see Figure 5–11 later in the chapter). Rockslides affect thousands of tons of material and normally follow the surfaces of joint, bedding, or fault planes that may be lubricated with wet clays. Rockslides are most typical of steep, possibly undercut cliffs, with deeply weathered rock or where resistant rocks are underlain by unstable shales or other weak materials. The trigger action can be provided by

ROTATIONAL SLUMP   PLANAR GLIDE

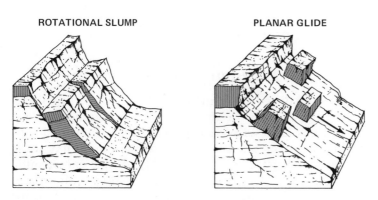

Figure 5–7. Slumping or Gliding of Undermined Blocks on a Steep Slope.

a protracted period of heavy rains or by earthquakes. In cold environments the spring thaw may have a similar effect.

Rockslides provide a potent geomorphic agent in mountainous terrain, and a single such slide may move more material in a matter of minutes than all other mass movements combined can transfer in a few centuries. Rockslides produce conspicuous, spoon-shaped scars on a hillslide, while the slide mass itself forms a hummocky surface on the foot-slope or beyond. Slopes prone to rockslides are marked by many concave scars with corresponding convex and irregular masses of debris below.

When bedrock is partly or largely unconsolidated or when large quantities of regolith are involved, the terms "debris slide," "avalanche," or "flow" are used, depending on the relative importance of dry flow or fluid deformation. As in the case of rockslides, the resulting deposits are intensively deformed, poorly bedded, and generally unsorted.

Catastrophic landslides, on the scale of several million tons of debris, involve complex movements as well as combinations of materials. In fact, the Vaiont Reservoir disaster, in northern Italy, was a combination rockslide and debris slide, one of the largest on record. About 300 million cubic yards (240 million cu m) of material slid into the reservoir on October 9, 1963. The dam was destroyed, and the surging waters killed 2,117 people in the valley below.

Dirty Snow Avalanches. Surprisingly enough, snow avalanches can also build up a peculiar variety of rock deposits. The term "dirty snow" is applied to snow that includes quantities of rock waste of all sizes. Snow avalanches in mountain country may pick up large masses of weathered rock and talus as they desend. This snow cascades down over existing drifts or patches of fairly "clean" snow. Later, as the dirty avalanche snow melts away, the rock waste appears and may form low, irregular ridges parallel to the cliff face but separated from it by a depression formerly filled with clean snow. These features are known as protalus ramparts or *nivation ridges* (Figure 5–8).

5-4. THE PERUVIAN CATASTROPHE OF MAY 1970

One of the worst natural disasters of the century struck in the Andes Mountains of north-central Peru on May 31, 1970. Triggered by a severe earthquake, billions of tons of debris were dislodged from unstable mountain slopes, setting off rockfalls, landslides, and dirty snow avalanches that killed at least 25,000 people. The major catastrophe occurred when a half-mile (800 m) ice overhang broke off Mount Huascarán, the second highest peak of the Americas. Falling free for 3,000 feet (900 m), the mass of ice and snow crashed onto a lower slope

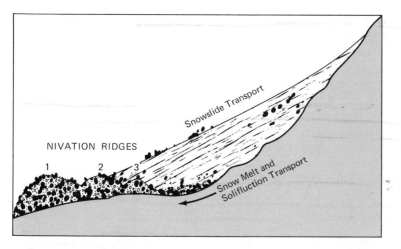

Figure 5–8. Successive Nivation Ridges (1, 2, 3) Laid down by Dirty Snow Avalanches. (Modified after Sharpe, 1960.)

at a height of 18,500 feet (5,500 m) and pulverized on impact (Clapperton and Hamilton, 1971). Frictional heat melted much of the ice and lubricated the avalanche which by now incorporated massive blocks of rock and loose debris. This highly fluid mass then slid from an elevation of 18,500 to 8,500 feet (5,500 to 2,500 m) over increasing gradients at an average speed of 300 miles (480 km) per hour. Charging over a 2.5-mile (4-km) front with a thickness of 150 to 500 feet (50 to 150 m), the debris flow completely wiped out the towns of Yungay and Ranrahirca. Along its way it ricocheted from side to side along the valley walls, leaving some vegetation in its path untouched—suggesting that the avalanche may have traveled over part of its course on a cushion of compressed air. In fact the town buildings may have been demolished by the turbulent air blast that preceded actual impact!

5-5. MASS MOVEMENTS ON INTERMEDIATE SLOPES OF LIMITED RELIEF

In areas of limited or moderate relief, slopes are commonly covered with a soil mantle or regolith. Consequently, on most intermediate and many steep slopes, mass movements involve soil or unconsolidated material. The resulting movements differ from those of rough country and compact bedrock, even though there are certain analogies. For one, they are much less obvious and usually require field experience to recognize. The mechanisms involved are also complicated and often controversial. And, above all, the sum total of these movements is both potent and effective in sculpturing hillslopes in most of the environments densely settled by man.

Soilfalls and Soil Slumping. A counterpart to rockfalls and talus is provided by the collapse of unconsolidated materials on near-vertical slopes. Such conditions may apply along the banks of streams and gullies, or in unconsolidated material along coastal cliffs that are undermined by wave action. Such *soilfalls*, together with soil slumping, are primarily responsible for the headward cutting of gullies and the lateral shift of stream meanders.

Soil slumping operates in analogy to block slumping but is confined to moderately steep slopes (20 to 34 degrees) with a grassy sod, a deep soil mantle or unconsolidated sediments, and a fair amount of soil moisture. Slumping can affect an entire hillside, favoring a slow, valleyward movement of the slope regolith. One of the most common expressions of soil slumping are steplike surfaces that follow the slope in linear or concentric arrangements. Individual steps may vary from 1 to 4 feet (30 to 120 cm) in width and 8 inches to 5 feet (20 to 150 cm) in height. They mark successive slump blocks that have slipped and, in the process, rotated backward so that the steps have a gentler dip than the general slope. Sod tears apart along the slip planes and subsequently serves to stabilize the block surfaces while accentuating the steplike effect. Such features can be best described as *slump scarplets*.

Slump scarplets are but one variety of the widespread phenomenon generally designated as *terracettes* and locally known as cattle terraces, cowtours, sheep tracks, or cat steps. Some terracettes can be attributed to overgrazing and repeated use of certain trails by pastured animals: Vegetation and soil are compressed and hooves cut into the sod, producing bare surfaces readily attacked by surface waters. Such tracks aid in the development of small slump planes. Some terracettes also reflect underlying, horizontal rock structures, while still others form as a result of solifluction (to be discussed shortly).

Rates of movement associated with soil slumping seldom are spectacular, although the attrition of pasture and soils can be destructive enough. In some instances, too, slump mechanisms are involved in the catastrophic landslips described as debris slides or earthflows.

Soil Creep. The most universal of all mass movements in soil and regolith on slopes of as little as 5 degrees is the surface phenomenon known as *soil creep*. Grain-by-grain downhill displacement of soil affects ground vegetation as well as man-made structures. Rates of movement are greatest near the surface and, depending in part on soil thickness, become ineffective at depths of anywhere between 2 inches and 3 feet (5 and 90 cm). In detail, much of the motion is produced by repeated movements of infinitesimal dimensions. Measurements on slopes suggest mean annual surface rates varying from 0.01 to 1.0 inches (0.025 to 2.50 cm).

Soil creep is effected through several different processes (Figure 5–9):

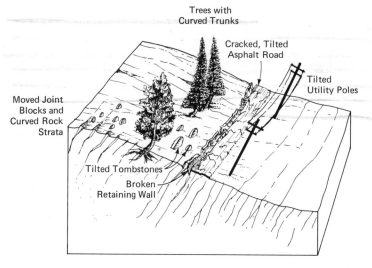

Figure 5–9. Common Effects of Soil Creep. (Modified after Sharpe, 1960.)

1. *Raindrop spatter:* On a bare slope, raindrops may produce a mudspatter that showers out tiny soil particles within a radius of up to a few inches. The muddy water flies farther downhill than uphill, due to the inclined surface, and there is net downslope transfer of minute dimensions.
2. *Moisture changes:* Alternate wetting and drying leads to repeated expansion of soil. Also, dehydration can form small cracks, with similar net downhill movements, on both opening and closing, or as they fill in with loose dirt.
3. *Frost heaving:* The freezing of soil water leads to soil expansion. As the soil thaws and contracts, gravity allows the soil particles to settle a little downslope. Rocks are also lifted by frost heaving or needle ice. Subsequent melting allows them to settle by gravity, with a slight shift downslope.
4. *Temperature changes:* As in the case of talus creep, temperature changes lead to minute expansion or contraction of soil particles. A net downhill component results, since greatest expansion is *toward* the downhill side and greatest retraction *from* the uphill side.
5. *Weathering:* As rocks disintegrate in the course of weathering, detached particles fall or roll downslope.
6. Last but not least, worms and other burrowing animals, as well as plant roots also favor net downslope displacement of disturbed soil particles.

Rates of soil creep are greatest in regions with frequent moisture changes or intensive frosts, and creep mechanisms grade over into solifluction in subpolar or mountain environments. Yet creep is still effective in humid tropical lands and even in arid climates. Annual measurements may be negligibly small, but the effects on tree trunks, walls, fences, or roads are often conspicuous (Figure 5–9).

Solifluction. Whereas soil creep involves particle-by-particle movement, that is, dry flow, *solifluction* is a matter of plastic flow by water-saturated soil or regolith. Unlike soil creep, solifluction frequently carries larger rocks in suspension while motion is not necessarily slowed down at depth. Instead, the entire regolith can ooze downhill. Rates of movement are greater than those of soil creep and may be measured in inches per year, and annual rates of 1 or 2 feet (30 to 60 cm) are not uncommon. Whereas creep is soldom found on slopes of less than 5 degrees, solifluction can be effective at gradients as low as 2 degrees.

Solifluction processes are most active in subpolar or high mountain environments, where the subsoil is seasonally or perennially frozen. Under these conditions water infiltration is slow while groundwater movement is limited, at least seasonally, to the topsoil. The resulting conditions of water supersaturation over an impermeable base are seldom matched in other environments. Two more factors aid solifluction in cold climates. Frost heaving and ice thrusting accelerate normal creep processes. At the same time, thawed ground is particularly erodible: Freezing reduces the cohesion of clayey soil, while the dilation of soil through ice expansion and frost heaving reduces compaction. However, solifluction also operates in warm, humid lands through saturation of very deep, clayey soils on slopes of 5 degrees or more.

Denudation by solifluction may affect entire hillsides. Characteristic deposits are usually restricted to lower slopes and most commonly are confined to partly stratified sheets that mantle or obscure existing topography. On some solifluction slopes with a complete vegetation mat, terracettes may result from the checking of a slow, subsurface flow of soil sludge. These terracettes may broadly follow the contours, or they may be marked by wavy protuberances, pointing downslope. When the vegetation mat is less continuous, linear ridges of grassy sod or mosses develop perpendicular to the contours on steep slopes. Bare soil is found between these vegetation ridges. On bare slopes the latter may be replaced by lines of stones (Figure 17–2). Such linear or geometric arrangements are called *patterned ground* and are peculiar to many polar and high mountain environments.

Despite the fundamental differences of flow mechanics between creep and solifluction, the two processes do intergrade and are often difficult to isolate. Similar transitions can be observed between solifluction and certain kinds of block streams, rapid earthflows ("solifluction lobes"), and mudflows. Solifluction overlaps with what are sometimes described as slow earthflows.

Rapid Earthflows. A rapid and localized form of solifluction is provided by certain types of *earthflows*. Within a matter of hours, large masses of water-saturated clayey soil or sediment will flow downslope. Concave scars mark the focus of erosion, while lobes of oozing soil sludge form shallow tongues downslope, commonly 1 to 6 feet (30 to 180 cm) deep and as much as 60 to 100 feet (20 to 30 m) across. Slump scarplets may appear on the upper part of these deposits, reflecting collapse and sliding of more compact materials at the head of the scar. Flow is always dominant lower down, where buckling or successive overthrusts of mud can form ridges concentric with the tongue of debris (Figure 5–10).

Rapid earthflows on a small or intermediate scale can be observed at all latitudes on vegetated slopes of as little as 2 degrees or as great as 35 degrees. In subpolar and high mountain regions they reflect local intensification of solifluction. In humid mid-latitudes they occur during the spring thaw or after long periods of abnormally wet weather and are then restricted to clayey, impermeable subsoils. In the humid tropics earthflows may even take place under closed forest or beneath the topsoil in areas of deep, plastic soil mantles.

Catastrophic earthflows, involving several million tons of material, require unusual bedrock conditions. The overall effects resemble those of landslides, although flows of supersaturated clayey material are the dominant mechanism; slope gradients may be unusually low (2 to 4 degrees).

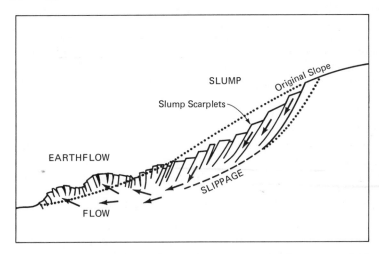

Figure 5–10. Rapid Earthflow Involving Both Flowage and Slumping. (Modified after Sharpe, 1960.)

Mudflows. The last major category of mass movements comprises *mudflows*. These are rapid flows with affinities to debris slides, rapid earthflows, and localized flood torrents. Mudflows result when masses of uncompacted silt are suddenly saturated with water. A liquid ooze, containing over 50 percent sediment, begins to flow downslope, following existing drainage lines with gradients of as little as 5 degrees. Ultimately, the viscous mass comes to rest at the foot of the slope, forming a lobate tongue or, if there are repeated surges of mud, a fanlike deposit, possibly ringed by low ridges. Although silt and clay-sized particles are dominant, mudflows can transport rock waste and even boulders.

In general, mudflows are confined to surfaces with little or no vegetation mat. For this reason they are most common in semiarid regions, in subarctic or high mountain country, and on volcanic slopes. In each case an abundant supply of suitable material is prerequisite. Silty residual products, wind-borne dust, fine-grained glacial debris, and volcanic ash are most common.

Mudflows play a role similar to that of rockslides, debris slides, and earthflows in producing rapid and often destructive slope changes. Velocities ranging from 0.5 to 14 feet (15 cm to 4.2 m) per second have been observed, and some mudflows have transported millions of tons of material. Such localized and sporadic changes probably result in more slope denudation than slow, cumulative forces working over long periods of time. However, none of these more spectacular movements is universal, and even where they do occur, they are restricted in area.

5-6. DISASTER IN LOS ANGELES HILLSIDE DEVELOPMENTS

In January 1969, following weeks of heavy rains, suburban developments on hillslopes around Los Angeles were struck by catastrophic mass movements. Hundreds of homes slipped downhill or were half-buried in masses of oozing mud. Property damage was severe, and there was some loss of life. Similar events have occurred before and since in southern California. In every case unstable slopes—both natural and man graded—were "developed," regardless of natural hazards such as steep gradients and inclined, uncohesive rocks liable to lubrication by water. The mass movements involved were slumping (Figure 5–11), solifluction, and mudflows, and the costly results show how effective such agencies can be.

In retrospect, most mass movements in soil and regolith are of primary importance in imparting particular forms to hillsides. To a very large extent, however, they work hand in hand with surface runoff. A fuller evaluation necessarily follows in the next chapter.

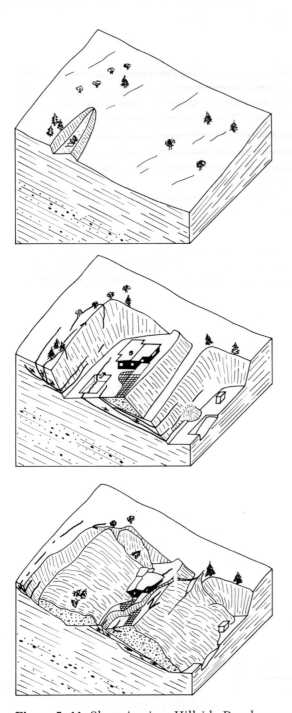

Figure 5–11. Slumping in a Hillside Development near Los Angeles. The clayey, inclined strata are prone to slumping, and in addition to rain, water is consistently added from cesspools, lawns, and, here, from a leaking swimming pool. (After R. H. Jahns, *Engineering and Science*, copyright 1958, California Institute of Technology.)

5-7. SELECTED REFERENCES

A. Terrain Representation, Aerial Photography, and Map Interpretation

Ahnert, Frank. An approach towards a descriptive classification of slopes. *Zeitschrift für Geomorphologie, Supplement*, Vol. 9, 1970, pp. 71–84.

Clark, J. I. Morphometry from maps. In G. H. Dury, ed. *Essays in Geomorphology*. London, Heinemann, 1966, pp. 235–274.

Hammond, E. H. Classes of land-surface form. *Map Supplement No. 4, Annals, Association of American Geographers*, Vol. 54, No. 1, 1964, with text pp. 11–18.

Murphy, R. E. Landforms of the world. *Map Supplement No. 9, Annals, Association of American Geographers*, Vol. 58, No. 1, 1968, with text pp. 198–200.

Ollier, C. D. Landform description without stage names. *Australian Geographical Studies*, Vol. 5, 1967, pp. 73–80.

Raisz, Erwin. *Principles of Cartography*. New York, McGraw-Hill, 1962, 315 pp.

Ray, R. G. Aerial photographs in geologic interpretation and mapping. *Professional Paper, U.S. Geological Survey*, No. 373, 1960, pp. 1–230.

Robinson, A. H. *Elements of Cartography*, 3rd ed. New York, Wiley, 1970, 375 pp. See Chap. 11.

Savigear, R. A. G. A technique of morphological mapping. *Annals, Association of American Geographers*, Vol. 55, 1965, pp. 514–538.

Shimer, J. A. *A Field Guide to Landforms in the United States*. New York, Macmillan, 1972, 272 pp.

Zakrzewska, Barbara. Trends and methods in landform geography. *Annals, Association of American Geographers*, Vol. 57, 1967, pp. 128–165.

B. Mass Movements

Bloom, A. L. *The Surface of the Earth*. Englewood Cliffs, N.J. Prentice-Hall, 1969, 152 pp. See Chap. 3.

Carson, M. A. *The Mechanics of Erosion*. London, Pion, 1971, 174 pp. See Chap. 4.

Clapperton, C. M., and Patrick Hamilton. Peru beneath its eternal threat. *Geographical Magazine*, Vol. 43, 1971, pp. 632–639.

Flawn, P. T. *Environmental Geology*. New York, Harper & Row, 1970, 313 pp. See Chap. 2.

Gray, D. H. Soil and the city. In T. R. Detwyler and M. G. Marcus, eds. *Urbanization and Environment*. Belmont, Ca., Duxbury, 1972, pp. 135–168.

Highway Research Board Special Report, *Landslides and Engineering Practice*. Washington, D.C., NAS-NRC, Publication No. 544, 1958, pp. 1–232.

Kirkby, M. J. Measurement and theory of soil creep. *Journal of Geology*, Vol. 75, 1967, pp. 359–378.

Schumm, S. A. Rates of surficial rock creep on hillslopes in western Colorado. *Science*, Vol. 155, 1967, pp. 560–561.

Sharpe, C. F. S. *Landslides and Related Phenomena: A Study of Mass Movements of Soil and Rock*, 2nd ed. Paterson, N.J., Pageant Books, 1960, 137 pp.

Chapter 6

HILLSLOPE SCULPTURE (II): BY SURFACE WATERS

Gravity movements and surface waters are the most effective agents in hillslope sculpture. Rainfall and melting snow provide water that, in part, is absorbed by the soil or added to the groundwater. This infiltrating water can perform chemical denudation, particularly in limestone, and ultimately may feed streams by seepage or spring flow. Surface runoff, much more important for slope evolution, takes the form of diffuse sheet wash and more concentrated rill wash. These two types of overland flow erode hillsides, particularly when there is no vegetation to intercept raindrops and neither ground cover nor humus to bind the soil, increase infiltration, and retard runoff. Overland flow can strip off the surface soil as well as cut into the soil profile by shallow rills or larger, master rills. Sediments removed from the crest- and mid-slopes are often laid down, temporarily, on the foot-slope before being carried into streams. Human interference by deforestation, overgrazing, cultivation, or construction greatly

*accelerates slope denudation and may possibly lead to
gully cutting adjacent to streams and back up onto the
slopes. The resulting "accelerated" soil erosion is not
only of economic significance but is a geomorphic
process now prevalent in most of the world's environments. In the end it affects the balance of weathering
and denudation. If weathering and soil development
match or temporarily outpace slope erosion, the hillside remains mantled in soil. Without cultivation or
other "disturbance," this applies in humid environments of mid-latitudes and the tropics. It favors erosion by soil creep and other mass movements on the
crest- and mid-slopes, with thickening of soil or sediment on the foot-slope. In this way the slope is
smoothed, and its gradient is reduced by downwearing.
However in arid or mountainous environments erosion usually outpaces weathering so that soil mantles
remain thin. Sheet and rill wash attack the mid- and
foot-slopes equally but remain less effective on the
slope crest. As a result, there is no accumulation of soil
or sediment on the foot-slope and mid-slopes wear
back evenly; slope inflections remain marked, and mid-
slope gradients do not decrease with time. This backwearing trend is currently favored by accelerated soil
erosion in all environments intensively used by man.*

6-1. RAINWATER INFILTRATION

The basic principles of hydrology were first expressed by Plato early in
the fourth century B.C.:

> The annual supply of rainfall was . . . received by the country, in all its
> abundance, into her bosom, where she stored it in her impervious potter's
> earth and so was able to discharge the drainage of the heights into
> the lowlands, in the form of springs and rivers with an abundant vol-
> ume . . . .[1]

Rainfall is initially absorbed by the soil. However, if the rain is either
intensive or persistent, the rate of *infiltration* will be too slow and sur-

---

[1] Plato, *Critias*, 111, E.

face *runoff* results. Ultimately, too, some of the rainwater will evaporate from the surface of the vegetation or from the ground. In effect,
precipitation may be disposed of by infiltration, runoff, or evaporation.

At first the water that infiltrates is absorbed as soil moisture. As
infiltration continues, free water gravitates down through the soil. If
there has been no rain for a long time, a great deal of water will be absorbed into the subsoil, and there may be insufficient free water to
pass on to the groundwater. If the subsoil is moist, a certain portion of
the rainwater is transmitted to the groundwater zone. From here the
water seeps slowly to adjacent valleys, emerging in the form of springs
or seeps that feed local streams (Figure 6–1).

The ability of the soil to absorb water is conditioned by several
factors:

1. *Vegetation Cover.* Woodland or tall grasses will slow down rain
impact by interception, so that water drips off the leaves or runs down
stems and tree trunks. A rooted sod is usually well aerated and rich in
humus, all of which soak up and hold considerable quantities of water.
By contrast, bare cultivated soil that has not been maintained or rock
surfaces have a low infiltration capacity.

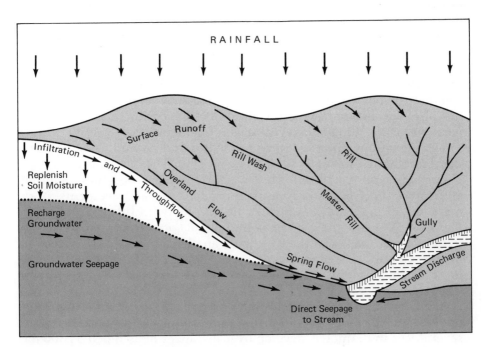

Figure 6–1. Infiltration, Surface Runoff, and Stream Flow.

2. *Soil Moisture.* Whereas a moist soil initially absorbs water rapidly, a dry, baked soil with little organic matter will repel water. However, moist soil is soon saturated with water, while dry soil can potentially absorb a great deal, although at a slower rate.

3. *Soil Texture and Structure.* Porous and permeable soils, with a high proportion of sand or gravel, absorb water far more rapidly than dense, clayey soils. The loose structure of sandy soils and the crumb structure of agricultural or woodland soils also favor greater absorption.

4. *Parent Material.* The nature of the subsurface plays an obvious role for infiltrating soil water. Unconsolidated sands or gravels are optimal, while clays and most forms of compact bedrock are unfavorable, depending on their permeability.

5. *Time.* Infiltration capacity in most soils rapidly decreases with time, even in moderately permeable material. Once the subsoil is saturated with water, subsequent percolation is limited to lateral seepage (throughflow) in the topsoil and slow, downward flow of free water to the groundwater zone.

Infiltration water plays a geomorphic role. Solubles at or below the surface are partly dissolved and either carried down into the soil or to the groundwater. These solubles are ultimately removed from the soil by lateral throughflow or from the groundwater reservoir to nearby streams. The solubles in question may be carbonates attacked directly by dilute carbonic acid, or they may be other minerals produced by hydrolysis or oxidation. In addition to leaching, eluviation of humus, oxides, colloidal silica, and clays will also transport small but cumulative amounts of material into the subsoil or groundwater. Consequently, infiltration waters not only permit and accelerate chemical weathering, but during the course of time they lead to a selective removal of subsoil materials in solution or suspension, effecting an imperceptible but nonetheless real form of denudation.

Under special circumstances soil water and groundwater may produce accelerated forms of subsurface erosion. The best known of these phenomena can be widely observed in soluble limestones. Solution and corrosion of exposed limestone produce jagged pavements, pockmarked with irregular cavities, or fluted rock with linear grooves (see Figure 3–5). Below the ground, solution and corrosion can widen joints to form deep fissures in the rock and can develop major cavities along bedding planes. Limestone caves, grottoes, and subterranean caverns form in this fashion. These *karst* features are described further in Chapter 15.

Underground erosion of a rapid type is peculiar to some soft shales or silts. Interlocking vertical or horizontal pipes or tunnels, from 1 to several feet in diameter, develop over distances of as much as 300 feet (90 m). The pipes are fed by miniature sinkholes that enlarge through mechanical abrasion and collapse of unconsolidated silt or clay, as rainwaters are funneled through. Known as *piping,* this phenomenon can often be observed on a limited scale along the margins of steep-walled stream banks; more severe forms develop within geological deposits, particularly in a semiarid climate. Piping is probably initiated by rodent burrows, by rainwater following tree roots underground, or by enlargement of dehydration cracks within silts rich in montmorillonitic clay. Another possible origin is differential permeability along the contact between different materials.

6-2. RUNOFF AND ERODIBILITY

When infiltration and throughflow are inadequate to cope with a continuing rainfall, water begins to accumulate in small surface depressions that ultimately overflow. A shallow skin of water moves over the surface, gradually gaining in volume and speed. Further down the slope this sheet wash begins to concentrate in minute channels known as *rills*. Both sheet wash and the more concentrated form of rill wash are included in what are best called surface or *overland flow*. This type of runoff, as distinguished from stream flow, is always a temporary and sporadic phenomenon, during and immediately after rainfall. *Stream flow*, on the other hand, is seasonal or permanent in humid environments because stream waters are only partly derived from surface runoff. Much of the time streams are maintained by throughflow, groundwater seepage, or spring flow into the stream channel.

Overland flow may take several forms (Horton, 1945; Strahler, 1952):

1. *Laminar Surface Flow.* A sheet of water, seldom more than a fraction of an inch deep, floods slowly over the crest-slope and any bare rock surfaces with gradients of less than 5 degrees and again on cliff faces of over 40 degrees. The capacity for mechanical erosion is limited, but ions, colloids, and clays are removed in solution or suspension. Pitting, fluting, and gooving of limestone are also possible. Laminar flow stops when the water thins to below the capillary control limit.

2. *Turbulent Overflow Flow.* Localized concentrations of water begin to develop on uneven surfaces, particularly as the runoff gains momentum on steeper mid-slope sections. Mechanical erosion now becomes possible.

3. *Surging Overland Flow.* Irregular surges of water pulsate through the rills on a hillside. The reasons for such surges can best be identified in the case of a plowed slope, where water is coursing down the furrows. Clods of soil cave in on the furrow as they are soaked, temporarily damming back water. When each mud dam is

breached, the gathered water surges out. Similarly, twigs, leaves, or crop debris can pile up in a furrow, creating small, temporary dams that are sooner or later washed out. Periodic surges of water can also be generated during intervals of especially heavy rain. Both types of surges are potent agents of erosion and transport, on mid-slopes in general and within rills in particular.

All in all, the intensity of overland flow reflects several factors:

1. Intensity and duration of precipitation
2. Rate of infiltration into the soil
3. Vegetation cover, soil mantle, and roughness of surface
4. Gradient and length of slope
5. Type of overland flow

The effectiveness of surface erosion in part depends on the intensity of overland flow. However, a second major factor is the erodibility of the surface under average conditions of slope and roughness. Resistance to erosion is largely determined by vegetation, the nature of the soil mantle, and the surface materials themselves.

It makes a great deal of difference whether rain falls on a bare plowed field or on a closed forest. In the case of a bare surface, raindrop impact sets off a chain of events leading to soil removal. A heavy drop hitting water-soaked soil raises a mudspatter, first helping to seal the surface and then mobilizing particles of dirt that are then frequently picked up by overland flow. Without a mat of vegetation, water courses unimpeded over the surface, free to pick up particles everywhere. A mat of grass, on the other hand, will slow down the water, which is forced to filter through a tangle of stems, grass blades, dead leaves, and root hairs. Meanwhile, infiltration is encouraged. There is some or even total interception of raindrops, while the rooting network serves to hold down the soil. The effectiveness of vegetation in protecting the soil varies: Forest with a grassy sod is optimal, and closed grassland with contiguous sod is next best. Sparse desert shrubs leave extensive areas of bare soil, as do many vegetable crops or widely spaced cotton plants. Most grains offer a fair amount of protection when present in mature stands, but all cropland is largely or entirely bare for all but a few months of the year. Consequently, cultivated land and deserts are most erodible on this count.

Recent experiments by Likens and others (1970) in a small watershed of New Hampshire show that the impacts of deforestation on runoff are even greater than expected. Summer discharge increased by 40 percent while almost 15 times as much dissolved inorganic material is now removed from the catchment. The latter includes a substantial proportion of soil nutrients. This confirms results from 219 rivers in Russia, reported on by Bochkov (1970). In the smaller Russian watersheds it was found that forest cover strongly decreases overland flow and runoff while increasing minimum stream flow during dry weather.

The topsoil also plays a role in cushioning overland flow. Quite apart from raindrop interception and the important holding function of the rooting network, the organic matter of the O2 and A horizons greatly increases infiltration capacity and permeability. This allows a substantial lateral movement of water through the soil, known as saturation *throughflow*. A further category of seepage, in the zone of aeration between the base of the soil and the groundwater table, can be called *interflow*. Together, throughflow and interflow constitute *quick* return flow, as opposed to *delayed* return flow from the groundwater. The total of such subsurface seepage takes care of most or all of the rainfall received in the case of well-vegetated slopes with low-density, organic soils.

In fact, the overland flow model developed here is not applicable to undisturbed woodland in mid-latitudes (Carson and Kirkby, 1972). In such environments the litter and fermentation horizons must be removed first, before overland flow and sheet wash can begin. The destruction of these crucial horizons by plowing will immediately enhance the erodibility of cultivated land (Figure 6–2). In practice, therefore, overland flow is the rule rather than the exception on the cultivated lands of humid mid-latitudes. It is also the norm in semiarid and subpolar environments with an incomplete mat of vegetation, as well

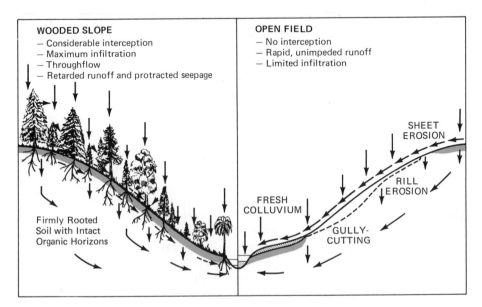

Figure 6–2. Runoff and Infiltration on Wooded and Cultivated Slopes.

as in the tropics where the majority of soils tend to be compact and clayey and insufficiently permeable to allow adequate throughflow. The true complexity of overland flow, quick return flow, and delayed return flow in providing variable sources for stream discharge can only be expressed in three dimensions: Depending on variables such as vegetation, soil type, bedrock, and inclination, their proportions vary along the length of the slope.

The substrate conditions both quick and delayed return flow, primarily in terms of the permeability of the soil horizons, the zone of aeration (if any), and the groundwater zone. The soil and the subsurface materials are critical in determining resistance to erosion. The presence of consolidated and intact bedrock in the subsoil sets an effective limit to accelerated erosion. This is not so in the case of unconsolidated parent materials or deeply rotted bedrock. Noncohesive soils, such as sands and sandy loams, are readily removed when exposed. Cohesive, clayey soils must, by contrast, be thoroughly soaked before they become erodible. From the point of view of texture, intermediate grade-size combinations of clay, silt, and sand are most resistant because they are highly compact, as well as cohesive and moderately permeable. Dense clays have a low infiltration capacity and therefore soak more readily at the surface. The presence of lime in the soil enhances cohesiveness, while increasing water content inevitably increases erodibility.

The sum total of the factors influencing erodibility is expressed in the universal soil-loss equation (UN Food and Agricultural Organization 1965):

$$E = R.K.L.S.C.P.$$

where

    E = average annual soil loss
    R = rainfall factor
    K = soil erodibility
    L = length-of-slope factor
    S = steepness-of-slope factor
    C = cropping-and-management factor
    P = conservation-practice factor

This equation incorporates a wealth of empirical observations made by the U.S. Department of Agriculture over many years. The formula can be easily applied for predictive purposes by use of appropriate tabular check lists. The factors themselves fall into two main categories, meteorologic influences (R, the capability of local rainfall to erode soil from an unprotected field) and physical properties of the land surface relevant to potential erosion. The universal soil-loss equation not only provides a good generalizing device but also offers opportunity to map areal variations in soil loss (see Figure 6–3).

6-3. RILLS AND GULLIES

Surface waters are sluiced away by drainage lines of various dimensions and permanence: rills, gullies, and streams. The smallest and least conspicuous of these is the *rill*, which can be a few feet wide and up to more than 1 or 2 feet (30 to 60 cm) deep. Well-defined banks and bed deposits are usually lacking, and except for plowed fields there is vegetation on the floor and banks of the channel. A *gully* is a channel of temporary drainage sufficiently well developed to show up on standard aerial photos. The banks are steep or vertical, frequently without vegetation, and characteristic water-laid deposits line the floor. Gullies commonly range from 3 to 50 feet (90 cm

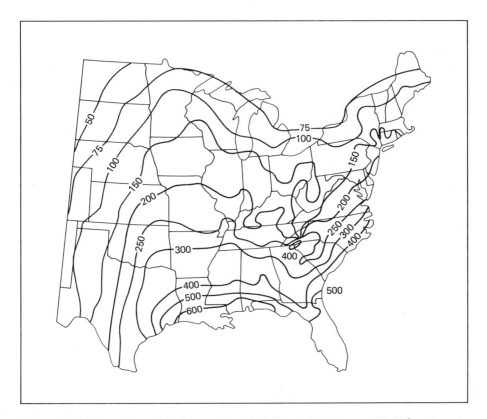

Figure 6–3. Mean Annual Values of Rainfall Erosivity Factor (R). (After American Society of Civil Engineers, 1970, with permission.)

to 15 m) in both width and depth. Finally, in humid environments a *stream* can be best distinguished from a gully by the nature of its water flow: The gully carries water only during and immediately after rains; the stream is still fed by seepage or springs long after rainfall has ceased. Consequently, a stream has perennial or seasonal flow in a humid climate, whereas a gully is ephemeral. In many ways, gullies are simply an extension of stream channels into deep alluvium, soil, or unconsolidated bedrock. In well-vegetated, humid areas they are symptomatic of increased storm flow, usually brought about by human activity.

Although it is sometimes difficult to distinguish among rills, gullies, and streams, particularly in arid lands, the distinctions are basically usable and important: Gullies and streams produce linear erosion or dissection, while rills form the link between dissection and denudation.

Observations made by Horton, a consulting engineer on practical problems of everyday erosion, showed that rills and gullies develop best on exposed slopes with gradients of 3 to 35 degrees. Here they may account for as much as 10 to 15 percent of the surface. Sheet wash also affects the same slopes, becoming an almost exclusive agent with either steeper or gentler inclinations. In other words, on a typical exposed slope, sheet wash dominates on the crest-slope, while rill wash characterizes the mid-slope (Figure 6–4). Further downhill, concentrated flow is restricted to a smaller number of master rills or gullies, while sheet wash is once more prominent over most of the foot-slope.

Rills inevitably develop as overland and throughflow converge in relation to surface irregularities. At first these rills are subparallel and run perpendicular to the contours (Figure 6–5). Gradually, however, they evolve into a dendritic system, converging on a master rill. This process is best illustrated by the rapid changes that can take place on a steep, cultivated slope that is not plowed for a few years. As the rills deepen, the shallow interfluves between them are lowered by raindrop erosion and sheet wash, while the muddy "banks" frequently cave in. Eventually, a smaller, higher-flowing rill breaches a low spot on the rain-soaked divide, and its water will be diverted into a deeper, larger rill. This is repeated until a complex hierarchy of rills develops. Finally, when two or more master rills converge, they may be large enough to work their way across the foot-slope to a stream. Possibly, too, they will produce a gully that gradually works its way back across the footslopes onto the hillside.

Since rills are a measure of erosional intensity, their size and spacing will reflect factors such as vegetation mat, soil mantle, gradient and length of slope, as well as intensity of precipitation. Smaller rills are not necessarily permanent features when surface vegetation is limited. In some areas frost action obliterates rills each winter. An-

Figure 6–4. Combined Rainsplash, Sheet Wash, and Rills Eroded This Culti-vated Field on Mid-Slope, Iatah County, Idaho. (Courtesy of USDA—Soil Conservation Service.)

Figure 6–5. Steep Rills, Master Rills, and Gullies Combine to Erode These Clay Badlands, near Cornelia, South Africa. (Karl W. Butzer)

other study on steep slopes showed that comparatively large rills form during high-intensity rains, only to be partly obscured by low-intensity rains and frost action. On cultivated land, contour plowing will obliterate all but the largest rills, whereas broad rills developed in grassy sod have great stability.

Gullies are almost irreversible. Gullying represents rapid headward erosion and deepening as channels work their way uphill into unconsolidated clays, silts, or sands. In many ways it is an accelerated form of rill cutting in a suitable parent material (Figure 6–6). Gully systems are usually dendritic, most commonly reaching out from a deeply incised stream. In other instances, however, independent gully systems develop on mid-slopes as a result of unchecked rill erosion.

Altogether overland flow produces little or no visible erosion on a crest-slope, but sheet wash and rill erosion are both effective over most of the mid-slope, which is an area of conspicuous erosion. Finally, there may be erosion *and* deposition on the foot-slope, since sediment is spread out in thin sheets of wash. This *colluvium* is transient, being repeatedly reworked and ultimately carried into the next stream (Figure 6–7). Colluvium is nonetheless a common feature where sheet and rill wash are active and in some environments can build up to a thickness of several feet. Only large rills or gullies cut right across the foot-slope and deposit their sediment directly into streams.

Figure 6–6. An Active Gully in Southeastern Oklahoma. (Karl W. Butzer)

Figure 6–7. The Colluvial Silt in the Foreground Was Deposited After a Single Rainstorm Stripped off the Cultivated Field Upslope, Posey County, Indiana. (Courtesy of USDA—Soil Conservation Service.)

6-4. BACKWEARING AND DOWNWEARING

On any slope a combination of mass movements, overland flow, and weathering is involved in hillslope development. This interplay of different forces is conditioned by climate, vegetation, lithology, structure, slope, relief, and human use of the land. Correspondingly, some particular set of processes will be dominant on a particular part of a hillslope. Despite these many variables, the fundamental criterion is whether or not the rate of erosion exceeds the rate of weathering on the mid-slope. Depending on the answer, two rather different trends of hillslope development will ensue.

When slopes are eroded as fast as or faster than soil and regolith can accumulate, mid-slope gradients remain constant with time. This assumes that the soil mantle on the mid-slope remains thin, and that there is no net accumulation of colluvium or other debris on the lower mid-slope or on the adjacent foot-slope. Such conditions are best met on steep slopes with inclinations of more than 35 degrees and where streams or other geomorphic forces are available to remove foot-slope accumulations or undercut the mid-slope. The effect is similar when the rock exposed on the crest-slope is more resistant than that below. Whatever the reasons, slopes remain steep and both the convex and concave slope inflections will probably be abrupt and angular. Under these circumstances mid-slopes retreat in parallel fashion. This is known as *backwearing* (Figure 6–8).

On the other hand, if the rate of weathering exceeds or keeps pace with the rate of erosion, deep soils develop. Mass movements such as creep or solifluction eventually become prominent on the convex slope, smoothing and reducing both angularity and slope. This is so because each creeping soil particle must move down the length of the slope behind the next downslope particle, leading to greater erosion near the top of the slope. Further downhill the soil profile inevitably thickens near the base of the mid-slope. This means that bedrock beneath the regolith of the concave slope is immune from erosion and, as a result, geomorphic processes will selectively wear down the convex slope. In the course of time mid-slope angles are reduced, slope

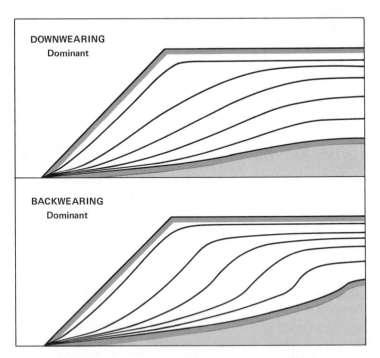

Figure 6–8. Computer-Simulated Models of Hillslope Development with Back-wearing or Downwearing Dominant. (Modified after Ahnert, 1971. With permission.)

inflections smoothed, and upland surfaces reduced in level. This is called *downwearing* (Figure 6–8).

Most leading geomorphologists today accept that slopes may retreat through backwearing, downwearing, or combinations of the two. In the past, however, individuals have championed a single possibility to the exclusion of all others. This led to the sharp controversies between the disciples of Davis and Walther Penck (see Chapter 1) that began during the 1920s and have only been muted during the past decade.

One of the most influential persons to help redirect this discussion has been Schumm (1966). He sought to avoid indirect arguments—based on almost imperceptible changes of hillslopes over the years—by direct observations of rapid, small-scale changes in soft materials. He studied spoil heaps of clayey sludge near Perth Amboy, New Jersey, and went on to study microslopes in the soft shales and siltstones of the Badlands National Monument, South Dakota. By careful measurements of vertical and horizontal changes over a period of years, Schumm was able to show that steep, bare hillslopes underwent parallel retreat without reduction of the mid-slope angle. Other slopes of low permeability were subject to sheet wash and also retreated in parallel planes. However, permeable hillslopes were prone to creep and, as a result of downwearing, showed a progressive reduction of slope length and inclination. These observations have greatly influenced modern geomorphological thinking, and most geomorphologists today favor an empirical approach to the question of backwearing versus downwearing.

6-5. COMPLEX HILLSLOPE DEVELOPMENT

Steep slopes with erosion in excess of weathering can be expected in several different settings:

1. Mountain slopes with lithosols
2. Desert hillslopes with little or no vegetation and soil
3. Semiarid hillslopes with effective denudation across the footslope
4. Cliffs or escarpments undermined by streams, waves, or ice or coincident with fault zones

Each of these can be discussed briefly.

Mountain slopes, such as those of the Rockies or Himalayas, whatever their origin, seldom accumulate deep soils or regolith. Rockfalls and talus creep are widespread, while block slumping, landslides, and the like, are commonplace. In addition to these mass movements, which often appear to be dominant on each slope segment, sheet wash may surge over convex slopes and flush through the talus and other accumulations on the foot-slope. Rill cutting may also be evident amid the talus cones at the base of the slope. In contrast to gentler slopes with soil, mechanical weathering by frost and pressure unloading will favor exfoliation, block wedging, and rock burst. The shape of the foot-slope varies according to its origin: stream incision, stream undercutting, or glacier activity.

In desert country, erosion may be slow, but soil development is slower still. As a result, mid-slopes are usually bare. First, in the case of sedimentaries, the more resistant rock types will form caprocks that protect underlying, weaker strata. Rapid recession of cliffs in weaker beds is checked by the caprock, so that the entire scarp retreats at about the same rate. The caprock breaks down as a result of weathering and undermining, whereas sheet wash, rill cutting, block gliding, and talus creep are more characteristic of the lower mid-slope. The details of scarp form depend on the nature of the caprock, the dip of the rock strata, spacing of bedding planes, and alternation of weaker and more resistant rocks (Figure 6–9). Fine materials on the slopes are washed away by sporadic rains or removed by the wind while rill erosion, sheet wash, and deflation manage to keep the foot-slope free of excessive accumulation. Second, in the case of massive igneous or metamorphic rocks, the crest-slope is less prominent, and differential

Figure 6–9. Backwearing in Horizontal Sandstones near Nile Valley, Nubia. Note rilled, alluvial veneers on rock-cut valley floor. (Karl W. Butzer)

Figure 6–10. Backwearing in Complex Igneous and Metamorphic Rocks, Red Sea Hills of Egypt. (Karl W. Butzer)

weathering will favor greater irregularity (Figure 6–10). Overall rock resistance and limited permeability lead to parallel slope retreat similar to that for sedimentary rocks.

Backwearing is equally prominent in semiarid or subhumid areas with a long dry season and fairly open vegetation. Of primary importance seems to be the presence of denudational agents (rill and sheet wash) that effectively halt accumulation in the foot-slope area. To an even greater degree this applies for steep cliffs maintained by direct undercutting or continued uplift.

Slopes with deep soils, where weathering equals or outpaces erosion, can generally be associated with two situations: (1) gentle and intermediate hillslopes in humid tropical environments where soil development is rapid and intensive and (2) gentle slopes in humid mid-latitudes where soil development is of moderate intensity.

In both cases, given a good cover of vegetation, rill and sheet wash are unimportant, but mass movements play a major role instead. Gravity movements that affect soil and regolith, rather than bedrock, are dominant. Many of these are slow but almost universal in their activity on a hillslope and, therefore, all the more efficient. Soil creep is probably the major agent at work on crest- and foot-slopes, with the possibility of rapid movements such as earthflows, soil slumping, or debris slides at mid-slope. When the rate of weathering exceeds that of

erosion, the increasing depth of soil and regolith leads to an acceleration of mass movements until a new balance is ultimately established.
With so much soil and regolith, any overland flow may actually diminish on long slopes of less than 15 degrees, through protracted infiltration. Consequently, unless the foot-slope is kept clear of sediment
by streams, all but the steepest hillslopes will be prone to some degree
or other of downwearing.

Accelerated soil erosion in the wake of cultivation over the past
few centuries or millennia has increasingly restricted those humid
environments where soil development matches slope denudation. The
stripping of established soils from crest- and mid-slopes by sheet wash,
rilling, and gully cutting tends to remove the prerequisites for downwearing. The temporary increase of colluvium on foot-slopes retards
backwearing, but any new or future balance of weathering and erosion
implies a thinner and possibly incomplete soil mantle on the upper
slope. As a result, wash will be favored over soil creep. Therefore, it is
reasonable to suppose that there will be shifts from downwearing to
backwearing dominant. In this way man exerts a greater impact on
geomorphology than the great climatic and environmental changes of
the recent geological past.

6-6. MAN AND ACCELERATED SOIL EROSION

The preceding sections discussed the major processes of hillslope
development. It remains to consider the role of that last-but-not-least
factor, man. *Man is one of the most important geomorphic agents.* Wherever man is present in sufficient numbers, he affects or upsets the natural balance of forces through deforestation, cultivation, animal
grazing, and other direct or indirect activities. Today croplands, pasture, and urban agglomerations dominate all but the Earth's deserts
and arctic wastes. Everywhere there is evidence of interference with
the natural equilibrium, so much so that in some environments it is
difficult or impossible for the geomorphologist to reconstruct the natural balance of forces.

The most conspicuous and widespread geomorphic impact of man
has been accelerated soil erosion, that is, the destruction of one of the
most vital elements to human existence on the planet. The concept of
soil erosion was first outlined by Plato in the fourth century B.C., and it
is difficult to improve on his description of the denudation of the hill
country around Athens:

> In consequence of the successive violent deluges . . . there has been a
> constant movement of soil away from high elevations; and, owing to the
> shelving relief of the coast, this soil, instead of laying down alluvium as it
> does elsewhere, has been perpetually deposited in the deep sea round the
> periphery of the country or, in other words, lost. . . . All the rich, soft soil

has moulted away, leaving a country of skin and bones [so that rainfall] is allowed to flow over the denuded surface [directly] into the sea. . . .[2]

Erosion of soils is an inevitable part of geomorphic activity, but when an established soil is removed faster than it forms, there is something unusual—a disturbance of the normal equilibrium between weathering and soil formation, erosion and deposition. As a result of topography or climate, lithosols or regosols may be a permanent feature in steep mountain country, the polar world, or in hyperarid deserts. However, when an existing soil of an "intermediate" environment begins to show net erosion, there are two possible explanations: a change of climate affecting the vegetation cover or interference by man through cultivation or construction activity. Practically all *accelerated soil erosion* today is a result of man's interference, even though the rate and the extent of erosion vary from place to place, depending on soil type, climate, the system of agriculture employed, or the intensity of urban or suburban disturbance.

The basic causes of man-induced soil erosion all stem from the removal of vegetation and the misuse of the land, which in turn lead to accelerated runoff and increased erodibility (see Figure 6–2):

1. Deforestation or removal of grassy vegetation is caused by lumbering, plowing, grazing, burning, or construction. Trees, shrubs, or grasses intercept rainfall, and their removal means direct raindrop impact on the soil as well as accelerated soil creep. Simultaneously, rill cutting is greatly facilitated by bare ground surfaces, while accelerated runoff from built-up areas with limited infiltration capacity favors stream entrenchment.

2. Plowing and severe overgrazing destroy the litter or fermentation horizons that constitute much of the organic mat. In both cases compaction results, and this further reduces infiltration capacity, increasing the volume and velocity of surface runoff, and exposes bare soil to alternating rain and drought. The last result in turn accelerates oxidation of organic matter and reduces the variety and number of soil microorganisms that generate beneficial humus. As a result, soil structure is modified, with heavy soils taking on forms that are less permeable and aerated and, therefore, more erodible and less fertile; light soils, on the other hand, lose their aggregation properties and become incohesive. Depletion of organic matter through soil overuse has similar effects on structure. At this stage of disturbance the probability and destructiveness of soil slumping, debris slides, earthflows, or deflation are greatly increased.

3. Bad cultivation practices are another factor in man-induced soil erosion. Plowing across the contours ("uphill and downdale") provides

---

[2] Plato, *Critias*, 111, D–E.

countless ready-made channels for water to rush through after each rainstorm. This allows appreciable sheet erosion on slopes of as little as 2 degrees and rill erosion on slopes of 5 degrees. Planting of open-row crops (e.g., corn, cotton, or tobacco as opposed to wheat or alfalfa) provides bare soil for ready attack by splash erosion and moving waters not only during the sowing season and after the harvest but at the height of the summer, when thundershowers may be particularly violent. Removal of natural vegetation along stream banks and head-waters will invite bank and headwater erosion. Cultivation of slopes steeper than 8 degrees is an invitation to disaster unless fields are carefully strip-cropped (alternating field strips with crops and hay) or artificial terraces are constructed as, for example, in parts of the Mediterranean world or in the wine regions of western Europe. Finally, overuse of the soil also leads to structural deterioration and increased soil erodibility.

The individual processes of accelerated erosion are broadly similar to those normally operating on gentle or intermediate slopes through a wide range of environments. But there are significant differences, since overland and rill flow become both viable and important. Also, gentler slopes are affected, and a greater variety of processes can be seen at work in any one area. Apart from the chemical erosion already referred to, the mechanical processes include the following:

1. *Sheet wash* is the gradual removal of topsoil over wide areas by raindrop spatter and overland flow during successive rainstorms. Sheet wash becomes conspicuous when light-colored B or C horizons are exposed on convexities and dark accumulations of A-horizon topsoil in concavities. Removal of the porous and permeable A horizon exposes the more compact and heavy subsoil, which greatly reduces infiltration and throughflow in favor of surface runoff.

2. *Rill wash* is the rapid removal of topsoil along plow furrows or natural lines of concentrated drainage. The effectiveness of rill wash is commonly obscured by plowing that erases temporary rills but that cannot restore lost topsoil.

3. *Gullying* is the rapid and catastrophic erosion of all soil horizons by deep channels that eat back from permanent drainage lines. Gullies deepen, widen, and cut headward after each rainstorm, mainly in unconsolidated parent materials. Headward erosion in silts may be aided by subsurface piping.

4. *Mass movements* include soil creep, soil slumping, earthflows, and debris slides. Such processes accelerate and intensify the impact of running water on slopes of over 5 degrees and can lead to catastrophic destruction of hillsides with clayey substrates. Bank slumping and collapse as well as soilfalls also aid in the growth of gullies.

5. *Deflation* or wind erosion occurs primarily with dry, incohesive soils. Although most common and effective in semiarid and coastal areas, deflation can attack exposed soil in humid lands during periods of drought.

The significance of human interference in geomorphic balances has been dramatically shown by Wolman (1967). Studying the Piedmont region of Maryland, he reconstructed the amount of sediment removed annually per unit surface since the beginning of European settlement. Initially, while all or most of the land was in forest, this sediment yield was probably under 25 tons per square mile (9 tons/sq km) (Figure 6–11). More intensive cropping since the 1820s brought the sediment yield to a first maximum of 600 tons per square mile (about 200 tons/sq km) about 1900. The 25-fold increase of denudation illustrates the impact of agricultural land use on a mid-latitude woodland. Sediment yield declined to 300 tons per square mile (about 100 tons/sq km) since the early 1930s as more and more farmland was put into grass or allowed to return to brush and forest while awaiting development on the fringes of the growing urban centers. The impact of urbanization was calamitous. Clearing, excavating, and grading for construction increased sediment yields to several thousand tons per square mile during a short interval of one to three years. After an entire drainage area is "developed," a new urban and surburban landscape of streets, rooftops, garden plots, gutters, and sewers increases runoff but cuts down denudation rates to values lower than those of the farming era. Clearly, cultiva-

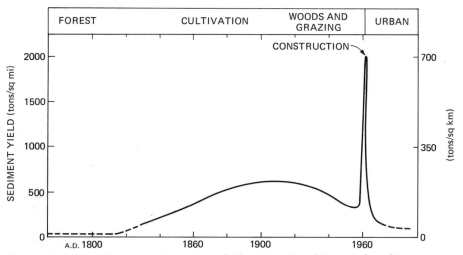

Figure 6–11. A Schematic Sequence of Changing Land Use and Sediment Yield in the Maryland Piedmont Region. (Modified after Wolman, 1967.)

tion and urbanization are two geomorphic factors that introduce radically different balances to the environment. It is in fact doubtful that any natural change or repeated catastrophes could exert a comparable effect over wide areas in less than a few millennia.

Can accelerated soil erosion be checked? The current difficulties in enforcing antipollution laws and passing even diluted legislation for conservation illustrate how the control of soil erosion is mainly a matter of politics and capital investment. From the viewpoint of practical measures, several options are possible:

1. On severely eroded surfaces (bad gullying and 75 percent of topsoil removed) reforestation or permanent grass is often recommended.
2. In areas where gullying has begun or is underway, construction of check dams or obstructions, favoring gully stabilization, are essential.
3. Where topsoil erosion is significant, contour plowing, strip cropping, and avoidance of open-line crops are recommended.

Contour plowing and strip cropping also are a "must" where gully erosion is active. Terracing may be necessary on steeper slopes, if used for crop cultivation rather than for woodland or hay.

Needless to say, the time to initiate erosion-preventive measures is before accelerated erosion begins. At present there seems to be little that can be done during the urbanization process, but in the case of cultivation the pressures are too great to ignore. Contour plowing should always be employed if slopes exceed 2 degrees, and on almost level land checkerboard plowing (each adjacent plot with furrows at right angles) is often recommended. Plowing as well as construction should stay away from the edge of steep bluffs, leaving areas of potential gullying or mass movements along river breaks in permanent grass or woodland. Alternation between open- and closed-grown crops, and even strip cropping, should be practiced on all slopes of 3 to 5 degrees or more. Woodland shelter belts or windbreaks may be necessary in areas with potential wind erosion, a practice pioneered in southern Russia.

Every form of cultivation or grazing activity in almost every environment produces some accelerated soil erosion. But with careful cultivation and pasturing of limited numbers of suitable livestock, soil erosion can be kept to a minimum and will not affect the soil resource over extended periods of time. Poor land management, on the other hand, can have catastrophic results within a very short time. So, for example, almost 30 percent of the land surface of Oklahoma was ruined in less than 30 years, and for the United States as a whole the agricultural resource base may have been cut by perhaps one-half during the past 150 years as a result of misuse. Seldom has so much been destroyed so fast and so efficiently. The tragedy is that this was done in callous disregard of the land, under no stress of population pressure.

Fortunately a highly successful soil conservation movement, begun in the 1920s, brought a turnabout in the waste of soil resources, and, during the past 40 years, steady improvement has been made. This was followed in the 1960s by a new ecological concern in urban society—for the conservation of wildlife, woodland, clean air, water, and soil. Whether or not cautious optimism proves to be justified, accelerated soil erosion has become and will remain a "natural" process in most world environments utilized by man. The effects on hillslope development are inescapable because everywhere the balance of weathering and erosion has shifted and even where new equilibria of soil development and denudation are established, soil mantles are invariably thinner. The dominant processes of slope development are modified accordingly.

6-7. SELECTED REFERENCES

A. Surface Runoff

Agricultural Research Service. *Present and Prospective Technology for Predicting Sediment Yields and Sources.* U.S. Dept. of Agriculture, ARS-S-40, 1975, 285 pp.

Carson, M. A., and M. J. Kirkby. *Hillslope Form and Process.* London and New York, Cambridge University Press, 1972, 476 pp.

Chorley, R. J., ed. *Water, Earth and Man.* London, Methuen, and New York, Barnes & Noble, 1969. See Chaps. 5.I–5.III (pp. 215–255) by M. J. Kirkby and M. A. Morgan.

Emmett, W. W. The hydraulics of overland flow on hillslopes. *U.S. Geological Survey Professional Paper 662A,* 1970, pp. 1–68.

Horton, R. E. Erosional development of streams and their drainage basins: hydrophysical approach to quantitative morphology. *Bulletin, Geological Society of America,* Vol. 56, 1945, pp. 275–370.

Parker, G. G. Piping, a geomorphic agent in landform development of the dry lands. *International Association of Scientific Hydrology,* Report 65, 1964, pp. 103–113.

Schumm, S. A. Evolution of drainage systems and slopes in Badlands at Perth Amboy, New Jersey, *Bulletin, Geological Society of America,* Vol. 67, 1956, pp. 597–646.

———. Erosion on miniature pediments in Badlands National Monument, South Dakota, Vol. 73, 1962, p. 719–724.

Strahler, A. N. Dynamic basis of geomorphology. *Bulletin, Geological Society of America,* Vol. 63, 1952, pp. 923–938.

B. Complex Hillslope Development

Ahnert, Frank. A general and comprehensive theoretical model of slope profile development. *University of Maryland Occasional Papers in Geography,* Vol.

1, 1971, pp. 1–95. See also further refinements of his computer program in *Geocom Programs*, No. 8, 1973, pp. 99–122.

Birot, Pierre, and Paul Macar, eds. International contributions to the morphology of slopes. *Zeitschrift für Geomorphologie*, Supplement, Vol. 1, 1960, pp. 1–240.

Birot, Pierre, Paul Macar, and Hans Mortensen, eds. *International Advancement in Research on Slope Morphology. Zeitschrift für Geomorphologie*, Supplement, Vol. 5, 1964, pp. 1–238.

Bunting, B. T. The role of seepage moisture in soil formation, slope development and stream initiation. *American Journal of Science*, Vol. 259, 1961, pp. 503–518.

———. *The Geography of Soil.* London, Hutchinson, and Chicago, Aldine, 1965, 213 pp. See Chap. 6.

Mortensen, Hans, ed. *New International Contributions to the Morphology of Slopes.* Nachrichten, Akademie der Wissenschaften in Göttingen. Math.-Phys. Klasse 1963, pp. 1–293.

Schumm, S. A. The development and evolution of hillslopes. *Journal of Geological Education*, Vol. 14, 1966, pp. 98–104.

———, and M. P. Mosley, eds. *Slope Morphology.* Stroudsburg, Pa., Dowden, Hutchinson and Ross, 1973, 468 pp.

C. Accelerated Soil Erosion

ASCE (American Society of Civil Engineers). Sediment sources and sediment yields. Proceedings, ASCE, Journal of Hydrology Division HY6, 1970, pp. 1283–1329.

Bochkov, A. P. Forest influence on river flows. *Nature and Resources* (UNESCO), Vol. 6, 1970, pp. 10–11.

Butzer, K. W. Accelerated soil erosion: a problem of man-land relationships. In Ian Manners and M. W. Mikesell, eds., *Perspectives on Environment.* Washington, D.C., Association of American Geographers, 1974, pp. 57–78.

FAO (UN Food and Agricultural Organization). Soil erosion by water: some measures for its control on cultivated land. FAO Developmental Paper 21, 1965.

Likens, G. E., et al. Effects of forest cutting and herbicide treatment on nutrient budgets in the Hubbard Brook watershed-ecosystem. *Ecological Monographs*, Vol. 40, 1970, pp. 23–47.

Meade, R. H., and S. W. Trimble. Changes in sediment loads in rivers of the Atlantic drainage of the United States since 1900. International Association of Scientific Hydrology, Publication 113, 1974, pp. 99–104.

Mrowka, J. P. Man's impact on stream regimen and quality. In Ian Manners and M. W. Mikesell, eds., *Perspectives on Environment.* Washington, D.C., Association of American Geographers, 1974, pp. 79–104.

Stallings, J. H., *Soil Conservation.* Englewood Cliffs, N.J., Prentice-Hall, 1957, 575 pp.

Wolman, M. G. A cycle of sedimentation and erosion in urban river channels. *Geografiska Annaler*, Vol. 49A, 1967, pp. 385–395.

Chapter 7

STREAM VALLEYS AND RIVER PLAINS (I): DYNAMICS

7-1. About stream flow

7-2. Stream velocity and turbulence

7-3. How streams transport sediment

7-4. Scour and fill in the stream channel

7-5. Mechanisms of stream erosion in alluvium

7-6. Mechanisms of stream erosion in bedrock

7-7. Depositional patterns on stream beds

7-8. Mechanisms of stream alluviation

Streams are potent geomorphic agents. Yet stream flow is variable during the course of the year, with moderate or high discharge reflecting on surface runoff after rainfalls or snow melts. The base flow provided in humid lands by groundwater seepage and springs is relatively ineffective in modeling a stream channel. However, a number of minor and major "flood" surges during a few months of the year are able to erode, transport, and deposit sediment, leading to repeated, short-term episodes of scour and fill within the channel. Sediment is quickly removed when a stream channel is deepened, widened, or extended in soft materials such as older stream beds, soil, or weak rocks. However, downcutting into compact rocks is a slow process that results from chemical weathering and from mechanical wear by sand and gravel. The sediment load itself is transported in solution, in suspension as a result of gentle agitation, in saltation or skipping along the channel floor, or by traction, that is, pushing or rolling along the immediate bed. Since ability to transfer a load depends on flow velocity and turbulence in the stream, sediment patterns vary from place to place and from time to time along the channel floor. Primarily since convergence increases water volume downstream, there is an increase of velocity and a greater capacity to carry large quantities of sediment. In addition, reduced turbulence downstream favors deposition of heavier materials that were eroded upstream on the interfluves. These basic mechanisms of stream behavior provide the dynamic background to analysis and explanation of both the depositional landforms of river plains (Chapter 8) and the erosional topography of the interfluves (Chapter 9).

7-1. ABOUT STREAM FLOW

To the ancient Egyptians, the Nile was a mysterious source of life identified with a male-female god intimately bound up with death and life, want and plenty. To the more practical Greeks, the Nile was a unique example of a river and its behavior. This was lucidly explained by the geographer-historian Diodorus of Sicily, writing about 50 B.C.:

> Beginning to rise at the summer solstice, the Nile increases in volume until the autumnal equinox . . .[1] day by day until it overflows practically all Egypt. Similarly it afterwards follows precisely the opposite course and for an equal length of time gradually falls each day, until it has returned to its former level. And since the land is a flat plain, while the cities, villages and farm houses lie on artificial mounds, the scene comes to resemble the Cyclades Islands. . . .[2] Every year continuous rains fall in the mountains of Ethiopia from the summer solstice to the autumnal equinox, so that it is quite reasonable that the Nile discharge should decrease in winter when it derives its natural supply of water solely from springs, but should increase its volume in the summer on account of the runoff which pours into it.[3]

These lines not only describe and explain the seasonal patterns of the Nile, but they also hold the key to several principles of hydrology.

Streams do not flow evenly during the course of the year. Instead,

[1] Book I, 36:2 *Diodorus of Sicily*. London, W. Heinemann, 1960. Translated by C. H. Oldfather.
[2] Book I, 36:7–8.
[3] Book I, 41:4–5.

discharge fluctuates between low and high water with great differences of level, volume, and velocity. The annual curve shows one or more well-defined maxima, commonly corresponding to periods of major rainfall and runoff. Where snowfall is heavy, thaws may also bring flood levels. However, within the probable pattern of peaks and lows defined by the discharge curve, the actual volume of stream water fluctuates rapidly in response to successive surges of high water (Figure 7–1). During periods of low water, when streams are fed by springs and seepage and so reduced to *base flow*, only a small fraction of the channel is occupied by water. By the time a flood is about to spill over the stream banks, discharge may be in the order of ten to several hundred times greater than the base flow. Even a great river such as the Nile carries almost 15 times as much water at the peak of the flood season than it does at low water. But the rise in level, when this water is spread thinly over a broad valley, was only about 25 to 28 feet (7.5 to 8.5 m) prior to inauguration of the High Dam at Aswan.

A major increase in volume produces a corresponding change in velocity, and careful on-the-spot studies by hydrologists of the U.S. Geological Survey have verified that geomorphic activity is greatly accelerated at such times. In fact, the morphology of stream channels—width, form, and gradient—is determined at times of moderate to high discharge. Relatively little change takes place during periods of base flow. When a stream is at the brink of overflowing its banks, discharge is *bankfull;* if the flood spills over, there is *overbank*

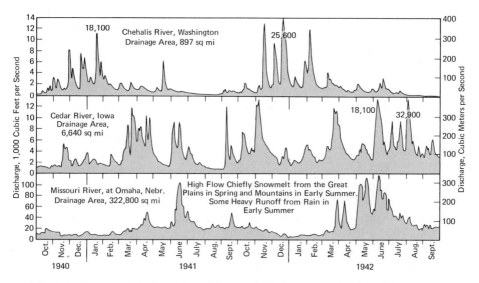

Figure 7–1. Daily Discharge of the Missouri, Cedar, and Chehalis Rivers. After Foster (1948). (Copyright 1948, The Macmillan Co., New York. Courtesy of Rose A. Foster.)

flow. A minor surge may produce one-half or three-fourths bankfull. On the average, the bankfull stage occurs 1.5 times each year in humid mid-latitudes, with the typical range of variation lying between 1 and 4 times. Once every 10 years or so a flood of exceptional severity will not only produce overbank discharge but will flood to the farthest perimeter of the river valley. Such catastrophic floods may produce visible changes and occasionally lead to severe destruction, such as the flood that killed over 125 people near Rapid City, South Dakota, in June 1972. But they are generally too infrequent to play a major role in modeling the stream channel.

Calculations by Leopold and others (1964) have shown that even bankfull floods do not occur often enough to do the bulk of sediment transport. Instead, it is the cumulative effect of frequent and more sustained flow at one-half or three-fourths bankfull that is responsible for most of the changes that take place. Altogether the duration of minor or major "flood" surges account for only a few months of the year, often for only a few weeks.

High water develops rapidly in small streams after any protracted, heavy rain. In small drainage basins, a day or more of heavy rain will lead, within 12 to 36 hours, to bankfull that may last from 1 to 3 days (Figure 7–2). In larger basins, measuring several hundred square miles (a few 1,000 sq km), the time lag will be greater and the flood stage more attenuated. Bankfull may be reached 2 to 4 days later on the largest stream of the drainage basin, since water must first build up and fill all the channels (*channel storage*). Correspondingly, high water is sustained over 2 to 5 days, but the increase of channel length and surface area smooth out the flood peak. In the case of very large drainage basins, measured in thousands or tens of thousands of square miles (several 10,000 sq km), the flood crest will pass through the largest stream 3 to 7 days later, with the flood levels persisting a week or more.

As Diodorus realized, long after surface runoff stops, streams persist as a result of groundwater seepage or direct spring influx. This base flow shows a smooth annual curve reflecting rainfall seasonality and rates of evapotranspiration. In higher latitudes groundwater is severely curtailed during the warmest months of the year, and base flow is at a minimum by the end of the summer. In low-latitude humid lands, base flow is more strongly influenced by the rainy season or seasons, while in arid lands the water table is low so that many streams flow only during and immediately after periods of rain. As a result of these climatic patterns of base flow and surface runoff, it is possible to distinguish *perennial, intermittent,* and *ephemeral* types. Most of the streams of Diodorus' native Sicily were intermittent, flowing more or less continuously during the rainy autumn, winter, and spring months but drying out in summer. It was in the deserts of

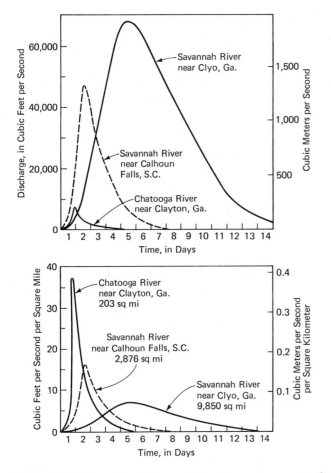

Figure 7–2. As a Flood Wave Moves Downstream, Both Discharge and Flood Duration Increase. This example from the Savannah River and one of its tributaries illustrates the effect of channel storage when shifting from small to larger and larger drainage basins. (Redrawn after Fig. 8, p. 39, in William G. Hoyt and Walter B. Langbein, *Floods*, copyright 1955 by Princeton University Press. Reprinted by permission of Princeton University Press.)

Egypt that Diodorus first saw the dry beds of ephemeral streams that carry water only during brief periods after irregular rains, often at intervals of many years.

Intermittent and ephemeral streams are commonly found in semi-arid and arid environments, some of which have *internal drainage*—that is, the surface waters are insufficient to reach the ocean and instead terminate in closed water bodies, such as the Caspian and Dead seas, or in the temporary lakes and shallow depressions that are common in deserts. Many of the largest streams that flow through desert environments derive most of their water from moister, distant highlands. Further downstream there is little or no influx from tributaries, and water volume shrinks through evaporation and seepage. Next to the Nile itself, other good examples of such *exotic rivers* are provided by the Tigris and Euphrates or the Colorado River. In each case the discharge is primarily derived from heavy rainfall or melting snows near the headwaters of the river.

7-2. STREAM VELOCITY AND TURBULENCE

Many recent advances in the interpretation of stream behavior have been a result of intensive gauging of stream flow, particularly since the 1940s. Current meters are lowered from a bridge or special cable car (Figure 7–3), and velocities are measured from all parts of the stream cross-section. Such readings can be evaluated in terms of mean velocities or discharge. In a general way the readings show that velocities are greatest near the center of a stream, just beneath the surface. Near the banks and along the bed, *current velocity* approaches zero (Figures 7–3 and 7–4).

The reasons for this distribution of velocity along any stream cross-section are complex but important. A great deal of the stream's

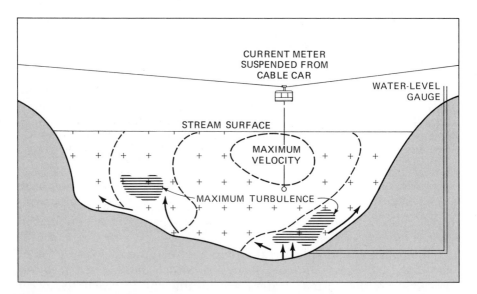

Figure 7–3. Stream Gauging. The current meter measures velocity at specific points. Since the channel cross-section is not quite symmetrical, maximum current velocity is found near the surface over the deepest water. Pockets of turbulence are found near the bed.

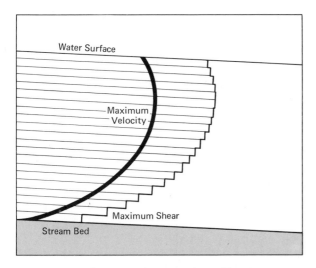

Figure 7–4. Theoretical Vertical Profile of Current Velocity. The layers of water shear one over the other, with maximum shear near the bed. (Modified after Morisawa, 1968.)

energy is used in overcoming external friction or *skin resistance* between the water and the *wetted perimeter,* that is, the banks and bed. Skin resistance varies according to how smooth the wetted perimeter is and how straight the channel runs. Another important variable is the *hydraulic radius* (R) defined as:

$$R = A/P,$$

where A is the cross-sectional area of water and P is the wetted perimeter. The hydraulic radius is small and skin resistance high in a broad, shallow stream; it is large with skin resistance at a minimum in a deep, narrow stream with a semicircular cross-section (Figure 7–5). Since the velocity is, in part, proportional to the square root of the hydraulic radius,[4] discharge is more rapid in deep semicircular channels than in shallow streams, all other conditions being equal. The energy used in overcoming skin resistance is externally dissipated, partly in eroding material or moving sand and pebbles on the stream bed.

A substantial part of the stream's energy is also used in overcoming *shear.* This resistance results from the adjustments of the

---

[4] The relationship is $V = C\sqrt{RS}$, where V is mean velocity, S is slope, and C is a constant. A closer and more practicable relationship is given by the Manning equation

$$V = \frac{1.49}{n} R^{2/3} S^{1/2}$$

where n is a roughness factor that is determined empirically, but which can be approximated from existing tables. On these relationships, see Morisawa (1968:35ff.) and Gregory and Walling (1973:125ff.).

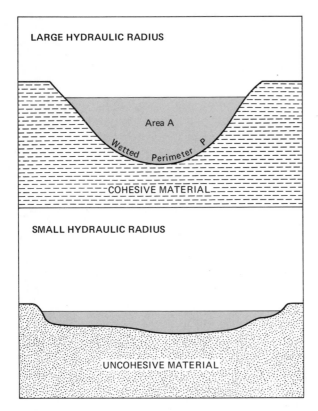

Figure 7–5. Streams with Large and Small Hydraulic Radii.

water to the changes in shape and roughness of the wetted perimeter (Figure 7–4). Shoals, bars, and pools on the stream bed require relative changes of velocity between the swiftly moving, undisturbed water above and the adjusting, retarded waters below. The energy employed in overcoming shear promotes turbulence and eddying motion within the stream. Unlike the laminar, streamline flow of the current in a deep river, *turbulent velocity* is difficult to measure effectively (see Figure 7–4), since the flow components are vertical and lateral, as well as downchannel. Although turbulence increases rapidly with current velocity, relative to current velocity, turbulence will be greatest in shallow streams with small hydraulic radius and significant shear across uneven channel beds.

Laminar, current velocity is a function of discharge, stream gradient, channel roughness, and hydraulic radius. Of these variables, recent work suggests that discharge may be most important. With over 6,000 current meters now in operation in the United States it is clear that, in humid environments, mean current velocity generally increases downstream, since discharge increases downstream as a result of: (1) increased discharge downstream, in response to tributary convergence; (2) decreased channel roughness; and (3) increased hydraulic radius, as stream depth and width increase, partly in response to increased volume. Stream gradient does not play an overriding role in determining current velocity, contrary to older theories.

This seemingly esoteric distinction of current and turbulent velocity in streams of large and small hydraulic radius is basic for the ability of a stream to erode and transport sediment.

7-3. HOW STREAMS TRANSPORT SEDIMENT

Coming right down to the essentials, a stream is only effective as a geomorphic agent when it has the ability to move sediment. This ability to carry a sediment load depends on the mechanisms involved. These break down into four types (see Figure 7–6):

1. *Solution.* Chemically dissolved materials, such as carbonates, sulfates, chlorides, and oxides, are carried almost indefinitely by streams. This *dissolved load* is subsequently precipitated in lakes or seas.

2. *Suspension.* Fine-grained particles with limited mass can be carried in mechanical suspension by agitated waters. As turbulence decreases, this *suspended load* is settled out, the coarsest particles first, the finest last. Clay or silt-sized material and colloids are primarily carried in suspension.

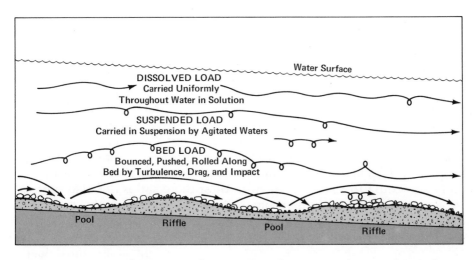

Figure 7–6. Types of Stream Sediment. (Based in part on Trewartha et al., 1967.)

3. *Saltation.* Intermediate-sized particles too heavy for suspended transport can bounce along the bed. Energy to leave the bed is supplied by hydrodynamic lift or by the impact of other, jumping sand grains. Particles rise as much as 1 to 4 feet (30 to 120 cm) and then travel downstream with the flat trajectory of a bullet. Unlike the suspended load, which may be destributed fairly uniformly throughout the water, material in saltation is restricted to the bottom of the stream. Sand-size particles are mainly involved.

4. *Traction.* Course sand and pebbles are normally too heavy for saltation. Instead they slide or roll along the bed as a result of the mechanical drag and push of turbulent stream waters. The materials moved by saltation and traction are called the *bed load*. Unlike the suspended and dissolved loads, the bed load is difficult to measure; the boundary between mobile and immobile bed load is impossible to define during bankfull stage, when the limits between bed and suspended load also become arbitrary.

A fair amount of numerical data is available regarding the types of stream transport typical for North America. Much of this is obtained by measurement and estimates of the load actually in transport at gauging stations along many of the major rivers. Another part of this information is derived from estimates of overall denudation of large watersheds. The dissolved load, which has been substantially increased as a result of human activities (accelerated flushing of natural plant nutrients and dissolved fertilizer from agricultural soils, industrial pollutants), comprises from less than 1 percent to over 99 percent of the total sediment (excluding bed load). These values indicate that an appreciable part of continental denudation is a direct result of chemical processes. The Mississippi River carries about 140 million tons of solubles into the Gulf of Mexico each year, and for the Earth as a whole Livingstone (1964) estimates that about 4,000 million tons of solubles are removed from the continents into the oceans. The proportion of dissolved load varies according to climate and lithology. It is minimal in arid regions with rocks that lack carbonates, and at its highest in humid lands with a high level of available solubles. No systematic information is available for the relative proportions of suspended and bed loads, but suspended sediments account for anywhere from about 50 to as much as 100 percent of the material transported mechanically. This ratio depends on local lithology and varies continually according to stream behavior. All of these ratios are also subject to very considerable fluctuations during the course of the year and from one year to another.

The ability of a stream to transport is best defined by two concepts: *Capacity* expresses the total potential weight of sediment load, which varies with particle size; *competence*, on the other hand, refers to the weight or size of the largest particles that the stream can possibly move along its bed. The capacity and the competence of stream transport depend on three major factors:

1. *Size of Material.* Since the specific gravity of most surface rocks and minerals varies within narrow limits (between 2.5 and 3.5), particle size is primarily responsible for particle weight. With identical stream velocities in an area of uniform bedrock, the distribution and proportions of the suspended and bed loads are determined by particle size, and to a lesser extent, shape.

2. *Turbulent Velocity.* It can be shown empirically that shear within a stream is closely related to the caliber of particles picked up, so that stream competence is proportional to turbulent velocity. Turbulent eddies maintain the upward force necessary to support sediment in suspension, while general turbulence will impart the energy to drag, push, and flip particles along the bed. Clearly, much of the energy used to overcome skin resistance will also be available to move the bed load. Consequently, stream competence will be relatively great in shallow, turbulent streams with a small hydraulic radius.

3. *Current Velocity.* Capacity, as opposed to competence, is proportional to the third power of mean current velocity. In other words, the total possible amount of sediment—of all sizes—in transport will be largely determined by the volume of discharge, being greater in large than in small streams.

In overview, two principles of stream behavior most relevant for erosion and deposition deserve to be restated. Firstly, geomorphic change is restricted to those periods of time when discharge and velocity are high. Secondly, turbulent and current velocity are distinct, the former closely linked to stream competence, the latter to stream capacity.

7-4. SCOUR AND FILL IN THE STREAM CHANNEL

For obvious reasons it is difficult to observe geomorphic changes within a flowing stream or river. There is little or no change for months, and then, rather suddenly, things happen during high water when the current is strong, the water deep and muddy. To circumvent these difficulties natural conditions have been partially recreated in the laboratory by running water down flumes filled with sand or a mixture of sand, silt, and clay. Even more instructive have been measurements made along the dry beds of ephemeral streams. Largely under the impetus of Luna Leopold, the U.S. Geological Survey has recorded stream-bed changes from storm to storm in New Mexico and several other states for many years. Individual pebbles were painted and numbered, and their positions measured after each period of discharge. Channel width, depth, and other features of the bed were also recorded by photography and by measurement with reference to fixed stakes or suspended chains. The following account of what happens on the stream floor is mainly based on this kind of experimental and observational evidence.

The bed of an average stream in humid mid-latitudes will consist of silt, sand, and pebbles. Most commonly the pebbles rest in and on the sand, while a thin film of silt and clay covers the whole bed. This is the case during low water. When a flood develops, stream velocity increases and turbulence becomes more and more evident. The silt layer is disrupted, churned up, and then dispersed into the water as suspended sediment. When the velocity and turbulence increase further, sand grains are set in motion by saltation and traction. Ripple patterns now develop in the surface sands of the stream bed. Then, as the flood surge hits, great masses of sand or mixed sand and silt are swept into the stream as the ripple patterns are destroyed. Isolated pebbles slide, roll, or skip along the bed in response to particularly strong eddies. Within a short time, several inches or more of the stream bed are removed and dispersed through the water while in temporary transport. The greatest amount of material is set in motion just as the surge arrives but before maximum discharge has been attained. Possibly this results from the maximum of turbulence that marks the approach of the flood surge. By the time the flood crest is reached, turbulence has decreased and given way to maximum current velocity.

While the stream is still at bankfull, the sand in saltation begins to settle out and pebble motion is sharply reduced. These first deposits rest on the scoured surface of the stream bed. They consist of the coarsest particles that have been moved by this particular flood. As discharge wanes, the last of the sands settle out, since the reduced turbulence is insufficient to support them. As the suspended silt and clay continue to pass by, the sands of the stream bed may once more be remodeled into ripple patterns. The suspended sediments settle out days or even weeks after the flood crest. They may be mixed with the sands or, most commonly, rest on top of the sands (Figure 7–6). This is the aspect of the stream bed that may persist, with little change, until the next flood.

The hypothetical description just given illustrates the processes of short-term erosion and deposition within a stream channel. Those changes that take place within the time span of a single flood are known as *scour* and *fill* respectively. Each flood is characterized by innumerable episodes of scour and fill along a particular stretch of the stream. And the state of the stream bed at a particular time and place may be quite ephemeral, the product of repeated scour and fill.

In the perspective of time, scour is concentrated during the period of increasing discharge, particularly as the flood surge approaches, while fill proceeds over days and weeks during the climax and waning stages of the flood. Little may happen for weeks or months thereafter, until the water level rises again.

7-5. MECHANISMS OF STREAM EROSION IN ALLUVIUM

Streams do more than shift sediment by repeated scour and fill along the bed. They actively erode by (1) channel deepening or by downcutting of the stream bed; (2) channel widening through bank caving or undercutting; and (3) channel extension, that is, headward or regressive erosion by streams and gullies (Figure 7–7). This erosion is of two basic types. Firstly, many valley floors over plains consist of stream-laid sediments or alluvium where streams shift their channels and readjust their beds by a little erosion and deposition. As a new channel is formed, an older channel is abandoned and filled, or, as one bank of a meandering stream collapses, the stream shifts in that direction and accretion follows on the opposite bank. Such erosion *in alluvium* is fundamental in the development of alluvial plains, being a mechanism whereby streams make short- or long-term adjustments. Secondly, streams cut their way into uplands where they usually erode into compact rocks. But the mechanisms and rates of erosion *in bedrock* are different, and the ultimate effect is to sculpture the interfluves.

Many small and most large streams flow between banks of the alluvium that fills the valley floor. In some such instances, the stream bed rests directly on the underlying bedrock, in which case there is bed-

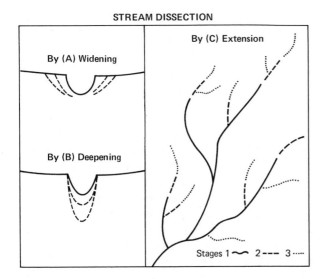

Figure 7–7. Three Methods of Channel Enlargement by Dissection.

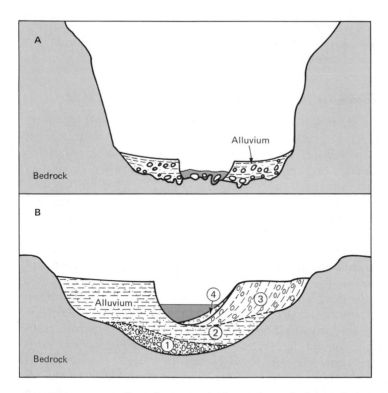

Figure 7–8. Examples of a Stream Channel Partly (A) and Completely (B) "Buffered" by Alluvium. Note the three generations of older alluvium (1, 2, and 3) as well as the active fill (4) in B.

rock dissection (Figure 7–8,A). However, when the alluvium is deep, there will be no change in the adjacent bedrock and stream erosion is largely confined to reworking of older stream deposits (Figure 7–8,B). Unlike in the case of compact bedrock, dissection of unconsolidated alluvium may proceed rapidly, and channel changes can often be discerned from year to year. Cutting in an established channel involves bed erosion, bank undermining, and bank collapse. Erosion of the alluvium on the stream bed involves lifting and pushing of loose particles, particularly by turbulent waters. Any cementing matrix in older alluvium under the bed or in the stream banks is partly dissolved, and individual particles are loosened by mechanical wear. Bank collapse occurs when the stream banks are undercut: At some critical angle the bank slope becomes unstable, and the upper part of the bank falls in or slides down (Figure 7–9,A), producing a gentler slope, until undermining begins anew. Bank collapse is further aided by freeze-thaw alternations and frost heaving in cool climates and by groundwater seepage toward the stream.

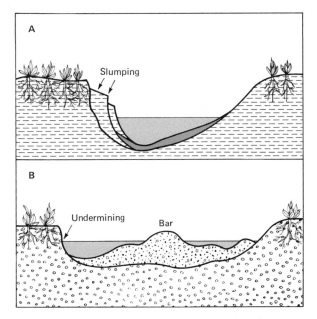

Figure 7–9. Bank Slumping in Silt (A) and General Bank Undermining in Loose Sand and Gravel (B).

Where stream banks consist of cohesive silt and clay and where a high proportion of the load is carried in suspension, channels maintain steep banks and deep channels. However, silty banks lose their cohesiveness as they become saturated at times of high water; as long as the water level remains high, hydrostatic pressure may support the bank; however, when the water drops, the bank may cave in. On the other hand, loose sands or gravels tend to fall into a stream, maintaining steep banks that are quite unstable. Erosion is rapid, and great quantities of bed load are immediately available (Figure 7–9,B). This favors channel widening instead of channel deepening, so that broad, shallow channels are formed. Such channels may be short-lived, since sand bars and gravel shoals develop, deflecting the stream or forcing the channel to subdivide.

New channels are cut into alluvium by gullying. Gullies are extended during periods of rain when surface runoff pours into a drainage line, undercutting the head and incising the floor. Piping accelerates headward erosion in some areas, while soilfalls can widen a gully at any time. Overflow channels or bifurcations can also be accelerated by gullying, when water spills over an unbreached body of older alluvium. A gully forms where the water plunges down and cuts back, until a major channel has been eroded (Figure 7–10).

Stream erosion in alluvium is usually matched by deposition of

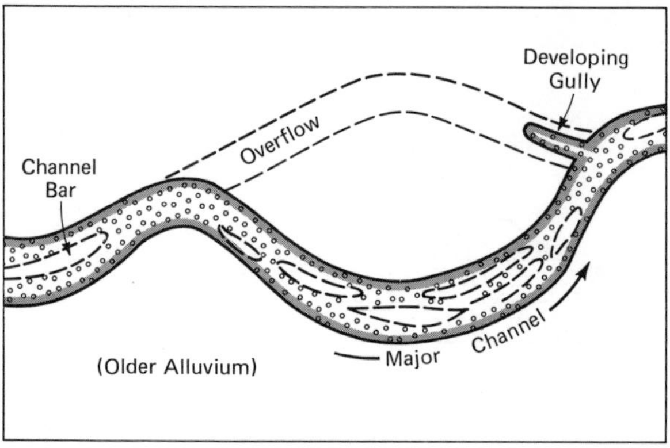

Figure 7–10. Map of a Developing Bifurcation with Overflow and Incipient Gully.

fresh sediment and is a normal part of the system of balances in most lowland streams. However, there may also be long-term changes leading to a general dissection of existing alluvium, with extensive gullying that may spread to adjacent hillslopes in softer rock types. Such disruptions of the normal state of affairs are most commonly triggered by a change in the hydrological balance, sometimes a result of man's activity, sometimes due to climate, and sometimes reflecting mechanical readjustments of channel gradients.

A striking example of such accelerated gullying is provided by Arizona and New Mexico, where ephemeral streams once flowed across broad, shallow draws that were grown over with grasses and shaded by trees. Even during the dry season, ponds were common along the stream beds, and descriptions of this kind were provided by the earliest American ranchers during the 1860s and 1870s. This part of the Southwest enjoyed two rainy seasons. The summer brought a number of severe thundershowers, while periods of light rains were vital for the vegetation, permitting a good grass cover and keeping the groundwater table relatively high (Figure 7–11, top).

However, during the 1880s a climatic change set in at about the same time that large herds of cattle and sheep were introduced. The beneficial winter rains declined. The summer rains persist as before—but these come as brief, violent downpours that flow off rapidly and evaporate without soaking the ground or recharging the groundwater. As a result of both the declining winter rains and the increasing grazing pressure, vegetation was gradually reduced, the groundwater table fell, and runoff was accelerated. As the water surged into the stream channels, it began to erode the older alluvium, cutting deep gullies into relatively steep alluvial surfaces. Shade trees died off, and grassland changed to desert shrub and brush country. This trend to

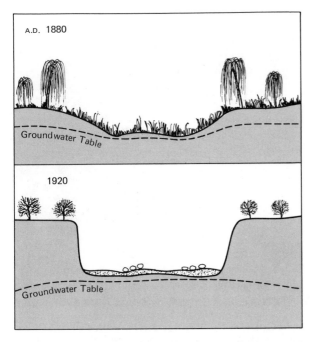

Figure 7–11. Stream Channel in New Mexico (A) 1880 with High Water Table and (B) 1920 with Low Water Table, after Gullying.

general dissection has created a countryside cut up by steep-sided channels and gullies, the floors of which are mainly bare and littered with sand and gravel (Figures 7–11, bottom, 7–12, 7–13); it has once more begun to build up, particularly in downstream valley segments. Similar examples of man-induced gullying can be cited from the Piedmont of the southeastern United States and from southern Africa.

7-6. MECHANISMS OF STREAM EROSION IN BEDROCK

Perhaps the earliest description we have of the force of a river cutting into rock is that of Diodorus for the rapids or cataract of the Nile at Aswan:

> This place has a steep gradient and is shut in by cliffs to form a narrow, rocky gorge along its entire length, full as well of great boulders that stand out of the water like pinnacles. And since the river diverges around these boulders with tremendous force and is often turned back, so that it hurtles in the opposite direction as a result of the obstructions, remarkable whirlpools are formed. Along the whole stretch, the middle space is also filled with foam produced by the swirling water, and strikes those who come near with great terror. Indeed the descent of the river is so swift and furious that it appears to the eye like the very rush of an arrow. At flood-

Figure 7–12. A Large Tributary Gully in the Orange River Valley of South Africa. Widespread gullying in South Africa began before 1900 for similar reasons to those in the American Southwest, namely, overgrazing and destruction of the grass cover, accompanying a change of rainfall seasonality. (Karl W. Butzer)

time of the Nile, when the jagged rocks are submerged and the entire rapids hidden by the large volume of water, some men sail down the cataract when the winds are against them; but no man can make his way up it, since the force of the river defeats every human device.[5]

Just how bedrock dissection proceeds has eluded observers until much more recent times, and some details remain controversial even today. The basic mechanisms can be outlined as follows (see Figure 7–14):

1. *Chemical Weathering.* Although the bulk of the solubles carried by streams is dissolved by groundwater, corrosion or hydrolysis is also effective along the wetted perimeter of the stream. Water is available so that the rate of chemical breakdown will depend on temperature and the dissolved chemicals of the water. Solubles, ions, and colloids can be removed rapidly. Attack is selective, affecting specific minerals or following joints and bedding planes. The net effect is often greater than that of all mechanical forces, commonly involving corrosion and fluting or pitting, while cracks are enlarged to prepare rock fragments for hydrodynamic rotation or lift. Chemical weathering is not confined to the wetted perimeter. Many European geomorphologists now believe that chemical decomposition is an important factor in breaking

[5] Book I, 32:8–10.

Figure 7–13. The Once Gullied Valley of the Middle Riet River in South Africa Restored to Conditions of the Nineteenth Century by a Flood-Control Dam Upstream. The water table is now high and vegetation abundant. (Karl W. Butzer)

down the bed load of perennial streams. Pebbles are "softened up," facilitating breakdown after wear, while lines of weakness in the rock are exploited to produce fractures. As a result, most stream beds in humid lands contain few boulders, and in tropical areas, where chemical attack is even more effective, quartz sand is corroded so that sand grains break up into smaller particles.

2. *Mechanical Abrasion.* For a long time most observers felt that stream downcutting was a matter of mechanical wear, performed primarily by the pebbles and sand of the bed load grinding and scouring on the wetted perimeter (Figure 7–15). Such abrasion is undoubtedly important, but it is substantially aided by chemical processes. Evidence to this effect is provided by the beds of desert streams, on which cobbles and boulders are preserved almost indefinitely in the absence of water and chemical weathering. A second fallacy about stream dissection is that pebbles are the best tools for stream bed scour. In some cases this is true, but pebbles tend to travel on top of the sand in a bed load of mixed calibration. As a result, the pebbles mainly collide with one another, leaving the sand to wear away at the intact rock below. The suspended load is nearly ineffective in abrasion of compact bedrock.

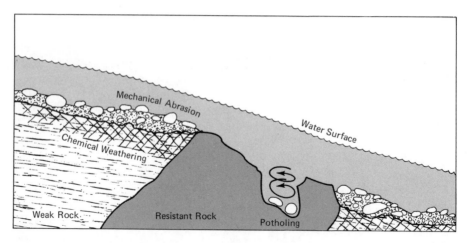

Figure 7–14. How a Mountain Stream Cuts into Bedrock.

3. *Potholing.* A special type of abrasion is peculiar to many streams with steep gradients and limited bed load. Pebbles or cobbles roll into shallow depressions on the bedrock floor of the stream, where they are subsequently churned about by stream turbulence. The

Figure 7–15. Cobble Gravels and Boulders Floor the Channels of the Steep Streams That Drain from the Serengeti Plains into the Olduvai Gorge of Tanzania, East Africa. (Karl W. Butzer)

pebbles grind away, carving out deep potholes (see Figure 7–14). A constant supply of fresh pebbles assures continued potholing.

4. *Rock Type and Sediment Supply.* The amount and type of sediment carried by a stream is strongly influenced by bedrock. Soluble rocks provide less bed load or suspended matter, since most of the rock is dissolved and removed chemically. This explains why streams in limestone country are normally poor in sediment. The same applies for quartzite and other highly resistant rock types that weather very slowly. Sandstones, on the other hand, provide abundant sand, while soft shales can deliver considerable quantities of silt and clay.

5. *Structure and Valley Morphology.* Differential erosion and other structural phenomena, such as inclined strata, can modify the cross-section of a stream valley or control the stream profile (Figure 7–16). Structural-lithological features such as bedding planes, jointing systems, and rock deformations may also be important in providing fragmented rock, which is more liable to erosion.

Whatever the stream velocity, bedrock dissection proceeds very slowly. Downcutting, undercutting, and headward erosion in compact bedrock can usually be estimated in inches per century or inches per millenium. Unless a large block happens to be dislodged, there will probably be no discernible change in a lifetime.

The immediate impact of bedrock dissection is that streams cut up their interfluves, regardless of whether the terrain is mountainous or constitutes a gently rolling hill country. Channels are deepened, widened, and extended, fingering deeper and deeper into the unconsumed interfluves. When cutting and deepening proceed rapidly, the

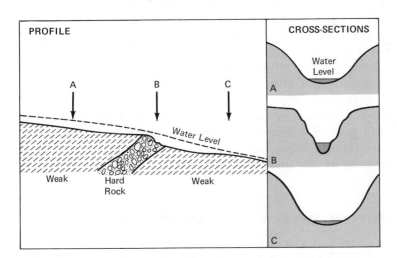

Figure 7–16. Resistant Rock Strata Interrupt Stream Profiles and Favor Valley Constrictions as Dissection Proceeds.

stream floor remains comparatively narrow and the valley sides rise steeply, giving the valley a V-shaped profile (Figure 7–17). However, when dissection is slower, the other agents of denudation are better able to adjust to the stream. As a result, overland flow, concentrated runoff, and mass movements wear back the valley sides, broaden the valley floor, and reduce the surfaces of the interfluves. In that case, comparatively broad valley floors will accompany the streams, even into the headwaters of a drainage basin. The resulting valley profiles will not be V-shaped (see section 9-2).

If we focus closer attention on the role played by the stream in valley cutting, we can see that it is easy to overemphasize the actual work done by the stream itself. The stream erodes its floor and its banks, within the confines of its valley. It seldom directly attacks the valley walls, except where a shifting channel is able to undercut a rock outcrop. Consequently, the greater part of the valley widening (as opposed to channel widening) can be attributed to mass movements, overland flow, and rill erosion. Streams are seldom directly responsible for hillslope development.

The greatest significance of streams for landform development among the interfluves is indirect. First of all, streams perform a vital function by removing the debris that is eroded directly by the stream channel as well as that brought into the stream by surface runoff and mass movements. In other words, streams serve as sewers that sluice away solubles, clay, silt, sand, and pebbles. By so removing the products of erosion, surface denudation is allowed to continue. If, on the other hand, total stream transport out of a drainage basin did not balance with the amount of material eroded by all agents, over geological time, denudation would gradually come to a halt. The valleys would fill up with debris, mantles of colluvium would accumulate on the lower hillsides, and erosion would be confined to upland surfaces. Topographic relief would be reduced, thus cutting down the potential energy of running waters. This brings up the second basic function of stream dissection, that of providing relief and potential energy for geomorphic agents. As the interfluves are denuded, streams frequently continue to cut down their channels into fresh bedrock. In this way the loss of potential energy due to denudation is partly or entirely compensated for. Consequently, streams essentially control rates of surface denudation by their influence on gradients and length of slopes.

7-7. DEPOSITIONAL PATTERNS ON STREAM BEDS

The stream channel is not only modeled by erosion but also by deposition. Sediment is shifted during each cycle of scour and fill, and stream cutting and shifting are usually matched by deposition of fresh

Figure 7–17. The Muddy Waters of the Orange River Swirl down Successive Steps of the Aughrabies Falls, South Africa, Cutting Deeply into Jointed Granite. Note high-water potholes. (Karl W. Butzer)

alluvium. Although most of these deposits are ephemeral and subject to continued change, there are a number of basic patterns and properties that persist through time and that are valid for all stream channels.

Within a *straight* stream segment, pebbles and other coarse particles are concentrated in the center of the channel, where the water is deepest, and velocity and competence are greatest. The finer materials are deposited near the foot of or up against the banks. In a curved stretch the coarsest load is concentrated in the deepest water along the undercut outer bank of every bend, while low ridges of finer sediment build up in the shallows near the inside bank. Sand and gravel *bars* are oriented with the channel axis and commonly form where bed load piles up behind an obstacle to form a tail pointing downstream. The original obstacle can be a large rock or the erosional hummocks left on the channel floor when localized scour temporarily deepens the channel to accommodate more discharge during bankfull stage.

Gentle undulations of the bed along the axis of the channel are longitudinally spaced at 5 to 10 times the channel width. The rises are called shoals, or *riffles*, the lower points, hollows, or *pools* (see Figure 7–6).

Pebbles are usually concentrated on the riffles. These undulations are accentuated immediately after high water and are probably due to selective scour as a flood surge develops; relief is reduced as the pools fill in with sediment during low water. Pool-and-riffle systems migrate downstream, but the constituent sands and pebbles move faster by jumping from one bank to the next during successive floods. Superimposed on the riffles and pools are ripple patterns transverse to the current (Figure 7–18), as well as miniature sand bars that run parallel. These irregularities reflect turbulent eddies that pick up sand grains, rolling them up and over the next ripple.

Where the bedrock of a drainage basis is homogeneous, the caliber of the coarsest pebbles will decrease downstream, while the actual bed sediment increases in quantity. This is partly related to a relative decrease in competence, despite increasing current velocity; at the same time, continued attrition of the coarser bed load by chemical weathering and mechanical abrasion slowly destroys pebbles and cobbles that entered the stream in the headwater region. If the drainage crosses different kinds of bedrock, the amount and the caliber of the load probably change from one tributary to another according to the porosity and permeability of the rocks to surface flow, their resistance to erosion, and the amount and type of load they provide. The ratio of suspended and bed loads can change rapidly above and below stream confluences, since some tributaries introduce more, others, less sediment. Bed load may even build up at river confluences when a through-stream lacks the competence to remove excessive amounts of coarse-grained sediment.

Figure 7–18. The Bed of the Upper Orange River in South Africa as Low Water Exposes Sand Bars and Shoals with Superimposed Ripples and Miniature Bars. Rapid sand accretion along this segment of the river is a result of accelerated erosion on the sandstone hillsides of Lesotho, 50 miles and more upstream. (Karl W. Butzer)

Apart from their forms and horizontal arrangements, the sediments themselves can be analyzed and the information used to decipher current processes or past history of a stream.

Sediment *texture* is similarly defined as for soils (see section 4-7), but simpler descriptive classes are normally used for stream deposits, depending on the relative importance of gravel, sand, silt, or clay-sized particles. Sandy or gravelly deposits are normally related to the bed load, whereas silts and clays are suspended sediments.

Sorting refers to the proportions of particular grade sizes present in a sediment. So, for example, if over 80 percent of a sample falls into one grade size, such as sand, the material is well sorted. Good sorting indicates uniform conditions of deposition, commonly with abundant water. Strong mixing of grade sizes is referred to as poor sorting and results from mixed deposition, such as bank collapse or a brief flood in a desert stream. As stream turbulence diminishes when a flood wanes, the coarser particles settle out first, the finest last; in that case the deposits of a single episode become progressively finer from bottom to top.

Stratification means the disposition of a sediment into well-defined beds or horizons (Figure 7–19,A). Many stream deposits show little or no horizonation although pebbles will usually rest on their flattest

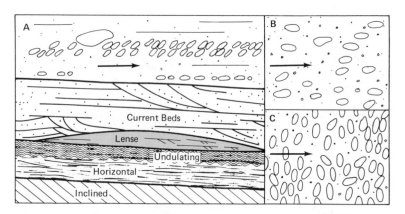

Figure 7–19. Examples of (A) Stream Bedding in Section and (B, C) Pebble Orientation in Plan View. Direction of current is from left.

surface. Some horizons are conspicuous on the basis of color, texture, or sorting. Thus, for example, a bed of yellow sands may rest on brown silt with an abrupt contact. The contact indicates that an interval of scour has separated the two beds. In general, thin beds up to 1 or 2 inches (2 to 5 cm) thick, and extending over greater distances, commonly record periods of moderate or quiet stream flow. More massive lenses of limited distribution reflect turbulent flow (Figure 7–19,A). *Horizontal* beds are, of course, the norm and indicate vertical accretion. *Inclined* beds, reflecting lateral accretion, are embanked against the side of a channel or laid down on the edge of deeper waters, either on a small scale on the channel floor or on a large scale where a stream enters a lake or the sea. *Current* beds consist of interfingering horizontal and inclined beds; they are common in turbulent streams where scour and fill alternate. *Undulating* beds are associated with former ripple patterns.

Whereas stratification refers to the vertical arrangement of a sediment, *orientation* is a matter of horizontal alignment. If pebbles are rolled during transport, they rotate on their longer axis, so that they are oriented perpendicular to the direction of flow (Figure 7–19,C); if they slide, they point nose downstream, that is, the long axis is parallel with the current (Figure 7–19,B). Both types of movement are always present, but one may be dominant. Rolling is most frequent on pebbly stream beds or in streams having very considerable turbulence; sliding is more common along sandy stream beds.

7-8. MECHANISMS OF STREAM ALLUVIATION

Scour-and-fill and channel deposits are processes or features found in almost any stream. However, experience shows that some streams are actively engaged in accumulating sediment along much of their courses. Similarly, the lower part of most stream systems is accompanied by broad expanses of recent deposits, whereas there seldom are massive stream sediments in the upper reaches of the drainage basin. The process of stream deposition is called *alluviation*. Whereas fill is a short-term process on a local scale, alluviation refers to a long-term phenomenon affecting a large part of the stream system. It involves a preponderance of fill over scour during long periods of time. In other words, scour and fill always go hand in hand, at all times, and along all parts of the stream course. But alluviation represents *net* fill, on a long-term basis, along part or much of the course.

In general, scour and erosion appear to be relatively more active than fill upstream, judging by the limited amount of alluvium in the stream bed and along its banks. On the other hand, the extent and the thickness of alluvium suggest that fill and deposition are more significant downstream. To express this composite picture of many floods over many years it can be generalized that:

1. Dissection tends to be dominant upstream.
2. Alluviation tends to be dominant downstream.
3. Denudation is effective to some degree or other over most of the drainage basin.

There are several reasons why alluviation is characteristic of the lower course of a stream.

1. Upstream, turbulence is at a maximum with steeper gradients, a small hydraulic radius, and greater channel roughness. This last property is a result of the gravel and cobbles of the stream bed, and of the stronger bedrock and structural control of channel width, depth, and profile. Consequently, erosion is active, with a maximum of stream competence.

2. Downstream, turbulent velocity and eddying motions are reduced with gentler gradients and reduced channel roughness, as the channel develops free of structural constraints within a body of alluvium. Consequently, stream competence tends to decrease with net deposition of increasingly fine particles. At the same time, the total load in transport is greater, since the greater volume of water increases current velocity and stream capacity.

3. Level and low-lying country is most extensive adjacent to the lower course of a stream. Such terrain is liable to flooding during overbank stage. Overbank waters are greatly retarded as they spread in thin sheets over a broad floodplain with an inefficient hydraulic radius and rapid dissipation of energy. Consequently, there is little current velocity or turbulence, and the suspended load is deposited in thin films of flood silts.

As a result of these factors, there is an overall trend to sediment accretion downstream, both in the channel and on the adjacent floodplain. The channel deposits consist primarily of bed load, the caliber of which varies from flood to flood, according to the peak competence reached. Beyond the stream margins a fairly uniform mantle of suspended matter eventually accumulates in those areas reached by overbank discharge.

Seen in a general perspective, alluviation downstream roughly balances erosion upstream. However, human activity or climatic change can suddenly increase the supply of sediment so that alluviation is accelerated. Changes of this kind took place in many parts of Italy where once wooded hillslopes were deforested in early Roman times and converted into vineyards, olive groves, and wheat fields. To hold the soil and to make cultivation simpler small stone ramparts were built around the edges of each small plot, creating artificial terraces. These terraced slopes functioned well as long as farmers took care of them. However, as the Roman state collapsed in the fourth and fifth centuries A.D., economic ruin became commonplace in the countryside. Farmlands were not properly tended, terraces washed out during violent rainstorms, and the soil behind them was often lost. This accelerated denudation of the hillslopes provided the streams with an excessive load to carry. Alluviation of almost catastrophic proportions followed in some stream valleys, with the result that fertile lowlands were buried by thick fills of sterile sands and gravel (Figure 7–20). Many Greek and Roman valley settlements are now often found under 20 to 30 feet (6 to 9 m) of silt and sand. Also, the rapid accretion of sediment at the mouths of rivers pushed the shoreline well seaward. Such siltation has noticeably modified the coastal configuration of the Aegean Sea, for example. Although an episode of "postclassical" alluviation was widespread in the Mediterranean world during the sixth to eleventh centuries A.D., and may well have been in-

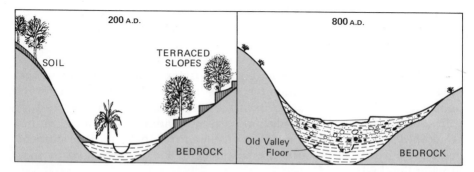

Figure 7–20. An Italian Valley Before and After Stream Alluviation, Following Soil Erosion in Late Roman Times.

fluenced by a period of heavier rainstorms, localized erosion and alluviation did take place repeatedly at earlier times in direct response to misuse of the land. Such examples provide striking illustrations of how human activity can affect the balance of natural forces.

Chapter 8

STREAM VALLEYS AND RIVER PLAINS (II): ALLUVIAL LANDFORMS

The depositional landforms created by streams are found in most riverine lowlands and are of prime value to land use as well as of great interest as geomorphic features. River floodplains of all kinds are character- ized by channels that either meander in sinuous loops or braid in interweaving, multiple channels. These dif- ferences of channel configuration reflect the function of the channel to carry a fluctuating volume of water and sediment as efficiently as possible. Floodplains also differ in section: Some are convex, due to raised banks that are higher than the surrounding flats, while most smaller or intermediate-sized examples are flat, rising gradually to the valley margins. Convex flood- plains commonly characterize the largest alluvial sur- faces; their special features are related to meandering streams, with a predominance of suspended sediments, and prone to overbank discharge. Where floodplains merge onto coastal plains or border on lakes, their

channels bifurcate and the waters diverge over broad, flat deltas. Where streams debouch from mountains onto plains or where steep-gradient streams merge with rivers of much lower gradients, there also is stream divergence with formation of convex alluvial fans. On account of a variety of external factors, including man, streams may readjust their floodplains by cutting to slightly lower levels–with narrower, lower-gradient floodplains–or by building up to slightly higher levels–with broader, higher-gradient floodplains. As a result, most alluvial plains are accompanied by terraces of older alluvium that mark former floodplains at different levels. Some such alluvial terraces are of historical age; others may be tens or hundreds of thousands of years old. Similar readjustments can leave tiny, misfit streams in large valleys or lead to the projection and entrenchment of meandering channels deep into bedrock. Alluvial terraces, misfit streams, and entrenched meanders provide the most striking examples of stream readjustments through time. Apart from their own importance, they serve to show that streams are dynamic and subject to change. At the same time, they provide a detailed record of the dissection and denudation that have sculptured the interfluves, creating the patterns of stream systems and landscape profiles outlined in Chapter 9.

Egypt is the gift of the river.
HERODOTUS, 2:5

8-1. THE CONFIGURATION OF FLOODPLAIN CHANNELS

There is a widespread misunderstanding about what Herodotus meant when he said that Egypt was a gift of the Nile. He did not link this statement with irrigation waters but with the alluvial origin of the valley and delta. The Greeks had long been aware of the depositional activity of streams, and Strabo, writing about 20 B.C., tells us that the legendary Trojan War hero Menelaus

had ascended the Nile as far as Ethiopia, and had heard about the inunda-
tions of the Nile and the quantity of alluvial soil which the river deposits
upon the country, and about the large extent of territory off its mouths
which the river had already added to the continent by silting.[1]

Strabo's *Geography* goes on to describe a variety of other stream-laid
surfaces in different parts of the Mediterranean world. We now call alluvial plains that are due primarily to channel or overbank deposition
floodplains. In the case of small streams, a floodplain may be only 100
yards across, and although bankfull may be attained several times a
year, there will probably be little overbank flow. The other extreme is
provided by the floodplain of a river such as the Nile, which, prior to
construction of flood-control and irrigation dams, enjoyed a flood
season of three months. The water began to rise in its channel during
July and spilled out over the floodplain during August. The flood crest
was reached in September, and only during October or November was
the floodplain drained once again.

Despite the contrast in size between a trout stream in Minnesota
or a great tropical river, the basic geomorphology of floodplains is
similar. The low-lying terrain is remarkably level, crisscrossed by
streams, and frequently dotted by lakes, ponds, or swamps. The
stream channels show predictable patterns, and any open water takes
one of a number of possible, characteristic shapes. The following section describes some of the more important geomorphic features of
floodplains, beginning with floodplain channels.

Floodplain streams show two possible patterns from the air: meandering or braided. A *meandering* stream winds across its floodplain in
a sinuous course, in a series of sprawling or contorted loops (Figure
8–1). Although there may be islands within the channel, the stream
course is unique and well defined. A *braided* stream, on the other hand,
has multiple channels that interweave as a result of repeated bifurcation and channel convergence (see Figure 7–9,B). There are remarkably few simple, straight channels in floodplain environments.

Why a stream meanders or braids is not completely understood.
But the function of a stream channel is to accommodate its discharge
and load. Width, depth, and velocity are simple power functions of the
discharge, such that the discharge is equal to the product of channel
width, mean depth, and mean velocity. Consequently, at any given
point of a stream, an increase in discharge requires an increase in
width, depth, and velocity. Since velocity is the most conservative
variable of the three, the major adjustments are accomplished by
changes in channel width and depth.

All the evidence indicates that the sinuous course of a meandering

[1] Book I, 2:23. *The Geography of Strabo.* London, W. Heinemann, 1960. Translated by H.
L. Jones.

stream responds to precise physical laws, no matter how random the meander swings may seem to be. Within a meandering channel the line of maximum depth and velocity crosses back and forth from outside bank to outside bank (Figure 8–2). As the water swings through a bend, it develops centrifugal force that raises the water level on the outer bend, thereby increasing stream velocity in the meander loops. Far from being sluggish or inefficient, a meandering channel has a greater capacity to carry sediment than a straight channel.

Experiments show that meanders develop rapidly from an initially straight channel. Uniform bends develop on uniform material and slope, and an increase in discharge or gradient is matched by an increase in the radius of curvature of the meander bends. In fine-grained material, channels are deep and narrow; coarse-grained materials are more erodible, and the channels that develop are broad and shallow, with steeper gradients. The results of these laboratory studies by J. F. Friedkin (see Schumm, 1972) correspond closely with those of

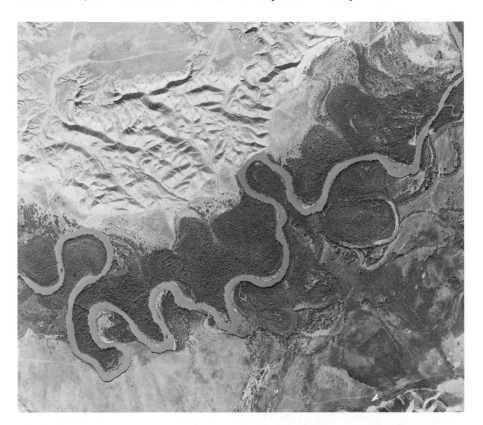

Figure 8–1. Fantastic Loops Characterize the Meander Belt of the Jordan River. Note the many abandoned meander scars visible on the margins of the fringing woodland. (Courtesy of Isaac Schattner)

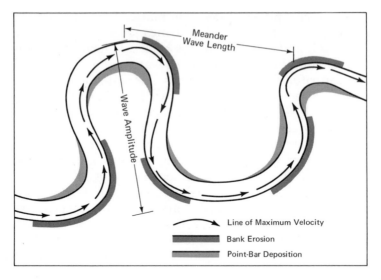

Figure 8–2. Meander Belt, Showing Line of Maximum Velocity and Areas of Active Bank Erosion and Point-Bar Deposition.

Schumm, derived from field analysis of several rivers of the Great Plains. Schumm (1963) found that meandering streams have deep, narrow channels, with high proportions of silt and clay in their banks and beds. On the other hand, broad, shallow stream segments are straighter and have a greater proportion of bed load. There is a close empirical relationship between meandering and a strong preponderance of suspended sediments.

Meander wavelength is proportional to channel width, so that the wavelength averages a little over 10 times the channel width. Wavelength is also proportional to the area of the drainage basin.

Meander patterns are unstable although their transformation is predictable. Over long periods of time meanders migrate downstream and tend to swing more widely, that is, to increase their wave amplitude (see Figure 8–2). Maximum velocity and erosion occur on the outside bank of the channel, a little downstream of the axis of the bend, so that the meander loop grows with a slight downstream vector component. After brief transport the eroded material is deposited on an inside bank of the channel in the form of low ridges known as *point bars* (Figure 8–3). As time goes on, the individual meanders shift downstream, but their wave amplitude cannot grow indefinitely without affecting the wavelength. Consequently, at some point the increasingly narrow neck of the meander is breached during flood stage, cutting off the meander loop (Figure 8–4). The ends of the *cut-off meander* may soon be blocked with bank sediments and the cut-off reduced to an ar-

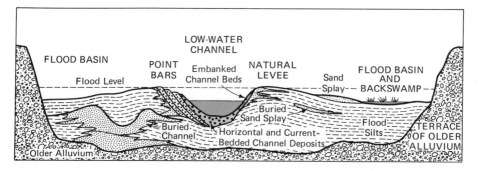

Figure 8–3. Section of a Convex Floodplain. Point-bar deposits are only found along inside meander bends and rarely opposite actively developing levees. Note considerable vertical exaggeration.

cuate lake called an *oxbow*. This gradually deteriorates into a swamp and ultimately dries out. These processes and features, as chaotic and desultory as they may seem, are all part of the basic dynamism of a meandering stream.

Braiding is confined to broad, shallow streams with great quantities of bed load, highly erodible banks, and steep gradients. Bifurcations develop around large mid-channel bars that grow downstream and increase in height. Such bars often become an interchannel island, the surface of which may be stabilized by vegetation. The new channels develop rapidly and can later bifurcate again around new multichannel bars. Not all bifurcations result from bar accretion, however. New channels are also cut into the floodplain during periods of overbank discharge, as described in section 7-5.

The horizontal patterns of a braided stream are far more unstable that those of a meander belt (see Figures 7–9,A and 7–9,B). This can be easily explained by the high erodibility of the bed and the banks, as well as by the limited relief of the channel, which is prone to major changes after each bankfull episode. The process is self-perpetuating, since the obstructing channel bars may consist of gravels with greater resistance to erosion than the more sandy material composing the channel banks. Consequently, the effects of channel widening are counteracted by a decrease in channel depth. This increases channel roughness and turbulence, providing the necessary competence to transport a large bed load.

8-2. CONVEX FLOODPLAINS

In describing a river plain it is common to use examples such as the Mississippi, the Amazon, the Nile, or the Mekong. These have very large floodplains and are conspicuous in that they are slightly convex

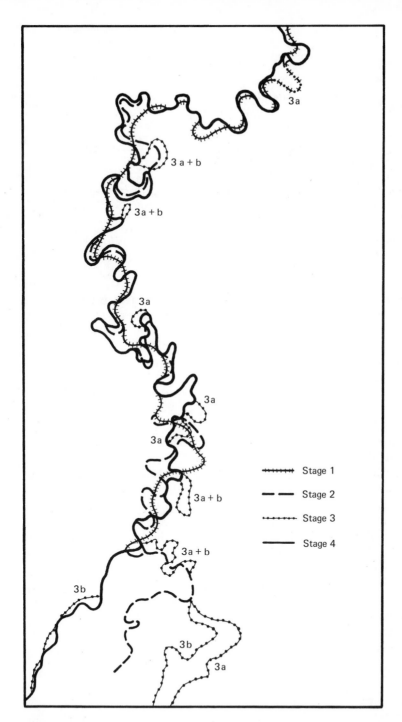

Figure 8–4. The Meander Belt of the Omo River in Southwestern Ethiopia Evolved in Four Stages During the Past 800 Years. (From Butzer, *Recent History of an Ethiopian Delta,* Copyright 1971.) See also Figure 8–6.

in sections—that is, the land slopes away gently from the river bank to the sides of the valley. In fact, the low-gradient floodplains of most large rivers are "convex," and they show a number of characteristic features illustrated by Figure 8–3:

1. The *low-water channel* forms a meander belt winding between banks of fairly fine-grained alluvium. Bed-load sediments range in texture from silt to sand, with horizontal, inclined, or current-bedded sediments.

2. *Point bars*, developed as a series of low levees on the inside meander bends, form arcuate or parallel ridges and swales. Since point bars include bed-load and suspended sediments deposited within the channel or its immediate banks, texture is highly variable. Bedding may be horizontal, inclined, current bedded, or undulating.

3. The *natural levees* rise above the river banks as low ridges that fall off gently away from the river. These berms are best developed on outside, undercut banks. They form during periods of overbank discharge, when silts and fine sands are deposited near the edge of the bank, where they are trapped by vegetation. Beds within the channel that are laid against the banks are inclined, while horizontal and cross-bedded sediments constitute the levee itself.

4. *Flood basins* lie beyond the channel and its levees and are repeatedly flooded with overbank discharge. In general, sediments consist of finely bedded, horizontal silt and clay, often rich in organic material. Specific features commonly present include the following.

    a. *Sand splays*, tongues of coarser sediment deposited where a levee has been breached during a violent flood. The surge of water from the channel into the flood basin has considerable turbulence and carries a great deal of sand.

    b. *Abandoned channels*, including cut-off meanders, and oxbow lakes. These relict features record shifts of the meander belt and are frequently prominent within the flood basins.

    c. *Backswamps*. The lowest parts of the flood basin may be near the groundwater table or even beneath low-water level in the channeled stream. As a result, seasonal or permanent swamps or lakes are found in the lowest parts of many flood basins. Backswamps are normally of circular or oval shape.

    d. *Gathering streams*. As the flood waters recede at the end of a high-water period, part of the overbank discharge drains back into the low-water channel. Shallow, gullylike channels are cut across the flood basins, sometimes running parallel to the river. Eventually, these gathering streams breach a low point in the natural levees and flow back into the low-water channel.

    e. Where levees have formed rapidly in response to accelerated sedimentation, valley-side tributaries can be blocked and forced to flow through flood basins until they can link up with a gathering stream. In some cases this pattern is permanent, with a meandering *yazoo* stream running subparallel to the main river across its floodplain.

The cross-sectional convexity of a floodplain is primarily due to the nature and texture of sedimentation. Bed-load and suspended sediments are deposited in the low-water channel and along its immediate banks, while suspended materials exclusively are deposited in the flood basins. For an identical unit of time, bed load accumulates far more rapidly than suspended sediment. In addition, however, deposition in channel vicinity takes place following every period of bankfull discharge, whereas overbank flow is less common. Consequently, sedimentation is more rapid in and near the channel than it is in the flood basins. As a result, the channel bed and levees build up at a faster rate, until the crests of the levees rise conspicuously above the general level of the alluvial flats. Depending on the size and dynamism of the floodplains, the relative relief between the levee crests and the floors of the basins may vary from 3 to 50 feet (1 to 15 m). A secondary factor, responsible for an accentuation of floodplain convexity, is the compaction and resulting subsidence of the fine-grained sediments that build up the flood basins.

Classical examples of convex floodplains, apart from the ones already mentioned, are provided by the Tigris and the Euphrates and above all by the alluvial plains of the Hwang and the Yangtze rivers in China and the Ganges and the Indus on the Indian subcontinent.

Floodplain levees have long provided optimal settlement sites on these river plains. The levee crests are only flooded briefly—and often incompletely—during exceptionally high floods, whereas the basins are poorly drained and may be inundated for weeks or months at a time. In response, the majority of the towns and villages are commonly located on the levees of active or abandoned channels. The natural levees of the lower Mississippi River are strengthened and heightened by embankments that form artificial levees. These dikes keep the waters out of most of the basins, except at times of exceptionally violent flooding. They also help to stabilize the meander belt. On some floodplains the channel has been artificially straightened to facilitate river transport. However, cutting through the meander loops disrupts the geomorphologic balance, often with undesirable results, since the shortened channels have steeper gradients, leading to gullying of the floodplain and a drop of groundwater table.

8-3. FLAT FLOODPLAINS

Despite the importance of "convex" floodplains, the great majority of river plains are flat or gently concave in cross-section. Natural levees are inconspicuous or absent, and the alluvial flats rise very gently to the outer margins of the floodplain. Except where small floodplains have experienced rapid sedimentation in response to accelerated soil erosion, most smaller streams have flat floodplains. This is probably best explained by a small floodplain area, liable to continued stream reworking.

However, most intermediate-sized and many major rivers also have flat floodplains. Such alluvial surfaces are primarily formed through lateral accretion, that is, bed-load sedimentation on inside meander bends (Figure 8–5). Coarse bed-load deposits are embanked against the inside bank during bankfull stage while finer, suspended sediments are laid down on the channel flood during low water. There may be some flood silts in the wake of overbank discharge, but these remain veneers because of repeated erosion. During the next flood surge, the suspended material is swept out of the channel and erosion on the outside bank is compensated by new bed-load accretion on the inside bank. In this way floodplain sedimentation is primarily a matter of inclined sands or gravels (Figure 8–5).

A second type of flat floodplain is related to alluviation by a braided stream. In fact, braided streams are almost exclusively associated with flat floodplains.

The fundamental genetic distinction between flat and convex floodplains seems to be that the alluvium of the flat variety consists primarily of bed load, whereas that of the convex type is predominantly a matter of flood silts and fine sands. This would explain why convex floodplains are most commonly associated with the low-gradient alluvial plains of large rivers that have a preponderance of suspended load. Corresponding with the origin of their alluvial surface, flat floodplains are not only distinctive in that they lack natural levees. Backswamps are absent, and they seldom have gathering

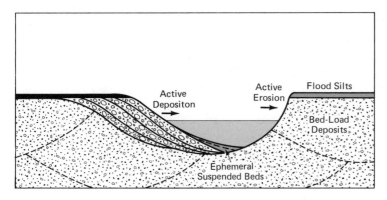

Figure 8–5. Section of a Flat Floodplain.

streams. Abandoned channels do occur but are commonly less conspicuous and lack the relative relief provided by natural levees. Further, the channels themselves are both wider and shallower, reflecting on the preponderance of bed load.

8-4. DELTAS

When a stream enters a horizontal body of water or a lower-gradient river, special floodplain types may develop. These include deltas and alluvial fans, two geomorphic features of considerable importance and widespread occurrence. Functionally, both deltas and alluvial fans serve and act as floodplains. Unlike floodplains, however, they are characterized by stream divergence, and whereas deltas have much lower gradients than those of the average floodplain, alluvial fans are remarkably steep. Accordingly, deltas are almost horizontal, while alluvial fans are prominently convex in longitudinal profile.

The name "delta" was coined by the Greeks, whose penchant for geometry led them to find a resemblance between that letter of the Greek alphabet and the Nile Delta, both of which look like an equilateral triangle. As Diodorus phrased it, almost as though he were writing a description for a geometry text:

> Now where the Nile in its course through Egypt divides into several streams it forms the region which is called from its shape the Delta. The two sides of the Delta are described by the outermost branches, while its base is formed by the sea which receives the discharge from the several outlets of the river. It empties into the sea by seven mouths. . . .[2]

A delta is simply an extension of a floodplain, with minimal gradients and built out into a body of water. At the point where a floodplain changes to a delta, there is a break of gradient, and the river channel bifurcates into two or more *distributary channels* (see Figure 8–6). In fact, channel divergence is one of the hallmarks of a delta. Each of the distributaries forms meander belts that are accompanied by natural levees and associated with flood basins that in turn are studded with abandoned distributary channels and natural levees. But, corresponding to the reduced gradients, levee-basin relief is considerably less than on the floodplain. Similarly, the relative area of backswamps is greatly increased, and flood-basin lakes may be linked to the open seas by interdistributary bays.

The greater part of a delta is a result of fluvial processes, deposited by running waters, and known as the *deltaic plain*. It contrasts with the *deltaic fringe*, which is a coastal environment where standing water and wave action assume considerable or even primary importance. Shallow bars of sand, silt, or clay are deposited rapidly in standing

[2] Book I, 33:5–7. *Diodorus of Sicily*. London, W. Heinemann, 1960. Translated by C. H. Oldfather.

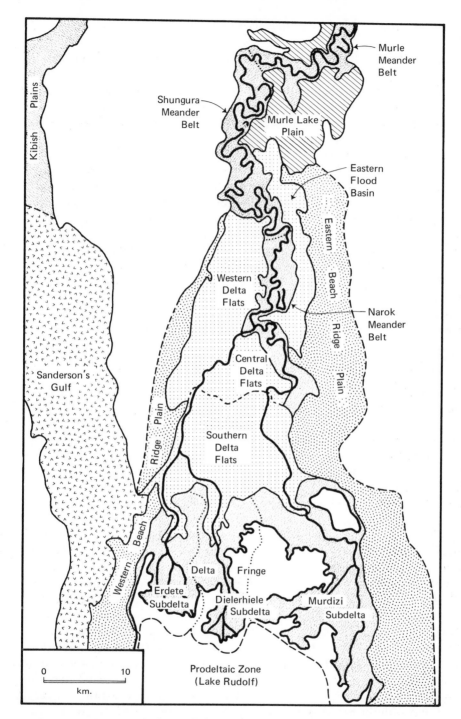

Figure 8–6. Geomorphology of the Omo Delta, a Complex of Units Developed in Response to Fluctuations of Lake Level Since the Twelfth Century A.D. The entire delta fringe was submerged by a sudden rise of Lake Rudolf in the 1960s. (From Butzer, *Recent History of an Ethiopian Delta*. Copyright 1971.)

waters within the distributary mouths. The levees extend out in the lake or seas as tongues of land, with interdistributary bays fingering their way up the adjacent basins. Lagoons and mudflats, flooded by sea water during storms or periods of high tide, form a further transitional belt between land and sea. Finally, a line of sandy beach ridges or barrier islands may form a discontinuous shoreline that connects the various distributary mouths.

Deltas can be classified on the basis of the associated body of water into lake, bay, inland sea, or marine deltas. More common, however, is a classification according to shape (see section 11-5). The Mississippi and the Omo deltas are examples of a *birdfoot* delta, with distributary channels protruding into the Gulf of Mexico or Lake Rudolf. The classic Nile Delta has an *arcuate* shoreline. Wave and current activity are stronger here, so that the distributary mouths do not develop as rapidly. Instead, they are linked by a well-developed series of beach barriers. Another type of delta can form within a river mouth. An example of such an *estuarine* delta is the confluence of the Sacramento and the San Joaquin rivers in the Central Valley of California.

8-5. ALLUVIAL FANS

When mountain streams debouch on flat plains, there is a sudden reduction of competence that produces fan-shaped spreads of alluvium (Figure 8–7). Such *alluvial fans* can also develop where high-competence tributaries converge with lower-competence streams. The resulting alluvial surfaces are convex or concavo-convex in profile and radiate out from the mouth of an incised valley or a tributary stream; they range in size from miniature forms, developed beneath low banks of erodible soil or sediment, to gigantic spreads covering hundreds of square miles (thousands of sq km).

Largest and most common are fans that reach down from the foothills of a mountain range. Upstream of the fan apex, the incised stream valley generally has a concave profile that becomes convex upon the actual fan surface where stream waters diverge into a plethora of distributary channels and rills. A study of alluvial fans in the Central Valley of California by Bull (1963) showed that there is no significant break of gradient at the apex of the fans. Instead, the loss of competence is due to a rapid loss of volume as the waters diverge into multiple channels. Fan alluvium is also far more porous and permeable than the bedrock valleys upstream, so that there is increased percolation of runoff. Gradients on the outer periphery of the fan decrease, and the profile becomes gently concave once more. Sediments here are finer and less permeable, and there is renewed stream convergence.

The morphology of alluvial fans is closely related to that of the incised drainage basin upstream. In California, Bull found that the size

Figure 8–7. Alluvial Fan Below a Fault Scarp on the Omo Valley of Southwest Ethiopia. (Karl W. Butzer)

of alluvial fans was directly proportional to the size of the drainage basin, while fan gradients were proportional to those of the incised valley upstream. However, the smaller the drainage basin, the steeper the fan.

Well-developed alluvial fans are most commonly found in arid or semiarid lands or where high mountains abut steeply on tectonic depressions or on the coast. When large, contiguous alluvial fans converge, they form the extensive *piedmont alluvial plains* that are so characteristic along the foot of desert mountains in the American Southwest (Figure 8–8). Bed-load sediments are most frequent although not exclusively associated with fan alluvium. However, stream confluence fans commonly form where tributaries with a preponderance of bed load join streams with a dominantly suspended load.

8-6. ALLUVIAL TERRACES

Geomorphologic studies of countless stream valleys have shown that the balance of erosion and deposition in a stream system does not remain constant through time. On some occasions dissection has been predominant throughout a stream system, while on others alluviation has proceeded right up into the headwaters. Another alternative is that the balance of erosion and deposition was reversed along a part of the

Figure 8–8. Coalescent Piedmont Alluvia in the Southeastern Sudan. The mudflats in foreground are seasonally flooded. (Karl W. Butzer)

stream course. Changes of climate or vegetation cover are most commonly responsible for such shifts in the geomorphologic balance. In some areas, however, tectonic deformation of a drainage basin may lead to alluviation or dissection. Similar effects can be produced in coastal areas, by ups and downs of ocean level. And last but not least, man has had a significant influence in historical times as *the* major agent of accelerated soil erosion, a process that requires extensive readjustments of the balance between erosion and deposition.

It is generally difficult to explain alluvial landforms adequately within a strictly modern context. The floodplains of most larger and many smaller rivers do not border on bedrock but on older, higher-lying alluvium (Figure 8–9). As a rule, older alluvial surfaces, no matter how fragmentarily preserved they may be, indicate that the floodplain was once both higher and wider than it is today. Older alluvial beds frequently consist of coarser materials than the modern floodplain and indicate that stream competence formerly was greater. Another common incongruity is that of the small stream flowing through a great valley that it could not possibly have excavated. These ancient alluvial features deserve attention too: They help explain contemporary forms, and both as surface materials and as landforms they are often more extensive than are their "modern" counterparts (Figure 8–10).

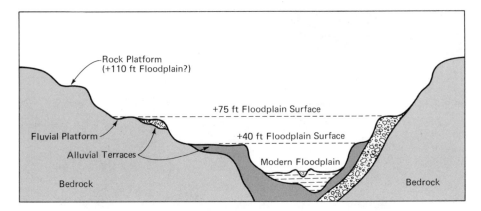

Figure 8–9. Alluvial Terraces Record Former, Higher Floodplains.

The width and the gradient of an active floodplain, as well as the texture of its alluvium, are all closely related to the geometry of the meander belt and to the competence and capacity of the stream. These in turn reflect a particular set of variables, such as size of drainage basin, available relief, lithology, structure, vegetation, hydrology, and climate. If any one of these given parameters is changed, there will be differences in stream velocity, turbulence, and sediment load, so that a new balance of forces must be established, often through complicated readjustments of floodplain geomorphology. Wavelength and amplitude of the meander belt change, commonly resulting in changes of the width and gradient of the active floodplain. Readjustments of this kind are normally achieved by accelerated alluviation or by a period of dissection, until a new balance prevails. There may also be more fundamental shifts from meandering to braided patterns or vice versa as a result of radical changes in the proportions of the suspended and bed loads.

Basin area and lithology normally remain constant. But the other factors are variable (see Table 8–1), and these are of considerable interest to the geomorphologist studying the history of a drainage basin or to the archeologist attempting to interpret an Early Man site:

1. Available relief may change through slow but protracted structural deformations, whereby the stream interfluves or the lower basin may be raised or depressed. In this way the potential energy of the basin is increased or decreased. If increased, the denudation and bedrock dissection would be accelerated, with an increase of stream turbulence and competence. If decreased, both total load and competence also decrease. Thus an increase in available relief would favor a broader, steeper floodplain downstream, with coarser alluvium (Table 8–1,A, Figure 8–11). Conversely, a decrease in available relief would

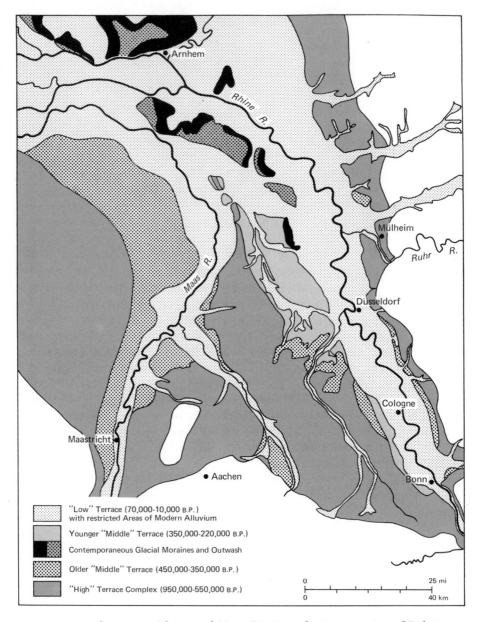

Figure 8–10. The Lower Rhine and Maas Basin at the Intersection of Belgium, Germany, and the Netherlands Was Built up of Alluvial Terraces Laid down by These Rivers During Approximately a Million Years. (Terrace distribution based on Kaiser, 1961, dating on Brunnacker, 1975.)

Table 8-1. POSSIBLE CHANGES IN THE BALANCE OF EROSION AND DEPOSITION OF A DRAINAGE BASIN

FACTOR	GEOMORPHIC RESPONSE		
	INTERFLUVES AND HILLSLOPES	UPSTREAM (CHANNEL)	DOWNSTREAM (FLOODPLAIN)
A. In response to tectonic deformation of the drainage basin			
Increased relief and potential energy	Accelerated denudation	Accelerated bedrock dissection	Accelerated bed-load alluviation with increase in floodplain width and slope
Decreased relief and potential energy	Some reduction of denudation	Reduced dissection	Initial floodplain dissection followed by reduction of sediment caliber and floodplain width/slope
B. In response to changing sea level, affecting lower floodplain only			
Increased gradient, with sea-level falling			Initial floodplain dissection followed by alluviation of suspended sediments on steeper, narrower floodplain
Decreased gradient with sea level rising			Accelerated alluviation of suspended sediments on a floodplain of reduced gradient
C. In response to changes of vegetation and hydrologic balance on the interfluves			
Increased or accelerated runoff	Accelerated denudation with foot-slope colluviation	Alluviation	Accelerated bed-load alluviation on a wider, steeper floodplain; braiding possible
Decreased or decelerated runoff	Less denudation	Dissection of alluvium	Initial floodplain dissection followed by alluviation of suspended sediments on a narrower floodplain of reduced gradient

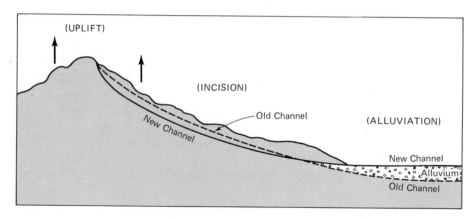

Figure 8–11. The Effects of Increased Potential Energy on Steam Profile. Uplift of the interfluves leads to intensified bedrock dissection upstream and accelerated alluviation downstream.

produce a lower-gradient floodplain with fine-grained alluvium (Table 8–1,A).

2. More rapid but no less effective are the changes in available relief that can affect the terminus of the stream: fluctuations in the level of the lake, sea, or ocean into which the stream debouches. If lake or sea level falls, the adjacent stream level must readjust. Such readjustments of the stream channel result in dissection of the floodplain, a process that is gradually transmitted upstream (Table 8–1,B, Figure 8–12,A). If lake or sea level rises, the channel compensates by accelerated deposition of fine-grained sediment (Table 8–1,B, Figure 8–12,B).

3. Changes in vegetation or climate affect the amount, rate, or annual distribution of surface runoff and stream discharge. This in turn will be reflected in the turbulence, current velocity, competence, and capacity of the stream during bankfull stage and, probably, by the frequency of bankfull stage.

    a. Accelerated runoff and denudation will lead to greater competence, transport, and, probably, capacity. Alluviation will proceed everywhere except in the interfluve zone. But even upstream there will be more sediment at any one time, although it may be only in the process of transfer, increment by increment, year by year. Thus the increased thickness of colluvial mantles in foot-slope areas and the increased thickness and breadth of channel and overbank alluvium all represent sediment that has been temporarily stranded between periods of surface runoff or turbulent discharge. Unlike the sediments of the floodplain downstream, which are mainly deposited on a permanent basis, the upstream sediments are constantly being replaced and fun-

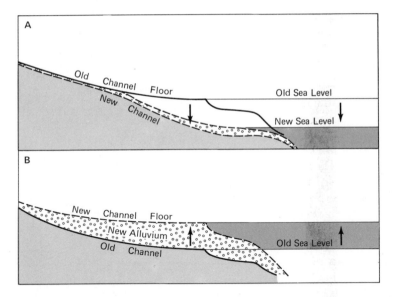

Figure 8–12. The Effects of Changing Sea Levels on Stream Profiles. A falling sea level leads to dissection of the lowermost valley (A); a rising sea level leads to accelerated alluviation in the lower valley (B).

neled on downvalley or downstream. Consequently, if the change of vegetation or climate is reversed, large quantities of sediment would be permanently abandoned upstream. This would give the impression that alluviation had been characteristic of most or all of the stream system (Table 8–1,C).

b. If an environmental change decreases or slows down surface runoff, then stream turbulence, transport, and capacity will probably be reduced. Stream activity upvalley may be restricted to dissection of older fill. Alluviation downstream would be slow and restricted to finer-grained sediments and to a smaller floodplain area. By implication there would initially be floodplain dissection (Table 8–1,C).

In this way environmental changes will normally affect the balance of alluviation and dissection similarly, that is, in the same direction, through most of the stream system. This is not a contradiction even though erosion and deposition must eventually balance. Erosion involves surface denudation or stream dissection or both. Consequently, alluviation may be balanced by denudation with little or no concomitant dissection.

In order to see the impact of these changes on floodplain geomorphology, a hypothetical sequence of alluviation, dissection, and renewed alluviation can be outlined (Figure 8–13):

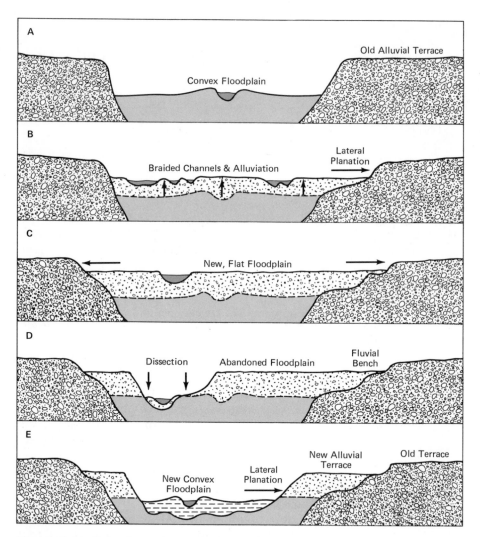

Figure 8–13. Hypothetical Development of an Alluvial Terrace in Five Stages. An initial convex floodplain (A) is buried by coarser sediments as a result of rapid alluviation by braiding channels (B). Eventually, a more stable, flat floodplain (C) is attained by alluviation and lateral planation. Another hydrological change leads to dissection of alluvium (D), abandoning the flat floodplain. A new, convex floodplain (E) is established by alluviation of fine sediments and lateral planation, approaching the conditions of stage (A).

Phase A. Initially, the floodplain is of moderate width, with a strong preponderance of suspended sediment.

Phase B. Accelerated bed-load alluviation, linked with readjustments of meander-belt geometry, channel form, and floodplain slope, begins. Rapid sedimentation in the channel bed leads to rapid shifts of the channel and possible braiding. As stream sinuosity increases, older alluvium on the floodplain periphery is undercut, thereby increasing the width of the active floodplain.

Phase C. A new balance is established with a broader and steeper floodplain than that of phase A, with greater proportions of bed load. This new floodplain has a higher elevation.

Phase D. A decrease of turbulence and bed load initiates a second period of readjustment, with deepening of the channel and temporary entrenching of the meander belt into the floodplain. Dissection stops when the floodplain gradient has been reduced and the meander belt geometry readjusted.

Phase E. A new floodplain is created within and at a lower elevation than the old one (phase C), as the meander belts swing freely and undercut the coarser sediments of phase C.

The final picture now is that of an active floodplain, surrounded by remnants of an older, higher-lying one. These flat surfaces of old alluvium have steep margins adjacent to the next lower alluvial surface and are therefore called *alluvial terraces* (see Figures 8–9 and 8–13). If there have been several cycles of alluviation and dissection, there may be several sets of alluvial terraces at different elevations and of different ages. The process of undercutting of older materials, as the floodplain is widened during a period of accelerated alluviation, is known as *lateral planation*. If lateral planation extends beyond the alluvial plain, *fluvial platforms* may be cut into the adjacent bedrock (Figure 8–14).

Laboratory studies by Schumm and Parker (1973) have shown that the shift from a braided channel and broad floodplain (c) to dissection and reestablishment of a narrower, lower river plain (d) need not necessarily be the result of a reversal of the external factors discussed previously. As upstream tributaries complete their readjustment to the original impetus, sediment loads are decreased and channel incision may set in by itself. This confirms the results of many field geomorphologists, that periods of alluviation are relatively brief compared with the total duration of anomalous external parameters such as a low sea level or climatic change. Overall, it needs to be emphasized that the model of alluvial terraces presented here is offered as a general framework only. Appropriate analytical frameworks must be devised for each specific investigation, in order to be sensitive to the multivariate nature of the accessible data. Sediment properties, sediment disposition, erosional interfaces, and accretionary forms all re-

Figure 8–14. The Fish River Canyon of Southwest Africa Records an Ancient Meander Train Entrenched over 1,200 Feet (400 m) into a High Plateau. (Karl W. Butzer)

quire careful, small-scale scrutiny. This information can then be evaluated in its regional context and in the light of any faunal, floral, and radiometric data. Ultimately, interpretation can only be made with a full appreciation of contemporary processes and balances and within the framework of a dynamic, interactive, paleosystem.

Most of the world's rivers and many smaller stream systems show at least one set of alluvial terraces, often extending well up into their headwaters. There may also be remnants of several higher alluvial terraces with associated fluvial platforms. The alluvial terraces are normally best developed in broad basins, whereas they will be narrow or replaced by rock-cut platforms in valley reaches constricted by bedrock. Similar cycles of deposition and erosion can also affect deltas and alluvial fans, creating sequences of analogous terraces.

8-7. MISFIT STREAMS AND ENTRENCHED MEANDERS

The changes of climate of the recent geological past were quite fundamental in some parts of the world, bringing about drastic changes in stream discharge. When volume of flow is reduced by 90 to even 99 percent, it is impossible for a stream to readjust its floodplain slope and width as it becomes impotent to dissect its own alluvium. Such streams are conspicuous by their insignificant size within broad, often nonfunctional floodplains or valleys. Meander amplitude is much smaller than floodplain width, and wavelength smaller than those recorded by bends once cut along the floodplain margins. The active floodplain further shows a veneer of fine-grained channel and flood silt, resting abruptly on top of coarse bed-load deposits. Streams of this kind have been called *misfit* or underfit, since they are now inadequate to account for their own valleys. Dury (1964) has shown that most major rivers in Europe and the United States are underfit along at least 50 percent of their length, as a result of reduced peak flow in postglacial times.

Another, radical form of adjustment may result from rapid and protracted tectonic deformation in a stream system. If the slope, that is, potential energy of a river is strongly increased over a long period of time, dissection will continue through the fill underlying the floodplain, down into the bedrock below. When this happens, the existing meander pattern is projected and incised into the bedrock. Such *entrenched meanders* are, of course, incapable of shifting downstream, although sinuosity may increase during entrenchment if the bedrock is weak (providing an excess of bed load) or if discharge is inadequate to mobilize the entire bed load during peak runoff (Shepherd and Schumm, 1974). Entrenched meanders are most commonly found where rapid uplift in the middle or lower course of an existing stream has been matched by rapid stream incision; they remain a museum record of long-vanished floodplain meanders. Although entrenched meanders are fairly common, among the most awe inspiring are the goosenecks of the San Juan River in Utah and the Fish River Canyon of Southwest Africa (Figure 8–14).

Chapter 9

STREAM VALLEYS AND RIVER PLAINS (III): GEOMETRY

9-1. Base level and stream profile

9-2. Valley profiles

9-3. Stream density and order

9-4. The drainage basin and general systems theory

9-5. Selected references

Erosional topography, as a counterpart to alluvial landforms, can be easily expressed in geometric terms. Firstly, there is the longitudinal profile of stream valleys, converging at tributary confluences, and projected ultimately to the sea. Secondly, there is the cross-section of stream-cut valleys. Thirdly, there is the horizontal arrangement and spacing of streams within the drainage system. Each of these coordinate systems can be expressed mathematically as well as in descriptive terms, and in many instances each can even be explained or predicted by mathematical techniques. Longitudinal profiles tend to be concave upward and exponential in form but interrupted by knickpoints that reflect on several distinct causes. Cross-sectional profiles take on any one of nine possible shapes, depending primarily on whether crest- and foot-slope inflections are smooth or angular. These possibilities are controlled by stream spacing, available relief, and by the dominance of backwearing or downwearing, as

*discussed in Chapter 6. Finally, horizontal arrange-
ments can be effectively approached from the perspective of stream hierarchies, by establishing stream
orders, or from the aspect of drainage texture, by determining stream spacing. Just as every functional feature of the landscape has its explanation, these generalized pictures of erosional topography on a broad
scale can be rationally expressed or even predicted.
And, at the highest level of abstraction, these parameters of the drainage basin can be related to an open
system that changes and adjusts through time in
response to its many controlling factors.*

9-1. BASE LEVEL AND STREAM PROFILE

In 1869 the one-armed Civil War veteran J. W. Powell led a boat expedition into the gorges of the Colorado River and through the abyss of
the Grand Canyon. Powell, who was working for the U.S. Geological
Survey, emerged not only as a great explorer but as one of the most
original geomorphologists of the nineteenth century. He was, and it
comes as no surprise, greatly impressed by the erosive potential of
running water, but he also recognized that stream cutting had a vertical limit:

> There is a limit to the effect of [dissection], for it should be observed no
> valley can be eroded below the level of the principal stream, which carries
> away the products of its surface degradation.
>
> . . .
>
> Let me state this in another way. We may consider the level of the sea to
> be a grand base level, below which the dry lands cannot be eroded; but we
> may also have, for local and temporary purposes, other base levels of ero-
> sion, which are the levels of the beds of the principal streams which carry
> away the products of erosion.[1]

Seen in a general context, each tributary stream projects its channel to
the level of the next larger stream valley, so that its erosional and depositional activities are adjusted to the confluence. In their turn, major
tributaries are projected to a through-river and, ultimately, to the sea.
The elevation to which a stream projects its channel is called *base
level*. Tributary confluences serve as *local* base levels for individual
stream segments, while the sea serves as the *ultimate* base level for a

---

[1] John Wesley Powell, *Exploration of the Colorado River of the West and Its Tributaries
(1869–72)*. Washington, D.C., U.S. Geological Survey, 1875, pp. 163, 203.

stream system. It is also common to consider lakes or hard rock barriers that interrupt a stream profile as *temporary* base levels.

The longitudinal profile of streams is typically concave upward and exponential in form (Figure 9–1,A), even though some floodplains and most deltas and alluvial fans exhibit marked local convexity (Figure 9–1,B). Stream profiles are controlled by the same variables that influence other aspects of stream behavior: size of drainage basin, available relief, lithology, structure, vegetation, and climate. The basic arrangement of the basin, its potential energy, and its base level provide the background; the pattern of runoff and discharge are determined by climate, vegetation, and groundwater geology; the amount and the type of sediment load are controlled by climate, vegetation, slope, and lithology; finally, the nature of local details or irregularities is modified or determined by lithology, structure, and above all, man's

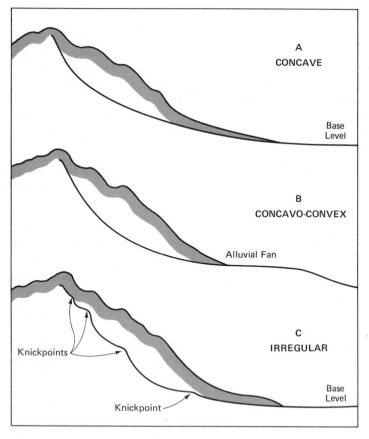

Figure 9–1. Stream Profiles and Base Level: (A) Concave, (B) Concavo-Convex, and (C) Interrupted by Knickpoints.

activities. Thus although mathematical curves can be fitted to stream profiles, the number of interdependent factors is so great that the constants used in such expressions are necessarily unique for each stream system.

Few stream profiles are smoothly concave (or concavo-convex). Instead, there generally are inflection points, known as *knicks* or *knickpoints*. These are most common upstream, where the stream bed is incised in bedrock. Knickpoints are most frequently associated with tributary confluences, with outcrops of hard rocks, or with valley constrictions that reflect structural controls (Figure 9–1,C). Several authors have been able to show that stream gradients vary in different types of rock, and alternating lithologies can create obvious discontinuities of profile, depending on the angle at which the beds are intersected. Flowing over a bed of resistant rocks, a stream will maintain low gradients; however, when the same stream cuts down through the resistant stratum into softer beds below, the gradient is abruptly steepened. Waterfalls or rapids may form at such breaks of stream profile.

Knickpoints are not entirely absent from the alluvial reaches of stream systems, where they form either through erosion or deposition. Local scour is possible when there is an excess of stream competence and transport ability relative to available load. Situations of this kind are common in ephemeral streams with silty alluvium and relatively steep channels; gullying proceeds headward from such knickpoints (Figure 9–2). Conversely, a sudden decrease of competence may lead to local fill. Increased deposition will subsequently proceed upstream as a result of the decreased gradient above the new knickpoint. Both of these categories of knickpoint in alluvial valleys reflect adjustments of stream profile that may only be temporary. Other knickpoints will

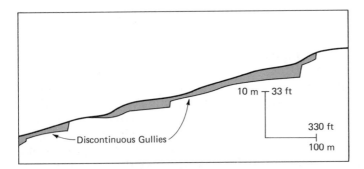

Figure 9–2. Severe Floods in Intermittent Streams Can Create Knickpoints in Fine-Grained Channel Alluvium by Cutting of Discontinuous Gullies (Gray) in Areas of Maximum Longitudinal Gradient. Mancos River, New Mexico. (Modified after Schumm and Hadley, 1957.)

form as a result of gross differences in bed-load caliber, usually near tributary confluences.

Once a stream system has adjusted to a particular pattern of vegetation and climate, on the one hand, and of lithology and structure, on the other, a balance of forces and processes becomes apparent. Denudation and dissection continue among the interfluves, concomitant with alluviation downstream of that part of the load that is not carried into lakes or seas. The boundary between areas of dominant erosion and areas of dominant deposition will remain more or less constant according to the external factors that determine the geomorphologic balance. Similarly, the relative rates of erosion and deposition, which also vary from one environment to another, will remain fairly constant under the given premises. Natural variability from one bankfull stage to another, or from one year to the next, is compensated by readjustment and self-regulation of the stream channels. Consequently, the functional role of any segment of the stream system—scour, transport, fill—remains identical over the centuries, even though the details of a bedrock channel or meander belt will change through time. In other words, geomorphologic evolution is normally accompanied by a degree of geomorphologic stability.

Change in one or more of the variables governing the existing balance will require major readjustments within the stream system, such as were described in section 8–6. If the external changes are significant, there may be a rupture of equilibrium, sufficiently drastic to leave its mark on the geomorphic landscape within a comparatively short time. Ultimately, however, a new form of *dynamic equilibrium* will be achieved, with a different rate and balance of erosion and deposition assuring constant adjustment (rather than constant form) (see Abrahams, 1968).

9-2. VALLEY PROFILES

Whereas longitudinal stream profiles may be concave, concavo-convex, or interrupted by knicks, transverse valley profiles can also take on a number of characteristic shapes. In section 5-1 a basic vocabulary was introduced, whereby a change of slope can be smooth or angular and a valley may have straight, concave, convex, or concavo-convex sides. In order to describe the transverse profiles of valleys in particular and of fluvial topography in general, Ollier (1967) devised a simple but useful classification. It is predicated on the fact that interfluves may be flat, angular, or rounded in general form, while valley bottoms may be rounded, flat, or V-shaped. This allows nine possible combinations:

Rounded interfluves with
$\left\{\begin{array}{l}\text{rounded valley bottom}\\\text{flat valley bottom}\\\text{V-shaped valley bottom}\end{array}\right.$

Flat interfluves with
$\left\{\begin{array}{l}\text{rounded valley bottom}\\\text{flat valley bottom}\\\text{V-shaped valley bottom}\end{array}\right.$

Angular interfluves with
$\left\{\begin{array}{l}\text{rounded valley bottom}\\\text{flat valley bottom}\\\text{V-shaped valley bottom}\end{array}\right.$

These combinations are illustrated graphically in Figure 9–3, where each profile type is coded.

With appropriate adjustments in the steepness of the slope facets, the curvature of the segments, and the vertical and horizontal spacing, the nine diagrammatic types of Figure 9–3 can account for the profiles of most erosional landscapes developed in homogeneous rocks. Further subdivisions can be made according to the nature of the slopes, which may be straight-sided (ss), concave (cv), convex (cx), or concavo-convex (cc). The classification does not, however, include flat plains, nor does it do justice to complex slopes, asymmetrical valleys, or prominent features related to lithological variations.

Ollier outlines several different sequences of landscape evolution in homogeneous rocks, depending primarily on available relief and drainage density:

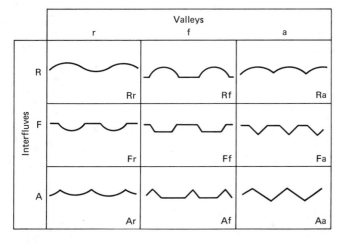

Figure 9–3. Classification of Erosional Topography Based on Profile of Valleys and Interfluves. After Ollier (1967). Copyright *Australian Geographical Studies.* The letters R and r denote rounded, F and f flat, A and a angular. Shape of interfluves (in capital letters) is given first, valley form (in lower-case letters) second.

1. Low Available Relief and Low Drainage Density. Dissection is impeded at first, producing V-shaped valleys cut into flat interfluves (type Fa, Figure 9–4,A). If we assume a dominance of backwearing (instead of downwearing), valley widening will soon begin to cut more or less flat-bottomed and straight-sided valleys (type Ff, Figure 9–4,B). Eventually, the interfluves will be reduced to conical residual hills (type Af, Figure 9–4,C). On the other hand, if downwearing is dominant, the profile will change from type Fa to a concavo-convex variety of type Ff, ultimately leaving dome-shaped residual masses (type Rf). In highly erodible materials, downwearing may also favor an initial profile of rounded interfluves with V-shaped valleys (type Ra).

2. Low Available Relief and High Drainage Density. Depending on how closely spaced the initial streams are, the valley sides will soon intersect, so that type Fa is rapidly replaced by angular interfluves with V-shaped valleys (type Aa, Figure 9–5,A). As valley widening replaces dissection, the ultimate residual forms will be of type Af (Figure 9–5,B) or Rf, depending on the ratio of backwearing to downwearing.

3. High Available Relief and Low Drainage Density. The evolutionary sequence is the same as for sequence 2, except that the Aa profile type persists for a very much longer period of time (Figure 9–6,A,B).

4. High Available Relief and High Drainage Density. The basic sequence is similar to that of sequence 3, but valley-side intersection almost immediately produces a profile of type Aa (Figure 9–7,A). Once

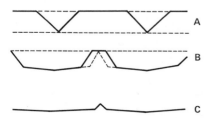

Figure 9–4. Landscape Evolution by Backwearing, Given Low Available Relief and Low Drainage Density. (After Ollier, 1967.)

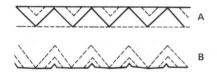

Figure 9–5. Landscape Evolution Given Low Available Relief and High Drainage Density. (After Ollier, 1967.)

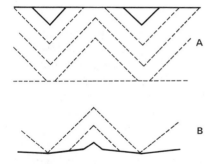

Figure 9–6. Landscape Evolution Given High Available Relief and Low Drainage Density. (After Ollier, 1967.)

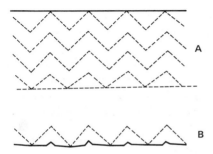

Figure 9–7. Landscape Evolution Given High Available Relief and High Drainage Density. (After Ollier, 1967.)

valley widening begins, the residual forms of type Af or Rf (Figure 9–7,B) will be cut more rapidly than in the case of sequence 3.

Sequences 1 and 2 are typical in the development of erosional plains. The initial and intermediate stages of evolution include irregular plains, hills, and low tablelands in various degrees of dissection. Sequences 3 and 4 are more characteristic for the erosional development of mountainous terrain. Sequence 3 begins with a tableland, until increasing dissection produces low or high mountains. Ultimately, under favorable circumstances, the mountain range may be reduced to a plain with hills. Sequence 4 does not include the initial tableland.

It can be seen that fluvial processes in the widest sense are able to model landscapes with any type of interfluve and with V-shaped or flat-bottomed valleys. Although rounded valley bottoms may be associated with running water, some of the most conspicuous concave valleys are a result of glacial erosion.

9-3. STREAM DENSITY AND ORDER

The basic horizontal geometry of drainage basins was expressed rather forcefully by Powell for the Colorado catchment: "Ten million cascade brooks unite to form ten thousand torrent creeks; ten thousand torrent creeks unite to form a hundred rivers beset with cataracts; a hundred roaring rivers unite to form the Colorado, which rolls, a mad, turbid stream, into the Gulf of California."[2] In order to have a set of systematic measures for drainage basins that allows meaningful comparisons, two parameters must be introduced.

The spacing of main streams and their tributaries, as well as the width of interfluves between valleys, is known as terrain *texture*. A useful measure of texture is obtained by calculating the cumulative length of drainage lines—both permanent and temporary—per unit area for a drainage basin. The index is defined by:

$$\text{Drainage density } (Dd) = \frac{\text{total length of streams (mi)}}{\text{area (sq mi)}}$$

This ratio can be converted to metric units (km/sq km) through multiplying by a factor of 0.62. A second measure of texture is obtained by counting the number of crenulations (see Figure 9–8) in that drainage basin contour that has the greatest number of crenulations. This value is then divided by the length of the perimeter of the drainage basin to obtain:

$$\text{Texture ratio } (T) = \frac{\text{number of contour crenulations}}{\text{length of perimeter (mi)}}$$

Again the result can be converted to metric units (crenulations/km) by applying a factor of 0.62. The two indexes are further explained and their relationships illustrated in Figure 9–8.

These indexes, developed by Strahler (1957), permit objective comparisons between different drainage basins and reflect primarily on rock type, vegetation, and climate. So, for example, the drainage density may range from less than 2 in areas of very *coarse* texture to 500 or more in areas of *ultrafine* texture. Most commonly these values range from between 2 and 40, with *medium* texture corresponding to a drainage density of 8 to 24. In the case of the texture ratio, coarse texture can be defined as less than 4, medium texture between 4 and 10, and fine texture over 10. Coarse texture is common in desert environments or where surface rocks are very permeable, such as on the Colorado Plateau (Figure 9–9) or the southern Great Plains. Fine texture, on the other hand, is more typical of moist environments or areas of soft, erodible rocks. A good example of ultrafine texture is the intri-

[2] Ibid., p. 4.

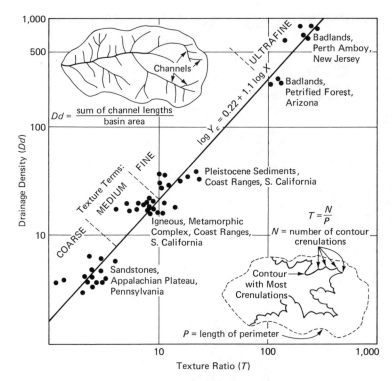

Figure 9–8. Drainage Density and Texture Ratio. (Redrawn from Strahler, 1957, with permission.)

cately gullied surface of the Badlands of South Dakota, developed in very soft but impermeable material. Gullying as a result of man-induced erosion can rapidly increase drainage density, while conservation measures can eliminate master rills and gullies, so as to reduce drainage density. In other words, man is another variable in controlling stream-basin geometry.

Although indexes of texture help express stream spacing, the relationship of minor tributaries to major streams in terms of relative frequency and size also requires definition. A common method of describing such a hierarchy of *stream orders* is illustrated by Figure 9–10. The fingertip tributaries along the watershed of a basin are first-order streams; two first-order streams converge to form a second-order stream, two second-order streams to form a third-order stream, and so on. In each case at least two streams of order n are required to form a stream of order $n + 1$. With such a system it is possible to describe headwater tributaries as *low-order* streams and downstream channels as *high-order* streams. It is also possible to talk about first-order or n-order drainage basins, and a particular basin can be described, for ex-

Figure 9–9. The Snow-Capped Colorado Plateau Has a Very Low Drainage Density, Despite Intensive Dissection Along the Margins of the V-Shaped Canyon of the Little Colorado River. (Karl W. Butzer)

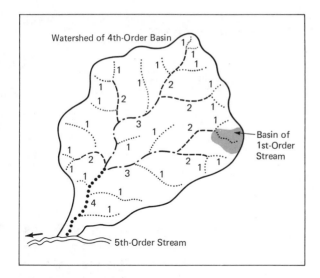

Figure 9–10. Method of Designating Stream Orders.

ample, as a fifth-order basin. The first-order streams of a drainage basin are the smallest gullies or channels having well-defined banks. Quite frequently, such first-order tributaries are intermittent streams of an ephemeral type, carrying water only after rains.

Two measures related to stream order can be calculated from good topographic maps.[3] The *bifurcation ratio* refers to the ratio of the number of stream segments of order n to those of the next highest order, $n + 1$. This dimensionless parameter indicates how many more streams of order n there are than of order $n + 1$. Since tributaries converge, there are fewer high-order streams than low-order streams. The *length ratio* is the ratio of the average length of streams of order n to the length of the streams of the next highest order, $n + 1$. Another measure that can sometimes be calculated from exceptionally good maps is the average channel gradient. The average vertical distance transversed by the longitudinal profile of streams of order n is divided by average horizontal distance.

Some of these values and ratios are shown in Table 9–1, as computed for a small drainage basin in southwestern Wisconsin. Parameters of this kind can be compared simply or in groups, between drainage basins of identical order; the differences can sometimes be explained by variations of lithology, vegetation, or climate.

[3] Field checks in many areas have shown that channels mapped as first order on U.S. Geological Survey 1:24,000 maps may in reality be second-order and, occasionally, even third-order channels. Consequently, for any such map exercises, stream channels must first be extended and multiplied whenever contour crenulations are V-shaped and then continued upstream until the contours are smoothed out.

Table 9-1. STREAM ORDERS AND CHANNEL SEGMENTS OF A SMALL
DRAINAGE BASIN IN SOUTHWESTERN WISCONSIN

STREAM ORDER	NUMBER OF CHANNELS	AVERAGE CHANNEL LENGTH (MI)	AVERAGE CHANNEL GRADIENT (FT/MI)	BIFURCATION RATIO	LENGTH RATIO
1	1210	0.16	97.5	3.18	1.07
2	381	0.15	71.1	4.43	0.32
3	86	0.47	55.7	3.74	0.55
4	23	0.85	70.6	4.60	0.53
5	1	1.60	90.0		

Data provided by Bruce Gladfelter.

The preceding discussion of fluvial geomorphology has suggested that there is an orderly development of stream systems. So, for example, the number of drainage basins in a fluvial system developed on a homogeneous surface is predictable or explicable from a hexagonal model that is used in central place theory (Woldenberg, 1969). Then, again, there is a predictable relationship between meander wavelength and drainage-basin area. In a more general fashion, several "laws" have been formulated to describe the normal development of drainage basins. These are called the *laws of drainage composition*, and, in order to emphasize their axiomatic nature, they are stated in nonmathematical language in Table 9–2. Some stream systems do not conform to the laws of stream numbers or stream gradients as a result of exceptional bedrock or structural controls. By and large, however, these laws help to formalize and express some of the fundamentals of drainage-basin geometry.

Table 9-2. LAWS OF DRAINAGE COMPOSITION

1. *Law of Stream Numbers.* The number of streams of a given order in a drainage basin decreases systematically with increasing stream order.

2. *Law of Stream Lengths.* The average length of streams of a given order in a drainage basin increases systematically with increasing stream order.

3. *Law of Basin Areas.* The average drainage-basin area of streams of a given order increases systematically with increasing stream order.

4. *Law of Channel Maintenance.* The average length of stream channels of a given order increases systematically with increasing average drainage-basin area of a given order.

5. *Law of Stream Gradients.* The average gradient of streams of a given order decreases systematically with increasing stream order.

6. *Law of Basin Relief.* The average relief of drainage basins of a given order increases systematically with increasing stream order.

After Horton, 1945, Schumm, 1956, and Morisawa, 1968.

9-4. THE DRAINAGE BASIN AND GENERAL SYSTEMS THEORY

Running water is one of the most universal geomorphic agents and, on a worldwide basis, the most effective. For these reasons fluvial processes and landforms are of critical importance for geomorphological theory. Since about 1945 the whole complex of problems ranging from hillslope development to the evolution of stream systems has become the focus of renewed interest and intensive study. At the highest level of generalization, all of these problems center on the geomorphology of the drainage basin.

In order to view these fundamental questions within a wider perspective, several geomorphologists have in recent years attempted to examine geomorphologic phenomena within the framework of systematic models. Chorley (1962) has explicitly applied general systems theory to the study of geomorphology. A *system* has been defined as a set of objects and the relationships between both the objects and their properties. A *closed* system has closed boundaries across which no import or export of energy or material can take place. An *open* system, on the other hand, has permeable boundaries and is maintained by a continuing supply and removal of energy and material. In geomorphology, free or potential energy is a function of available relief, that is, of the elevation of water and sediment above base level. Such energy is progressively lost as relief is reduced by erosion. The degree to which energy in a system cannot be transformed into mechanical work is described by the term *entropy*, so that an increase in entropy implies a trend toward minimum potential energy (Leopold and Langbein, 1962).

One of the traditional approaches to geomorphology, as represented by Davis and his followers, has some (limited) analogies to closed-system thinking. In Davis' view, landscape evolution begins with tectonic uplift, at which time a certain measure of potential energy is imparted to the system. As erosion proceeds from the dissection of rugged mountains to the planation of extensive lowlands, there is an inexorable loss of energy through time, until dissection and denudation have reduced available relief to a minimum. There are no points of equilibrium between the initial condition of maximum energy and the terminal one of maximum entropy. The sequence of events is inevitable, so that the assemblage of landforms at any one time is a function of the initial conditions and the time that has since elapsed.

Although this simplistic approach was popular between 1900 and 1945, the growth of empirical information has encouraged the development of more realistic concepts. The use of a more sophisticated, open-system model can give better credit to the interactive processes that model a drainage basin and its landform elements. Water is continually being added to the basin from the atmosphere and removed by stream discharge and evaporation. Since this flux of mass and energy is not uniform, except perhaps as a statistical mean, there necessarily is continuous interaction with the geomorphic elements of the basin. In this way there is repeated self-adjustment of form and process within the system on a short-term basis.

Changes in the supply of mass and energy from outside, be they climatic or tectonic, will lead to major readjustments of the system on a long-term basis. Thus, as developed by Schumm and Lichty (1965), the analogy allows for the effects of repeated climatic variation and crustal instability, elements that could not be integrated, except as "accidents," into the Davisian system. Unfortunately, the drainage-basin analogy is less satisfactory when the progressive loss of dissolved, suspended, and bed sediments from the stream system is considered.

Although there is no simple way for the geomorphologist to apply or test this or other system models, such a brief excursion into the philosophy of an open system has advantages for the student. It stimulates an awareness of the relationship between form and process and of the multivariate nature of geomorphic phenomena. At the same time, it emphasizes that both process and form can change with time, that there is no such thing as a unidirectional course of events, and that initial conditions often are of subordinate importance in determining the constellation of landforms at any one time.

9-5. SELECTED REFERENCES

A. General

Chorley, R. J., ed. *Water, Earth and Man.* London, Methuen, and New York, Bares & Noble, 1969, 588 pp.
———, A. J. Dunn, and R. P. Beckinsale. *The History of the Study of Landforms.* Vol. 1: *Geomorphology Before Davis.* London, Methuen, 1964, 700 pp. See Chaps. 25 and 26.
Gregory, K. J., and D. E. Walling. *Drainage Basin Form and Process: A Geomorphological Approach.* London, Edward Arnold, and New York, Wiley, 1973, 456 pp.
Leopold, L. B., G. B. Wolman, and J. P. Miller. *Fluvial Processes in Geomorphology.* San Francisco, Freeman, 1964, 522 pp.
Morisawa, Marie. *Streams: Their Dynamics and Morphology.* New York, McGraw-Hill, 1968, 175 pp.
Russell, R. J. *River Plains and Sea Coasts.* Berkeley, Calif. University of California Press, 1968, 173 pp.
Schumm, S. A., ed. *River Morphology.* Stroudsburg, Pa., Dowden, Hutchinson and Ross, 1972, 444 pp.

B. Stream Dynamics

Barry, R. G. The world hydrological cycle. In Chorley, ed., 1969, *supra,* pp. 11–30.

Beckinsale, R. P. River regimes. In Chorley, ed., 1969, *supra*, pp. 455–472.

Butzer, K. W. Fine grained alluvial fills in the Vaal and Orange Basins of South Africa. *Proceedings, Association of American Geographers*, Vol. 13, 1971, pp. 41–48.

———. Accelerated soil erosion. In Ian Manners and M. W. Mikesell, eds., *Perspectives on Environment*. Washington, D.C., Association of American Geographers, 1974, pp. 57–78.

Carson, M. A. *The Mechanics of Erosion*. London, Pion, 174 pp. See Chap. 2.

Chow, V. T., ed., Handbook of Applied Hydrology: A Compendium of Applied Hydrology. New York, McGraw-Hill, 1964.

Foster, E. E. *Rainfall and Runoff*. New York, Macmillan, 1948, 487 pp.

Friedkin, J. F. A laboratory study of the meandering of alluvial rivers. In Schumm, ed., 1972, *supra*, Chap. 13.

Gregory, K. J., and D. E. Walling. *Drainage Basin Form and Process*. London, Edward Arnold, and New York, Wiley, 1973, 456 pp.

Guy, H. P. Fluvial sediment concepts. *U.S. Geological Survey, Techniques of Water-Resources Investigation*. Book 3, Chap. C1, 1970, pp. 1–55.

Hastings, J. R., and R. M. Turner. *The Changing Mile: An Ecological Study of Vegetation Change with Time in the Lower Mile of an Arid and Semi-Arid Region*. Tucson, Ariz., University of Arizona Press, 1965, 317 pp.

Hoyt, W. G., and Langbein, W. B. *Floods*. Princeton, N.J., Princeton University Press, 1955, 469 pp.

Livingstone, Daniel A. Chemical composition of rivers and lakes. *U.S. Geological Survey, Professional Paper* 440 G, 1964, pp. 1–64.

Meade, R. H. Errors in using modern stream-load data to estimate natural rates of denudation. *Bulletin, Geological Society of America*, Vol. 80, 1969, pp. 1265–1274.

Picard, M. D., and L. R. High. *Sedimentary Structures of Ephemeral Streams*. New York, Elsevier, 1973, 224 pp.

Strahler, A. N. *Physical Geography*, 3rd ed. New York, Wiley, 1969, 733 pp. See Chap. 25.

Trewartha, G. T., A. H. Robinson, and E. H. Hammond. *Elements of Geography*, 5th ed. New York, McGraw-Hill, 1967. See Chap. 13.

C. Stream Channels and Floodplains

Bull, W. B. Geomorphology of segmented alluvial fans in western Fresno County, California. *U.S. Geological Survey, Professional Paper* 352E, 1963, pp. 89–128.

Butzer, K. W. *Recent History of an Ethiopian Delta. Research Papers, University of Chicago, Department of Geography*, No. 136, 1971, pp. 1–184.

Dury, G. H. Relation of morphometry to runoff frequency. In Chorley, ed., 1969, *supra*, pp. 419–430.

Ireland, H. A., C. F. S. Sharpe, and D. H. Eargle. Principles of gully erosion in the Piedmont of South Carolina. *U.S. Department of Agriculture, Technical Bulletin 633*, 1939, pp. 1–143.

Leopold, L. B., and W. B. Langbein. River meanders. *Scientific American Offprint* No. 869 (June 1966), 11 pp.

Schumm, S. A. The shape of alluvial channels in relation to sediment type. *U.S. Geological Survey, Professional Paper* 352B, 1960, pp. 17–30.

――――. Sinuosity of alluvial rivers on the Great Plains. *Bulletin, Geological Society of America*, Vol. 74, 1963, pp. 1089–1100.

――――, and H. R. Khan. Experimental study of channel patterns. *Bulletin, Geological Society of America*, Vol. 83, 1972, pp. 1755–1770.

Shepherd, R. G., and S. A. Schumm. Experimental study of river incision. *Bulletin, Geological Society of America*, Vol. 85, 1974, pp. 257–268.

Trimble, S. W. Landscapes of Wisconsin. *Annals, Association of American Geographers*, Vol. 66, 1976, in print.

D. Alluvial Terraces

Brunnacker, Karl. The mid-Pleistocene of the Rhine Basin. In K. W. Butzer and G. L. Isaac, eds., *After the Australopithecines*. The Hague, Mouton, and Chicago, Aldine, 1975, pp. 189–224.

Butzer, K. W., and C. L. Hansen. *Desert and River in Nubia: Geomorphology and Prehistoric Environments of the Aswan*. Madison, Wis. University of Wisconsin Press, 1968, 562 pp.

Dury, G. H. Subsurface explorations and chronology of underfit streams. *U.S. Geological Survey, Professional Paper* 452B, 1964, pp. 1–56.

――――, ed. *Rivers and River Terraces*. New York, Macmillan, 1970, 283 pp.

Gladfelter, B. G. Meseta and campiña landforms in central Spain. *Research Papers, University of Chicago, Department of Geography*, No. 130, 1971, pp. 1–204.

Helgren, D. M. Rivers of diamonds: an alluvial history of the lower Vaal Basin, South Africa. *Research Papers, University of Chicago, Department of Geography*, 1976, in press.

Kaiser, K. H. Gliederung und Formenschatz des Pliozäns und Quartärs am Mittel- und Niederrhein. In K. Kayser and T. Kraus, eds., *Köln und die Rheinlande*. Wiesbaden, Steiner, 1961, pp. 236–278.

Schumm, S. A. River adjustment to altered hydrologic regimen—Murrumbidgee River and paleochannels, Australia. *U.S. Geological Survey, Professional Paper* 598, 1968, pp. 1–65.

――――. Fluvial paleochannels. In J. K. Rigby and W. K. Hamblin, eds., *Recognition of ancient sedimentary environments*. Society for Economic Paleontologists and Mineralogists, Special Publication 16, 1972, pp. 98–107.

――――, and R. S. Parker. Implications of complex response of drainage for Quaternary alluvial stratigraphy. *Nature*, Vol. 243, 1973, pp. 99–100.

E. Drainage Basins

Abrahams, A. D. Distinguishing between the concepts of steady state and dynamic equilibrium in geomorphology. *Earth Science Journal*, Vol. 2, 1968, pp. 160–166.

Ahnert, Frank. Functional relationships between denudation, relief, and uplift in large mid-latitude drainage basins. *American Journal of Science*, Vol. 268, 1970, pp. 243–263.

――――. An approach towards a descriptive classification of slope. *Zeitschrift für Geomorphologie, Supplement* Vol. 9, 1970, pp. 71–84.

Chorley, R. J. Geomorphology and general systems theory. *U.S. Geological Survey, Professional Paper* 500B, 1962, pp. 1–10.

————. *Spatial Analysis in Geomorphology.* New York, Harper & Row, 1972, 393 pp.

————, and B. A. Kennedy. *Physical Geography: A Systems Approach.* London, Prentice-Hall International, 1971, 370 pp.

Doornkamp, J. C., and C. A. M. King. *Numerical Analysis in Geomorphology.* London, Edward Arnold, 1971, 372 pp. See Chaps. 1 and 2.

Dury, G. H. The concept of grade. In G. H. Dury, ed. *Essays in Geomorphology.* London, Heinemann, 1966, pp. 211–234.

Leopold, L. B., and W. B. Langbein. The concept of entropy in landscape evolution. *U.S. Geological Survey, Professional Paper* 500A, 1962, pp. 1–20.

Ollier, C. D. Landform description without stage names. *Australian Geographical Studies,* Vol. 5, 1967, pp. 73–80.

Schumm, S. A. Evolution of drainage systems and slopes in Badlands at Perth Amboy, New Jersey. *Bulletin, Geological Society of America,* Vol. 67, 1956, pp. 597–646.

————, and R. F. Hadley. Arroyos and the semi-arid cycle of erosion. *American Journal of Science,* Vol. 255, 1957, pp. 161–174.

————, and R. W. Lichty. Time, space and causality in geomorphology. *American Journal of Science,* Vol. 263, 1965, pp. 110–119.

Speight, J. G. A parametric approach to landform regions. In E. H. Brown and R. S. Waters, eds. *Progress in Geomorphology.* Institute of British Geographers Spec. Publ. 7, 1974, pp. 213–230.

Strahler, A. N. Equilibrium theory of erosional slopes approached by frequency distribution analysis. In Schumm, ed., 1972, *supra,* Chap. 12.

————. Quantitative analysis of watershed geomorphology. *Transactions, American Geophysical Union,* Vol. 38, 1957, pp. 913–920.

Woldenberg, M. J. Spatial order in fluvial systems: Horton's laws derived from mixed hexagonal hierarchies of drainage basin areas. *Bulletin, Geological Society of America,* Vol. 80, 1969, pp. 97–112.

Chapter 10

GLACIAL LANDFORMS

Glaciers today cover 10 percent of the earth's surface but in the recent geological past extended over an additional 23 percent. Since glaciers are the most powerful geomorphic agent, they have left a legacy of "relict" glacial landforms across large parts of the continents. Glaciers may form in highlands, in which case they carve out rock basins and U-shaped valleys, adorned with a wide variety of ice-laid and meltwater deposits. Unlike these mountain ice streams, which follow existing drainage lines, icecaps such as those of Greenland and Antarctica move radially outward from a central focus. Continental glaciers of this type covered much of North America and Eurasia as recently as 10,000 years ago, scouring extensive erosional plains near the former ice centers and leaving broad depositional surfaces near the former ice margins. The characteristic geomorphic features imparted by glaciation withstand remodeling by running water and gravity agencies for tens of thousands of years, remaining remarkably fresh and distinctive. In other words, the

*geomorphology of the continents cannot be explained
by contemporary processes alone. Instead, we must
turn back to the Pleistocene Ice Age, which brought
continental glaciers to Antarctica by 6 million years
ago and which led to alternating glaciation and deglaciation of large segments of the mid-latitude continents between 1 million and 10,000 years ago. In fact,
the present is part of an "interglacial," somewhat
cooler than the normal planetary climate of most of
the geological past—when glaciers were totally absent
on Earth. This historical setting to Pleistocene glaciation is outlined in Chapter 18, together with other imprints of glacial-age climate on mid-latitude environments.*

**RADIO ANNOUNCER: The unprece-
dented cold weather of this summer has
produced a condition that has not yet
been satisfactorily explained. There is a
report that a wall of ice is moving southward across these counties. The disruption of communications by the cold wave
now crossing the country has rendered
exact information difficult. Little cre-
dence is given to the rumor that the ice
had pushed the Cathedral of Montreal as
far as St. Albans, Vermont. For further
information see your daily papers.**

THORTON WILDER, *The Skin of Our Teeth*
(Act I)

10-1. WHY WORRY ABOUT GLACIERS?

Glaciers do not form during a single summer, but moving ice is the
single most powerful agency of the gradational forces. Although glaciers move slowly, their crushing impact allows them to gouge out
valleys or bulldoze broad surfaces. As a result, landscapes can be profoundly altered in a comparatively brief time—so much so, that a surface eroded by running water for 10 million years can potentially be
changed almost beyond recognition by 10,000 years of glaciation.

Glaciers do not play an important role on the major continents
today. They are found in some high mountain areas, where they some-

times are of little more than scenic interest, although the significance of summer meltwaters in maintaining river flow is often overlooked. Elsewhere, they are confined to remote landmasses such as Greenland and Antarctica. Altogether only about 10 percent of the Earth's land surface is now glaciated. If this were the extent of glacial landforms, glaciation would obviously be of little more than regional importance. But this is not so.

Glaciation is so effective as a geomorphic process that its impact is clearly evident 10,000 or even 100,000 years later. It is so evident, in fact, that ancient glacial landforms can continue to dominate the countryside. Except for a mantle of soil and vegetation, the vestiges of glaciation remain fresh over many millennia and the subsequent work of running water and mass movements seems almost incidental. This legacy of glacial landforms assumes great importance because almost 30 percent of Earth's land surface was under ice as recently as 15,000 years ago. At the height of the ice age fully 33 percent of the land was glaciated. In other words, in addition to the domain of contemporary glaciers, glacial features are paramount over almost a quarter of the continental surfaces.

Landforms that were modeled or sculptured in the past by geomorphic forces different than those prevailing at present are called *relict*. Contemporary processes either are far less effective or have not had sufficient time to remodel the landscape into one completely in tune with modern factors. Relict landforms of different ages are prominent in many regions. Needless to say, they are all undergoing some degree of change. But the imprint of past events still remains unmistakable.

Of all relict landforms, glacial features are the most striking. Consequently, in turning attention to the work of moving ice we are seeking to understand a wide range of glacial landforms that characterizes vast areas of the continents. The field of *glaciology* is concerned with the mechanics of ice movement, whereas *glacial geomorphology* is devoted to the study of glacial erosion and deposition and of the landforms so created. Glacial geomorphology became a subject of almost general fascination during the mid-nineteenth century, when it was first discovered that much of Europe and North America had been glaciated during the Pleistocene Ice Age (see Table 10–1). In fact, the enthusiasm for unraveling the history and impact of the ice age was so great that most European geomorphologists during the half-century preceding World War II concentrated their work on glacial geomorphology. As a result of these studies, and more recent work on glacial processes, a great deal is now known about the subject matter. Yet the diversity of features is no less than that of any other branch of geomorphology, and there is an intrinsic difficulty in reconciling subglacial processes that can only be observed indirectly with past erosional or

Table 10–1. PLANETARY GLACIER AND TEMPERATURE TRENDS: PLIOCENE—PLEISTOCENE—HOLOCENE

	DATES[a]	PLANETARY TRENDS	COMMENTS
	Since A.D. 1895	Minor warming	Mountain glaciers retreating
	A.D. 1500–1895	"Little Ice Age"	Mountain glaciers readvance
	A.D. 1000–1300	Warm spell	Mountain glaciers retreat
	1000 B.C.	Major cooling trend begins	Mountain glaciers readvance
HOLOCENE	3000 B.C.	Warming trend begins	Mountain glaciers retreat
	4000 B.C.	Major cooling trend begins	Mountain glaciers readvance
	5000 B.C.	Maximum Holocene warmth. Final disappearance of continental glacier in Canada before 4000 B.C.	
	8300 B.C.	Major warming trend begins, with disappearance of continental glacier in Scandinavia about 7000 B.C. Temporary readvances of mountain glaciers about 8000 and again 6000 B.C.	
LATE PLEISTOCENE	75,000 B.P.	Last (Würm or Wisconsin) glacial begins. Formation of new continental glaciers in Canada and Scandinavia. After expansion into the United States and across the Baltic Sea, by 60,000 B.P., a temperate oscillation with limited deglaciation after 40,000 B.P. was followed by maximum glaciation (Newer Drift) about 25,000 B.P. The North American glacier reached south to latitude 39° N, the European to 52° N. Rapid deglaciation 15,000–11,000 B.P.	
	125,000 B.P.	Last (Sangamon) interglacial begins. Climate and vegetation similar to that characteristic during Holocene, interrupted by colder intervals with expanded glaciers about 115,000 and again 95,000 B.P.	
MIDDLE PLEISTOCENE	700,000–125,000 B.P.	Alternating glacials and interglacials. Including at least 4 and probably 7 intervals of major, Northern Hemisphere glaciation (Older Drift) (traditional names and interregional correlations uncertain) that brought severe cold to mid-latitudes.	

(Continued)

Table 10-1 (*Continued*)

	DATES[a]	PLANETARY TRENDS	COMMENTS
EARLY PLEISTOCENE	900,000–700,000 B.P.	Beginning of glacial-interglacial cycles, accompanied by extensive continental glaciation during 3 or 4 cold intervals, each lasting over 50,000 years.	
	1.8–0.9 million B.P.	Moderately warm but punctuated by at least 4 episodes of cold climate, lasting more than 75,000 years, and probably producing extensive mountain and piedmont glaciation. Gradual eradication of tropical and subtropical floras in mid-latitudes.	
PLIOCENE	5.0–1.8 million B.P.	Warmer than at present in mid-latitudes but including long-term colder oscillations leading to extensive glaciation in Alaska, the Sierra Nevada, Iceland, and the southern Andes. Continental glaciers already established in Antarctica and on Greenland.	

[a] Calendar years (B.C./A.D.) are given for the Holocene; radiometric or paleomagnetic dates (B.P. = Before Present) for earlier phases.
Based on Butzer 1971, 1974, 1976.

depositional forms. The result is that many details as well as a substantial number of the minor landforms are controversial and some appear to have had several possible modes of origin.

10-2. TYPES OF GLACIERS

Glaciers may fall into either of two broad categories, *land* and *sea* ice.

The sea glaciers include *shelf ice* attached to coasts and *pack ice*, floating in compact masses on open oceans. The two best examples of shelf ice are found in great bays opening up the Antarctic continent. Of these, the Ross Ice Shelf has been intensively studied during the past 15 years or so. With an area of about 200,000 square miles (500,000 sq km) it has an average thickness of over 2,000 feet (600 m). Pack ice permanently covers the Arctic Ocean and, during winter months, forms a broad belt about 200 to 1,000 miles (500 to 1,600 km) wide, fringing Antarctica. However, pack ice averages only 20 feet (6 m) thick. Since sea ice has limited geomorphic significance, the rest of this chapter is devoted to land glaciers.

Land glaciers can be divided into *ice streams* and *icecaps*. The former follow well-defined channels, moving downvalley like ordinary streams of water; they include mountain, valley, and piedmont glaciers (see Figure 10–1). The latter take the form of extensive sheets that expand in all directions more or less independently of topography. These include the coalescent icecaps that constitute much of the Antarctic continent, the smaller icecaps of Greenland (Figure 10–2), and

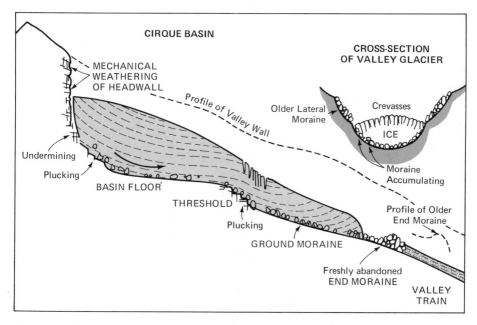

Figure 10–1. Schematic Diagram Showing Features of a Mountain Ice Stream.

several Arctic islands, as well as radially expanding ice domes found on certain Alaskan and Norwegian ranges. Highland ice streams range in size from as little as an acre (0.4 hectare) to a few hundred square miles (250 sq km) and seldom exceed 1,000 to 2,000 feet (300 to 600 m) in thickness. Continental-scale icecaps, on the other hand, are measured in millions of square miles and have an average thickness of over 6,500 feet (2,000 m).

Whereas there is only one ice sheet of continental dimensions today, that of Antarctica, there were two additional continental glaciers during the Pleistocene Ice Age. One of these covered northern Europe; the other spread over Canada into the northern United States (Figures 10–3 and 10–4). As late as 15,000 years ago the southern edge of the European glacier ran from Copenhagen through Berlin and Warsaw to Moscow. The North American counterpart extended south to a line running between Seattle, Chicago, Pittsburgh, and New York City. As recently as 10,000 years ago almost all of Canada and Scandinavia was still under ice, while Paleoindian hunters stalked mammoth, musk ox, and seals next to the ice front at Detroit and Montreal. The last traces of these continental glaciers melted away only as late as 4000 B.C., at a time when sizable villages of prehistoric farmers already dotted much of the Mediterranean world and central Europe and only shortly before the dawn of the historical era in Egypt and Mesopotamia.

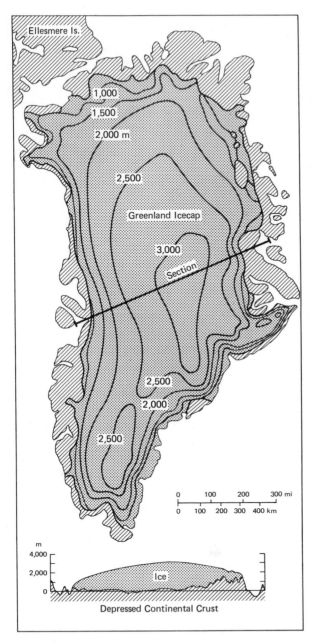

Figure 10–2. The Greenland Ice Sheet. (Based on Albert Bauer, Synthèse glaciologique, in *Contribution à la connaissance de l'inlandsis du Groenland*. Paris, Expéditions Polaires Françaises, 1954.)

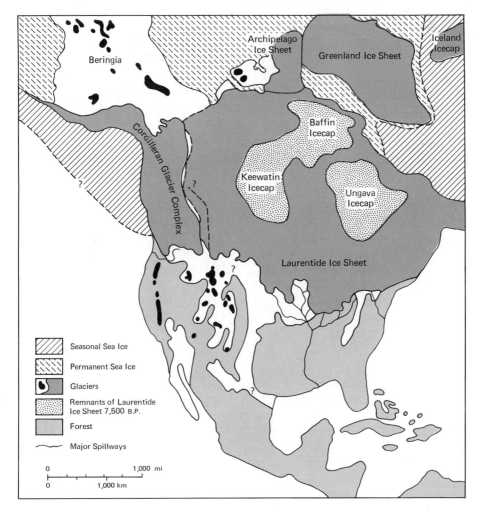

Figure 10–3. North America About 20,000 Years Ago. Shorelines were found 400 to 500 feet (120 to 150 m) below present sea level. (From Butzer, *Environment and Archeology*. Copyright © 1971 by Aldine Publishing Co., Chicago.)

10-3. SNOW ACCUMULATION AND ICE FORMATION

How does a glacier develop? This is best illustrated by the example of a high mountain range that receives abundant snowfall each winter. Snow accumulation is greatest in topographic hollows and on slopes prone to snowdrifting in the lee of snow-bringing winds. Avalanching also favors accumulation beneath steep hillsides. Shade slopes reduce the amount of snow that melts as a result of radiation. Together these different factors produce considerable variation in the snow depth on

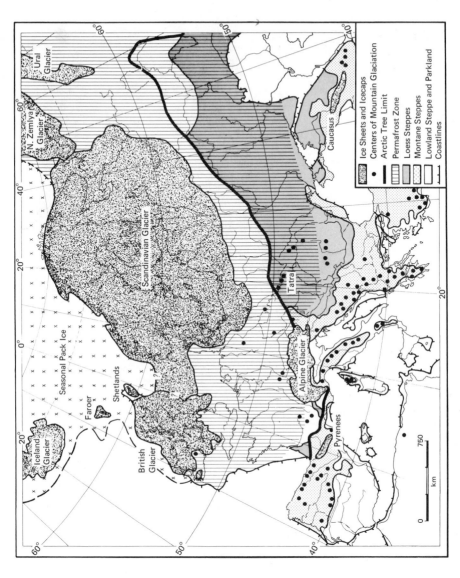

Figure 10–4. Europe About 20,000 Years Ago. (Modified after Butzer, *Environment and Archeology.* Copyright © 1971 by Aldine Publishing Co., Chicago.)

mountain country: Exposed slopes and precipices may be continually blown free of snow, whereas accumulations of 30 feet (9 m) and more may be found in hollows or beneath steep slopes. Consequently, long after the winter snow mantle has disappeared elsewhere, patches of snow remain in scattered, shaded spots or places or major accumulation. These snow patches may persist until the very end of the warm season and, after a particularly snowy winter or cool, cloudy summer, some of the largest patches remain until the next winter.

The snow in a perennial snow patch undergoes considerable change during the course of a year, partly as a result of compaction under pressure of its own weight, partly as a result of repeated melting and refreezing of some of the many minute ice crystals, at temperatures just at or below the freezing point. Compaction gradually eliminates air spaces from the snow, while recrystallization produces larger and larger ice crystals. An analogy is provided by ice cream in a freezer: Partial melting and refreezing also produce coarse ice crystals. The metamorphosis of snow in a perennial snow patch increases the density from between 0.01 and 0.16 to 0.6, after about a year. This new product is called *firn*. If a large snow patch persists for many years, the bottom layers can compact to a density of 0.85 or 0.9, that is, all of the air pockets are closed. This is true *ice*.

When snowfall is particularly heavy over a period of a decade or two or when summers remain abnormally cool or cloudy, snow patches increase in number and size and may become perennial.

Ice, however, is a quite different substance from snow: It is plastic and can be deformed under pressure. A thick mass of ice, resting on a hillside, is under considerable internal pressure due to its own weight, while at the same time gravity favors downhill deformation. Plastic flow results, and ice begins to move downslope. When ice thickness reaches 100 to 200 feet (30 to 60 m) or more and lateral movement occurs, this is in fact a *glacier*. Rates of glacier flow depend on several factors: ice thickness, ice temperature, and on slope and bed configuration. The greater the ice mass, the faster the flow. Similarly, ice near the freezing point flows more rapidly than very cold ice, which is more brittle. And, obviously, steep gradients favor any form of mass movement. Consequently, movements of small glaciers may amount to as little as a few feet per year, whereas the ice in intermediate-sized glaciers can move a few hundred yards or meters a year. Some exceptionally large, "galloping" glaciers in Alaska and Greenland can flow a few miles per annum, although they do not continually move that fast.

This sequence of development from a snow patch to a glacier has only been presented as an example of the underlying processes, and fresh glaciers only form in the wake of climatic changes. Instead, within a particular highland there normally is a balance between snow accumulation and snow melting or wastage. Above a certain limit, more snow falls than melts during the average summer. This permits local development of perennial snow fields and glaciers. Below this line, which fluctuates each year, all the snow that falls annually will melt. Glaciers from above the line may protrude below this limit, but their ice is constantly supplied from higher up the mountainside; it will continue to advance until the rate of local melting is greater than replenishment from upvalley. Glacier tongues commonly advance and recede a little in the wake of each year's winter-to-summer seasonal change, but in most instances the glacier terminus remains fairly stable as long as there is no climatic variation. All in all, glaciers exhibit a flexible balance between snowfall, temperature, and radiation, modified in detail by topography. This dynamic equilibrium means that the higher parts of the glaciers have net snow accumulation, the lower parts net wastage. The irregular dividing line between these two altitudinal zones is called the annual *snowline*.

The snowline is not a simple thing. Depending on exposure, irregularity, and steepness of mountain slopes, it moves either up or down. It is lower on poleward, shady slopes but higher on the sunny, equatorward flanks of mountain ranges. And from year to year it shifts upglacier or downglacier depending on the balance between accumulation and wastage.

Climates at the snowline are quite different from region to region. In low latitudes, air temperatures regularly oscillate across the 32° F (0° C) mark between day and night, but since there is little seasonal change, mean annual temperatures average out remarkably close to 32° F (0° C) in some mountains of the tropics. In mid-latitudes, mean summer temperatures near the snowline more frequently range from 32 to 40° F (0 to 4.5° C), although mean annual temperatures lie well below the freezing point. But these are *air* temperatures—only the surface of the snow or ice is exposed to such daily and seasonal changes, and it responds by repeated melting and refreezing of ice crystals.

10-4. NIVATION AND THE DEVELOPMENT OF MOUNTAIN GLACIERS

Glaciers are not born overnight, and even in high mountains with ample snow, glaciers do not form unless there are suitable hollows in which snow can pile up to sufficient depths. Such hollows are commonly although not necessarily prepared by snow-patch erosion or *nivation*.

Snow patches and firn tend to form under cliffs, in shallow dimples of rolling mountain topography, or in valley heads. Depressions of this type may be gradually enlarged by nivation and can ultimately become the feeding basin of a small glacier. Nivation is the result of several processes. Snow meltwaters fill any joints or cracks in the adjacent rock surfaces, favoring ice wedging during long, hard freezes. In general, a great deal of meltwater is available during periods of daily or seasonal thaw, thus accelerating all forms of frost weathering. Downstream of the snow field there will be rill wash by meltwaters, while solifluction and other mass movements sludge unconsolidated materials downhill. Finally, the comparatively high carbon-dioxide content of snow waters also favors fluting and other forms of corrosion in calcareous rocks.

Nivation can enlarge the hollows in which snow patches form. In other words, this almost inconspicuous process frequently is an important prerequisite for ultimate glacier development. As time goes on, specific deposits may accumulate around the lower margins of snow patches, either as a result of solifluction or of snow and debris avalanching (nivation ridges, see Figure 5–8).

If a snow patch develops into a glacier, the nivational hollow rapidly takes on more characteristic forms. These are only fully exposed if and when the glacier disappears centuries or millennia later. The peculiar erosional features (Figures 10–5 to 10–7) include:

1. A rock-cut *basin*, with a more or less smoothly sloping floor, normally with gentler gradients than valleys above or below, and

Figure 10–5. A Small, Abandoned Cirque in the Spanish Pyrenees. It shows the typical frost-shattered headwall and the rock threshold; the basin itself is veneered with talus and till. (Karl W. Butzer)

Figure 10–6. Multiple Pleistocene Cirques Have Eroded but not Fully Consumed This Highland in Iceland. (Karl W. Butzer)

commonly with a swampy depression or small lake (*tarn*) in its center
2. A steep or vertical *headwall*, usually a few hundred yards or meters high, rising behind the rear half of the basin
3. A low, rock *threshold* at the exit of the basin, demarcating it from the adjacent valley below

Altogether these glacier basins are called *cirques*, after the French word for amphitheater.

How do cirques form? A combination of glacial erosion and accelerated frost weathering appears to be responsible. Ice, moving over and out of the cirque basin, scours and grinds down rocks of the basin floor. Similarly, ice freezes to and plucks away rocks from the sides of the basin (see Figure 10–1). But the ice can neither scour nor pluck at the headwall back of the cirque. However, cirques form primarily by headward erosion rather than by deepening. Consequently, processes other than moving ice must be involved in the enlargement of cirques. Frost shattering is highly effective in the harsh microclimate of the cirque basin, leading to rapid slope retreat of exposed rock faces. In the case of many glaciers this process is accelerated by melting snow waters that stream over the headwall in early summer, running down into large cracks that often form between the head of the glacier and

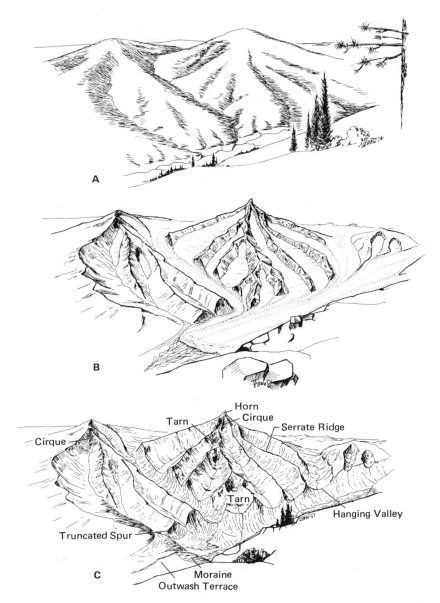

Figure 10–7. Evolution of Alpine Glacial Features Before (A), During (B), and After (C) Glaciation. After Hamblin and Howard, *Physical Geology Laboratory Manual.* Copyright © 1967 by Burgess Publishing Co., Minneapolis.

the base of the headwall. Here repeated freeze and thaw would also favor development of an almost vertical headwall that retreats by backwearing. Unloading effects, due to release of pressure on the face of the headwall, obviously favor these convergent processes of weathering. Whatever the primary agent, rockfalls, talus, avalanches, and other mass movements transfer the debris from the cirque walls onto the ice itself, from where it is transported downvalley.

When glaciers first form in a highland, the fresh cirques resemble scoops taken out of the mountainside. With time or after repeated glaciation, these rock basins are enlarged. Sooner or later, each valley head coincides with a cirque. Ultimately, headward erosion can consume and undercut mountain peaks from different sides. Some cirques coalesce while steep, jagged ridges are sculptured between others. These sawtooth, or *serrate*, ridges commonly lead up to striking, angular summits called *horns*—of which the Swiss Matterhorn is the type example (Figure 10–7). The resulting *alpine topography* is the trademark of mountain glaciation.

10-5. VALLEY GLACIERS AND VALLEY TRAINS

Glacier tongues that advance beyond the cirque threshold follow, enlarge, and modify existing stream valleys. Even after the position of the glacier tongue has been stabilized, ice continues to move along the valley from the cirque basin, feeding the lower parts of the glacier. In other words, even when the glacier itself does not move, the ice within it flows internally. Rock debris plucked from or fallen down from the walls is buried in snow and disappears within the ice. This waste travels much like the load of a normal stream, ultimately emerging on top of the ice tongue (see Figure 10–1) in the zone of net wastage where velocity approaches nil.

Equally as effective as in the cirque itself are plucking and frost shattering of the valley walls. However, much of the erosive potential of a valley glacier is due to the *scouring* effect of the debris carried in the lowest layers of the ice. This load may scratch, groove, gouge, polish, and crush the bedrock with which it comes into contact.

Glacially eroded valleys are semicircular or U-shaped in cross-section (Figures 10–8 and 10–9), the valley sides unusually steep, the floor comparatively flat, and both the depth and the width of the valley exceptional. Further characteristics of such *trough valleys* include irregular longitudinal profiles, interrupted by alternating basins and rock risers, due to selective gouging of weaker rock strata, so that previous lithological or structural influences are exaggerated. After deglaciation such basins may hold long, narrow *finger lakes* (Figure 10–9). Equally characteristic are some of the erosional details normally to be seen on the valley floors: glacial scratches and grooves in

Figure 10–8. Looking Down a U-Shaped Trough Valley in Iceland. Small tributary glaciers once scooped out parts of the smooth interfluve. The insignificant modern stream is braided. (Karl W. Butzer)

Figure 10–9. Trough Valley with Finger Lake Dammed Back Behind a Till-Mantled Riser, in the Spanish Pyrenees. (Karl W. Butzer)

bedrock and smooth, glacially polished rock knobs that resemble mid-channel bars in a stream. A final criterion of glacial erosion is given by *hanging valleys*. Both glaciated and unglaciated tributary valleys that feed a glacial trough empty on top of the principal ice mass and lack erosive potential to match the deep scour of the main glacier. Consequently, as the ice disappears, unglaciated side valleys are left truncated well above the main valley floor. These streams subsequently cascade over rocky ledges, frequently giving rise to scenic waterfalls.

Any material eroded or transported by and directly or indirectly deposited via a glacier is known as *drift*. This is normally subdivided into *till* or *moraine*, deposited by the ice itself, and *outwash*, deposited by the meltwaters. Till may accumulate (1) along the lateral margins of a glacier, as *lateral moraines;* (2) in front of the glacier tongue, as an *end moraine;* (3) between two converging glaciers, as a *medial* or *inter-lobate moraine;* or (4) beneath the glacier as a *ground moraine* (see Figure 10–1, and Figure 10–14 in section 10-7). The origins of till are complex. Much of the material is a result of glacial scour or plucking. The remainder, originally produced by mechanical weathering, is introduced into the glacier by mass movements. Although the grade size of till depends in great part on mineralogy, resistance, and jointing of the bedrock, there is little sorting according to size, and subrounded, coarse-grade particles are prominent. Another till characteristic is that the material is essentially bedded at random, just as it was carried. There is no stratification, and rocks are oriented and inclined at all angles.

The thickness and the irregularity of till deposits increase down-valley and, in part, reflect the original size and velocity of the glacier. Typical end moraines, recording standstills of a few decades or centuries, may be about 50 to 100 feet (15 to 30 m) high. Less resistant rock types also supply far more material, while older till is commonly reworked by glaciers. Depending on the thickness of a former glacier, lateral moraines may extend as much as 2,000 or 3,000 feet (600 to 900 m) or more vertically up a valley side.

Summer meltwaters gather into watercourses that move over, through, or under the ice. Heavily laden with detritus, these streams normally emerge in front of the glacier tongue where a further load of morainic debris may be picked up. Except where glacier tongues reach well below the snowline, discharge is essentially limited to the summer season and it fluctuates strongly between day and night. Braiding is characteristic of meltwater streams at the height of the summer, and channels are relatively shallow, normally between 1 and 5 feet (30 and 150 cm) in depth. Most outwash deposits consist of unco-hesive sands and gravel, although the suspended sediments normally swept far downstream may also be deposited in nearby lakes. Both sorting and stratification are pronounced (Figure 10–12), unlike in glacial till. The suite of outwash, constricted between valley walls down-stream of the glacier terminus, is known as a *valley train*. After degla-ciation the valley train is commonly dissected and reduced to one or more sets of outwash terraces (Figure 10–13).

10-6. ICE FIELDS AND PIEDMONT GLACIERS

Two types of highland glaciation fall beyond the scale of normal mountain and valley glaciers. These are *highland ice fields*, found on plateaus or broad mountain crests, and *piedmont glaciers*, which develop where large valley glaciers debouch in front of a mountain range.

Highland ice fields are particularly well developed in parts of Alaska, Norway, and the southern Andes—areas where snowfall is exceptionally heavy. Mountain glaciers grow until the cirques have al-most completely consumed the serrate ridges that delimit the different watersheds, leaving only isolated *nunataks* to rise above a smoothly domed expanse of ice. When glacier thickness is sufficiently great to outweigh the influence of the subglacial topography, ice movement be-comes radial, as in an icecap. Around its periphery the vast ice field dissipates into a number of valley glaciers (Figures 10–10 and 10–11).

Figure 10–10. Ice Streams Draining the Greenland Ice Sheet, Southeast Greenland. The foreground has been largely consumed by tributary cirque glaciers. (Karl W. Butzer)

Figure 10–11. Ice Tongues Converging and Calving in the North Atlantic. Note the dark lines of lateral and medial moraine near the confluence (southeastern Greenland). The icebergs in the foreground are as much as several hundred yards or more across. (Karl W. Butzer)

Piedmont glaciers are found well below snowline elevation where large valley glaciers expand beyond the edge of an adjacent plain, spreading out onto the piedmont like an alluvial fan. A ring of end moraines can develop around the piedmont glacier, with meltwater streams radiating out in fan form but ultimately converging into one or more major streams. As a result till and outwash surfaces can be as extensive as piedmont alluvial plains. During the ice age, piedmont glaciers and their related outwash forms were particularly well developed along the northern and southern foothills of the Alps. In fact, it was here that the relative chronology of successive glacial episodes that span the past million years or so was first worked out by Albrecht Penck shortly after 1900.

10-7. CONTINENTAL GLACIATION: ICE-SCOURED PLAINS

Icecaps of subcontinental dimensions are responsible for three major types of geomorphic environment:

1. *Ice-scoured plains*, primarily located in the central zone of former glaciation, in Canada and Scandinavia largely coincident with ancient crystalline rocks

Figure 10–12. A Diagonal Line Divides Chaotic Till Above from Well-Stratified, Inclined Outwash Below, Spanish Pyrenees. (Karl W. Butzer)

Figure 10–13. Pleistocene Glaciers from the San Juan Mountains of Colorado (Skyline) Deposited the Prominent Morainic Ridge (Center). Meltwaters laid down the now-dissected outwash terrace that stretches along the modern floodplain. (Karl W. Butzer)

2. *Till plains,* primarily the peripheral zone of former glaciation, in North America and Europe coincident with sedimentary rocks
3. *Outwash plains,* forming broad belts of country at and beyond the former termini of the continental glaciers (Figure 10–14)

These three settings will be considered in turn, discussing their Pleistocene formation and describing them as they appear now—about 10 to 15 millennia after deglaciation.

The ice-scoured plains of the Northern Hemisphere continents were once covered by about 3,000 to 9,000 feet (900 to 2,700 m) of ice, very much like Greenland and Antarctica are today. Although there was little ice movement near the center of the icecaps, radial flow velocities increased rapidly away from the focal points. As a result, most of the underlying rock surfaces were intensively scoured. The overall effects can be enumerated:

1. Soils and regolith that existed prior to glaciation were stripped off and almost entirely removed. Since soils form slowly on crystalline rocks in cool environments, bare and stony surfaces are not unusual today (see Figures 10–15 and 10–16); the new soils that have developed since deglaciation usually are thin.

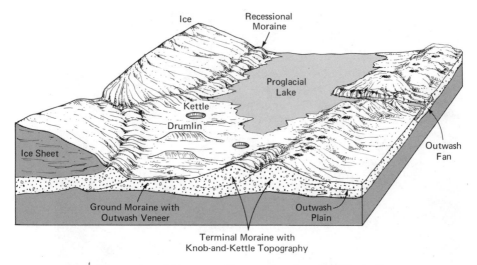

Figure 10–14. Depositional Features Due to Continental Glaciation.

2. Existing roughlands were considerably smoothed as a result of scour and plucking. Hills were truncated or rounded off and rough surfaces ground down and polished, sometimes into longitudinal ridges of bedrock that point in the direction of flow, with gentle slopes facing upstream and a rough, plucked surface, with a tail of till, downstream. The overall effect is an open, rolling topography with appreciable relief but a general dominance of smooth slopes, locally interrupted by rounded but steep-sided rock knobs.

3. The ice selectively grooves or gouges out rock basins along preexisting valleys and into surfaces underlain by inclined, weak rocks. Distinct linear trends, corresponding to flow streamlines, may be apparent; dimensions of such hollows are variable, but relief may exceed 300 ft (90 m) and some giant grooves are several miles long. After deglaciation these rock basins were converted into lakes, many of which have now been filled in with dead vegetation or sediment and reduced to swamps or bogs. Innumerable lakes and bogs are consequently characteristic of ice-scoured plains.

4. Depositional features, including thin mantles of ground moraine, as well as end moraines and outwash from the final stages of deglaciation, occur on a local scale. Immense boulders were often abandoned by the ice on ridges and slopes. Such *perched rocks* are another characteristic of glaciation.

5. The overall topography created by glacial scour in crystalline rocks appears chaotic, and the drainage patterns are unsystematic. The countless lakes and swamps fill in very slowly with mineral sediment but may fill in rapidly with organic material as bogs are formed.

Figure 10–15. Glacial Scour Produced by Convergent Valley Glaciers in Northern Iceland. Ice movement was diagonally upward. (Karl W. Butzer)

Figure 10–16. Scratched and Polished Glacial Pavement from the Permocarboniferous Glaciation, Some 225,000 Million Years Ago, Now Being Exhumed by the Riet River near Kimberley, South Africa. During the past few millennia Bushmen engraved drawings on the surface. (Karl W. Butzer)

At the same time, they do not drain rapidly, since they occupy basins with resistant rock thresholds: One basin overflows into another, leading to a maze of interconnecting drainage lines. Because water is detained by the countless retaining basins, stream storm flows tend to be small, thus giving the streams little opportunity to cut. Streams are poorly graded, so that waterfalls and rapids or cataracts form wherever water descends over a rock barrier.

This, then, is the environment of the ice-scoured plains—vast wildernesses of spruce, fir, and pine forests, studded with millions of lakes (over 2 million in the province of Quebec!). Lumber is abundant and trout fishing excellent, but often the area is of little other economic value to anyone other than miners and fur trappers.

10-8. CONTINENTAL GLACIATION: TILL PLAINS

Till plains have an appearance quite different from that of ice-scoured plains. Comparatively soft sedimentary rocks, which happen to be dominant in these areas, supplied large quantities of ready detritus for the glacier to remodel. As a result, depositional rather than erosional features are characteristic. It is fortuitous that the margins of the ice age glaciers of Europe and North America coincided with belts of sedimentary rocks, but it is hardly accidental that the till plains coincide with such bedrock formations.

Continental ice sheets created the till plains of southern Canada, the northeastern and north-central United States, north-central Europe, and northern Russia. Consequently, the landscapes of these diverse regions have much in common (see Figure 10–14):

1. The preexisting topography was partially buried in drift as valleys were filled in and hills partially masked with deposits. Average thickness of ground moraine probably ranges between 75 and 250 feet (23 and 75 m). Part of this was already deposited beneath the active glacier; the remainder melted out during deglaciation. Simultaneously, prominent bedrock features were subject to erosion and either truncated or streamlined and polished by ice scour. Local relief was thereby reduced and today seldom exceeds 100 ft (30 m) in areas of undifferentiated ground moraine.

2. A variety of special forms, including end moraines, drumlins, eskers, and kames, were deposited. These features are described in more detail later. Such depositional landforms have more relief in regions of a stony till than in areas of clayey till, a condition that mainly reflects the underlying bedrock. Therefore, clayey till plains are smoothly rolling and almost monotonously flat, whereas stony till maintains more prominent landforms and local relief may exceed 150 feet (45 m).

3. Till surfaces often constitute poorly drained plains with unsystematic patterns, widespread swamplands, and a large number of lakes and ponds. Unlike the ice-scoured plains, however, drainage improves steadily as dissection cuts drainage outlets in soft materials and the many depressions fill with sediment and organic material. Lakes and swamps in the till plain are favored by clayey, impervious subsoils; they normally belong to one of three types. Some were dammed back behind end moraines, since the till disrupted or obstructed previous drainage. Others formed as a result of differential compaction, where particularly deep till mantles filled older valleys. In fact, most of the larger lakes are due to compaction and sag of unconsolidated sediments. Finally, a great number of ponds and small lakes can be attributed to the gradual melting of great masses of "dead" ice—once buried in the till and left behind as the glaciers waned and retreated. The most common, smaller *kettles* measure about 100 to 300 feet (30 to 90 m) across, although many are considerably larger.

4. In most areas the glaciers of the last ice age did not extend as far as those of earlier glaciations, so that there may be several generations of relict glacial landforms. Those of the most recent glaciation (about 70,000 to 10,000 years ago) are conveniently called "Newer

Drift" and distinguished from "Older Drift"—forms and deposits due to one of several older glaciations (between 130,000 and perhaps 1 million years ago, Table 10–1) (and see Chapter 18). Depositional landforms of the Older Drift have been more intensively eroded and subject to deep soil development. Perhaps the most striking difference is that the Older Drift commonly exhibits normal dendritic patterns and there are almost no lakes, whereas the Newer Drift shows considerable evidence of a once unsystematic drainage.

End moraines constitute the most impressive landforms in till: great arcs of hummocky ridges that may undulate from one horizon to the other. The major ice fronts are recorded by multiple ridges with a local relief of 100 to 200 feet (30 to 60 m) and a total width ranging from several hundred yards or meters to as much as 2 or 3 miles (3 or 5 km). End moraines normally consist of coarser materials than ground moraine and show considerably more variation of relief. Erratic boulders, derived from outcrops of crystalline rocks by long-distance transport, are common at the surface. Kettles commonly occur as swarms of small depressions, and together with multiple moraines these favor an irregular *knob-and-kettle topography*. Although some end moraines are built up by traction, as ice pushes and bull-dozes its way across beds, most of the material accumulates gradually during periods of glacial standstill. The detritus that melts out of the glacier tongue is washed down, rolled, or sludged together along the perimeter of the ice.

Several types of end moraines are distinguished according to their relative positions. Continental glaciers commonly expand with a series of prominent tongues or lobes. A type of medial moraine, called *interlobate*, develops along the contact between two such lobes. The end moraines that accumulated during the farthest advance of a particular glacial stage are called *terminal*, as opposed to the *recessional* moraines of successive standstills and minor readvances during deglaciation.

Drumlins are elongated, streamlined hills that occur in swarms on the surface of some till plains. The upstream end of a drumlin is steep and fairly blunt, while the downstream terminus is smoother and the lee taillike in shape. Oriented in the direction of ice movement these features are commonly 100 to 150 feet (30 to 45 m) high and a few hundred yards or meters to a mile or so (2 km) in length. Some drumlins consist entirely of till; these may be the result of reshaping the ground moraine beneath the moving ice. Other drumlins have rock cores with a cap and tail of till; these probably result from reduced ice velocity in the rear of large bedrock obstacles.

Eskers are steep-sided, sinuous ridges of stratified sand and gravel. They may extend subcontinuously over distances of several hundred yards or meters to 100 miles (160 km) or more. Eskers consist of stream beds, originally laid down within large cracks of the glacier or deposited by meltwater streams moving through or beneath the ice. The fact that eskers often run uphill as well as downhill is not unreasonable: Subglacial meltwater streams flow under great hydrostatic pressure, so that the stream beds need not be horizontal. After deglaciation, eskers may remain behind as striking glaciofluvial landforms rising above till plains.

A number of miscellaneous deposits of mixed stream origin are called *kames*. In each case the main diagnostic feature is a hummocky or irregular mass of stratified but contorted sand or gravel. Some of these are morainic wash that originally accumulated within crevices of the glacier. Others formed in small pools between the glacier tongue and the end moraine; beds of this kind are frequently found intermixed with typical moraines.

10-9. CONTINENTAL GLACIATION: OUTWASH AND LOESS PLAINS

The end moraines of the former European and North American glaciers terminate abruptly on the edge of broad *outwash plains.* Numerous small meltwater streams once spread out sands and gravels here, ahead of the former ice front. Other streams deposited alluvial fans that merged and coalesced, while many of the largest breached the end moraines via a glacial gateway, building up massive fans crisscrossed by a maze of shallow, braided streams (see Figure 10–14). As in the case of valley trains, the outwash is well sorted and conspicuously stratified, consisting of sands and gravels. The surface is flat, dipping gently away from the ice terminus. The width of such glaciofluvial spreads may be 35 miles (56 km) or more.

Proglacial lakes sometimes formed in shallow depressions adjacent to the ice front, commonly in areas of older till now subject to compaction and sag (see Figure 18–7). In some regions large lakes were dammed up between the end moraines and former watersheds. This was the case in the north-central United States, where a succession of extensive proglacial lakes formed in the Midwest during the early stages of deglaciation. Although they are commonly tilted today, as a result of crustal readjustments, these lacustrine deposits were once absolutely horizontal. They are well stratified and consist mainly of well-sorted fine sands. Locally, deltaic beds may be found where meltwater streams once plunged directly into standing waters. Breaching of dammed lakes of this type occasionally sent catastrophic floods sweeping over the outwash plains downstream.

The waters of the outwash plains gradually gathered in broad sweeps, following existing river valleys or finding new outlets to the sea. Such outlets were cut across Poland and Germany, draining the meltwaters westward to the North Sea (Figure 18–6). In the United

States similar *glacial spillways* were formed across the Great Lakes-Ohio River watershed (Figure 18–7). Such spillways were up to 25 miles (40 km) wide; they are now swampy and often followed by insignificant watercourses that frequently qualify as misfit streams. Elsewhere, the meltwaters followed existing rivers, spreading masses of outwash well beyond the confines of the outwash plain. Subsequent dissection has reduced these sands and gravels to alluvial terraces, such as those found along the Rhine, the Ohio, and the middle Mississippi. The coarse grade of such deposits still bears mute testimony to the torrential summer floods that once raced down these valleys.

The significance of Ice Age glaciers even extends beyond the former meltwater streams. In particular, the spillways carried great masses of suspended material that was left stranded on broad floodplains in later summer, as ablation came to an end. Once dry this silt was swept up by windstorms and blown many miles beyond the valley margins, mantling the uplands with dust-sized particles. Such accumulations range in thickness from a few feet to several tens of feet (1 to 10 m and more). Known as *loess*, such mantles of wind-borne silt cover large parts of the Midwest as well as broad sweeps of country in central and eastern Europe (see Figures 18–7 and 10–4). Like the till plains and old proglacial lake floors, they provided a highly fertile parent material for subsequent soil development and now constitute some of the most productive agricultural lands of mid-latitudes.

In other words the once destructive glaciers have left a legacy of impressive and highly varied landforms across large parts of the United States and Europe, even well beyond the margins of the former ice. These same landscapes are now not only more interesting but form one of the prime agricultural resource bases for many of the Earth's billions of hungry mouths. In this sense the glaciers have been a blessing in disguise. The history of the ice ages and their landscape heritage are amplified in Chapter 18.

10-10. SELECTED REFERENCES

Butzer, K. W. *Environment and Archeology: An Ecological Approach to Prehistory*. Chicago, Aldine-Atherton, 1971, 703 pp.
———. Geological and ecological perspectives on the Middle Pleistocene. *Quaternary Research*, Vol. 4, 1974, pp. 136–148.
———. Pleistocene climates. In R. C. West, ed., *Ecology of the Pleistocene. Geoscience and Man*, Vol. 13, 1976, in press.
Charlesworth, J. K. *The Quaternary Era*, 2 vols. London, Edward Arnold, 1957, 1700 pp.
Davies, J. L. *Landforms of Cold Climates*. Canberra, Australian National University Press, and Cambridge, Mass., M.I.T. Press, 1969, 200 pp.
Embleton, Clifford, and C. A. M. King. *Glacial Geomorphology*, New York, St. Martin, and London, Edward Arnold, 1975, 640 pp.

Evans, I. S. Geomorphology and morphometry of glacial and nival areas. In R. J. Chorley, ed., *Water, Earth and Man.* New York, Barnes & Noble, 1969, pp. 369–380.

Fahnestock, K. Morphology and hydrology of a glacial stream—White River, Mount Rainier, Washington. *U.S. Geological Survey Professional Paper* 422-A, 1963, pp. 1–70.

Flint, R. F. *Glacial and Quaternary Geology.* New York, Wiley, 1971, 892 pp.

Giovinetto, M. B. The drainage basins of Antarctica: accumulation. *American Geophysical Union, Antarctic Research Series,* Vol. 2, 1964, pp. 127–155.

Hobbs, W. H. The cycle of glaciation. In Edward Derbyshire, ed., *Climatic Geomorphology.* New York, Barnes & Noble, and London, Macmillan, 1973, pp. 76–90.

Patterson, W. S. B. *The Physics of Glaciers.* New York, Pergamon, 1969.

Price, R. J. *Glacial and Fluvioglacial Landforms.* Edinburgh, Oliver and Boyd, 1972, 242 pp.

Robin, G. Q. The ice of the Antarctic. *Scientific American Offprint* No. 861 (September 1962), 15 pp.

Rooney, J. Economic implications of snow and ice. In R. J. Chorley, ed., *Water, Earth and Man.* London, Methuen, and New York, Barnes & Noble, 1969, pp. 389–401.

Schultz, Gwen. *Ice Age Lost.* Garden City, N.Y., Doubleday, 1974, 342 pp.

Thorn, C. E. A model of nivation processes. *Proceedings, Association of American Geographers* 7, 1975, pp. 243–246.

Tricart, Jean. *Geomorphology of Cold Environments.* Translated from the French by E. Watson. New York, St. Martin, 1970, 320 pp.

Von Klebelsberg, R. *Handbuch der Gletscherkunde und Glazialgeologie,* 2 vols. Vienna, Springer, 1948–1949, 403 and 621 pp. (Vol. 1, topical; Vol. 2, regional).

West, R. G. *Pleistocene Geology and Biology.* London, Longmans, 1968, 377 pp.

Woldstedt, Paul. *Das Eiszeitalter,* 3 vols. Stuttgart, Enke, 1954–1965, 374, 438, and 328 pp. (Vol. 1, topical; Vols. 2 and 3, regional).

Wright, H. E., and G. Frey, eds. *The Quaternary of the United States.* Princeton, N.J., Princeton University Press, 1965, 922 pp.

For the detailed topographic and geological maps essential to illustrating glacial landforms the reader is referred to:

Hamblin, W. K., and J. D. Howard. *Physical Geology: Laboratory,* 4th ed. Minneapolis, Burgess, 1975, 233 pp.

Hidore, J. J., and M. C. Roberts. *Physical Geography: A Laboratory Manual.* Minneapolis, Burgess, 1974, 203 pp.

Chapter 11

COASTAL LANDFORMS (I): MODELED BY WAVE ACTION

Earth's coastlines provide the interface between land and water. Although the area they cover is small, their linear extent is enormous and the variety of their landforms is considerable. The majority of seacoasts were sculptured by wave action. Many are dominated by cliffs and other erosional features, although submerged longshore bars and depositional platforms are also found not too far from the watermark. Other coasts are primarily depositional, with beach ridges or shingle beaches and coastal dunes, alternating with shallow rock outcrops cut by nips. Many shorelines represent complex zones of sediment accumulation, in the form of offshore or bay bars, spits, cuspate bars or tombolos, linked to the continent by mudflats or lagoons that are colonized by salt marsh or mangrove swamps. Finally, many tropical or subtropical coasts are fringed by reefs of organic origin, primarily coral.

These key types comprise the majority but by no means all of the Earth's coasts. The remainder now are or were in the past sculptured by terrestrial processes, and wave action has only achieved superficial modification. The legacies of past sea levels, both higher and lower than that of the present, are discussed in Chapter 12, together with the extensive occurrences of high-lying, abandoned shoreline forms.

> **Age by age, the sea here gives battle to the land.**
>
> HENRY BESTON, *The Outermost House*

11-1. SHORELINES AND COASTS

Wave action is the most striking geomorphic force on the shores of lakes, inland seas, and bays and particularly along the margins of the great oceans. Each year catastrophic storms bring home the fact that wind-lashed waves can be an uncontrolled force of great potency. On August 18, 1969, hurricane Camille smashed into the coast of Mississippi from the Gulf of Mexico. Storm waves washed out road embankments, crumbled brick and concrete structures, and reduced the shore zone to a shambles. Another storm, remembered by the Dutch with horror, brought masses of water through the dikes of the Netherlands during a period of phenomenally high tides, February 1, 1953. Vast stretches of fertile agricultural lands were submerged by salt water, and it will take decades to reclaim and restore them to their old productivity. Such events focus attention on the ability of waves to sculpture Earth's coasts.

Coastal scenery is a world by itself. The seamen of merchant vessels plying the oceans have known the peculiar appeal of far-flung coasts for centuries, shores that have only been discovered by vacationers, sun seekers, and surfers during the past decade or two. Waves beating at the foot of rocky cliffs over the ages have produced picturesque and often bizarre coasts that are among the few remaining unspoiled landscapes. The great sandy barriers of other, low-lying coasts often remain equally undisturbed—reminders of how many grimy and polluted industrial portsides may once have looked. And there still are the exotic coral coasts that only some have had the good fortune to visit. Traveling to different parts of the world, the coastal geomorphologist can sample not only some of the finest scenery but also learn to appreciate that the coast itself is an environment. Coastal processes are unique, and the constellation of landforms is distinctive.

The terminology applied to the coastal sphere also differs from that of other geomorphologic settings. Some of the general definitions illustrate that this is the meeting place of land and sea but that it belongs wholly to neither (see Figure 11–1):

> *Shoreline:* the two-dimensional line of contact between land and water. The ephemeral position of the watermark oscillates as the waves roll in and out, and the tides rush and ebb.
>
> *Shore:* the narrow strip of land between the low-tide watermark and the highest points reached by storm waves and high tides. It is sometimes dry, sometimes underwater, a zone of periodic submergence.
>
> *Sea level:* the average elevation of the watermark between the extremes of mean low and mean high tides.
>
> *Coast:* a general, loosely defined term for the land bordering the shore. From the geomorphologist's point of view, it should include all those land surfaces modeled conspicuously by waves—now as well as in the recent past.
>
> *Continental shelf:* the gently sloping, submerged margins of the continents.

11-2. WAVES, CURRENTS, AND TIDES

Waves are wind-driven oscillations of surface water. An oscillating motion travels through the water, with little or no transfer of the water molecules themselves—until the wave breaks at the shore. The crest and the trough of a moderate-sized wave on the open sea approximate a sine curve. The distance from one crest to the next is the *wavelength,*

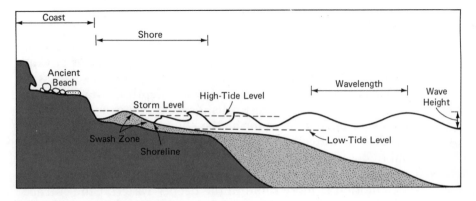

Figure 11–1. Coastal Elements.

the difference in level between crest and trough the *wave height* (Figure 11–1).

Two mechanisms combine to produce waves. Firstly, the wind has a *pushing* effect on the windward slope of each wave crest, driving the wave forward. Secondly, the air moving over the surface exerts a *frictional drag*. In detail, each molecule in the disturbed surface water is revolving in a circular orbit—in a forward direction on the wave crest, in a backward direction in the trough. At times of high wind velocity, the frictional drag of the wind accelerates the forward movements on each crest. As a result, the water molecules travel forward a little faster on the crests than they move back in the trough. A net flow of surface water, at speeds of up to 2 miles per hour (3.5 km/hr), is possible at such times. As a combination of these factors, wave velocity is commonly greater than the actual wind speed.

Wave height is controlled by (1) wind velocity, (2) duration of wind blowing from one direction, and (3) the available fetch, that is, the expanse of open water in which waves are generated. Thus, for example, wave height is greatest during protracted storms over large, open bodies of water. The lack of fetch, in fact, normally inhibits wave development on lakes, as well as in bays or straits. Stormy seas commonly take 12 to 24 hours to develop, with the basic potential of wave activity determined by wind velocity. A 10-mile-per-hour (16-km/hr) breeze can generate wave heights of as much as 2 feet (60 cm), a 25-mile-per-hour (40-km/hr) wind of up to 15 feet (4.5 m), and a 50-mile-per-hour (80-km/hr) gale a potential 60 feet (18 m). In other words, wave height increases geometrically with wind speed.

When waves move beyond their region of origin, they are transformed from short, relatively high, *free* waves into widely spaced waves of very regular form known as *swell*. This effect frequently indicates the approach of a storm at sea, and served as a crucial weather gauge for sailing vessels prior to radio communications. Swell also develops as a storm dies down. In either case, the swell may not conform in direction with light, local winds, although frictional air resistance gradually reduces wave energy.

As waves approach the coast, wave patterns and geometry are modified by the configuration of the shoreline and by changes of water depth. Decreasing water depth leads to a sharp increase in wave height and steepness. When the ratio of water depth to wave height is in the order of 1.1 to 1.5, the crest rolls forward and over as a *breaker*. This produces an uprush of water at the shore—the *swash*—followed by a return flow of water—the *backwash*.

In front of coastal promontories, factors such as shallow water, increased friction, and changing flow patterns retard wave motion. By comparison, motion is more rapid in the deeper water at the entrance of bays. As a result, wave patterns are distorted or refracted so that they break approximately parallel to the irregularities of the shoreline. This means that the promontories are subject to intensified wave attack (Figure 11–2). Refraction also occurs when the wind blows at an angle to a straight shoreline. The waves approach obliquely, but the changing water depth causes them to veer so that they break almost parallel to the shoreline.

The ideal surfing wave is produced when waves break on a straight shoreline, without refraction, and over a smoothly sloping surface. As these *plunging waves* move in perpendicular to the shore, they roll in orbital form and fall vertically forward. Refracting waves roll in an elliptical orbit and spill violently in a foamy swash (*spilling waves*) (Figure 11–2). Yet another type is found where inshore waters are so deep that waves do not break. Under these circumstances, waves are reflected back from the watermark and may produce an immobile gridlike interference pattern with incoming waves. *Standing waves* are then found at fixed points.

An important consequence of refracting waves is that they favor a net movement of water parallel to the shoreline. Additionally, during stormy weather, a slow shoreward drift of water is set in motion by intense frictional drag. Sea level at the shore rises temporarily as the water builds up under wind pressure. This in turn produces a *longshore current* parallel to the coast, in an oblique downwind direction (Figure 11–3).

Most coastal regions bordering on oceans further experience a periodic and predictable up and down of the coastal waters. These high and low *tides* are a response to the gravitational attraction of moon and sun. The moon is fairly small but also rather close to Earth and plays the leading role in this interaction of forces. The sun, despite its immense size, is at a great distance and exerts barely half the force of the moon.

As Earth rotates, the oceanic water masses closest to the moon are attracted to our satellite and "bulge up" minutely. The Earth rotates

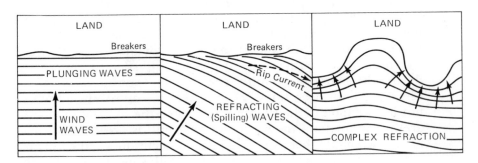

Figure 11–2. Plunging and Refracting Waves.

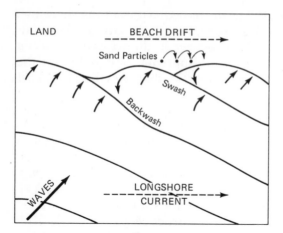

Figure 11–3. Longshore Current Drift.

once, with reference to the moon (which is not motionless but slowly revolving around the Earth), with a period of 24 hours and 50 minutes. This means a 50-minute delay in the tidal cycle from one day to the next. The tidal oscillation so produced travels around the world as a minute ripple of the great oceans. As the high point approaches, the tide *flows* in; afterward it *ebbs* out.

The details of tidal patterns vary from place to place and from day to day, with respect to both their timing and their day-to-day amplitude. Some areas only have one daily tide; others have two distinct tidal cycles per day. Some oceanic coasts experience little tidal variation; others are affected by considerable tides. Finally, during the course of the 27.5-day lunar month, tidal amplitudes vary everywhere, within a considerable range.

The factors responsible for the details of tidal variation are threefold:

1. The moon's orbit is slightly elliptical, so that when the moon is closest, tides are 15 to 20 percent higher than average; when the moon is farthest away, tides may be 30 percent lower than normal. In addition to this 27.5-day cycle, the angle of declination of the moon with respect to the equator changes, so that the Northern and Southern hemispheres are affected differently during the course of a similar cycle.
2. The sun's impact on tides is insufficient to control their timing, but it does modify their amplitude. When the sun and the moon are in conjunction, their gravitational pull is greater and tides are 20 percent higher. Such *spring tides* occur during the full and new moon phases. When the sun and the moon are at right angles with respect to the Earth, the gravitational effects are

partly canceled out, and such *neap* tides are 20 percent below average.

3. The complex configuration of the Earth's continents and oceans complicates matters immensely, so that tidal oscillations bear in strongly on some coasts but remain hardly perceptible on others. Typical amplitudes along open oceanic shorelines are in the order of 5 to 10 feet (1.5 to 3 m), although tidal effects may be much greater in small bays and inlets.

Despite the intricacies of tidal patterns, both spatially and temporally, tidal cycles and amplitudes can be calculated and predicted for any one point. Such data are critical for harbor installations, as well as for watercraft traveling in shallow coastal waters or in coastal rivers affected by tides. What cannot be predicted is the conjunction of exceptionally high tides with major storms.

11-3. EROSION AND DEPOSITION AT THE SHORE

Wave energy is largely expended at the watermark and immediately below it, as waves break and the swash rushes in at a velocity as much as that of the free waves at sea. Wave erosion of the bottom of a sea or lake is proportional to the speed of water movement: When water depth equals wavelength, water movement at the bottom is only 0.2 percent of that of the surface water; when water depth is one-half the wavelength, movement at the bottom is reduced to 4 percent of that of the surface. Consequently, the effective depth of wave erosion is very limited and strictly confined to inshore waters. Thus, for example, if maximum wave height during stormy seas is 20 feet (6 m), there can be little wave erosion beneath about 10 to 15 feet (3 to 4.5 m) below low-tide level.

The mechanics of wave erosion can be outlined as follows:

1. Wave impact against rocks or cliffs at the watermark exerts great pressures, particularly when water or trapped air is suddenly compressed in fissures or joints by shocklike impacts. The wedging effect pries rocks apart, while suction subsequently drags out loose material.

2. The swash hurls pebbles and cobbles against shore rocks, with a battering effect. Projecting corners are knocked off; loosened, well-jointed rock masses are dislodged; and rock is ground down.

3. Sand and rock are swept back and forth over the beach or rock bench by the alternating swash and backwash. This brings about mechanical abrasion beneath the watermark, as far down as there is significant wave agitation.

4. Chemical solution and weathering at the watermark or in the spray zone combine with mechanical attack, particularly in warm

environments. Breakdown of most rocks is accelerated, and, in the case of limestone, corrosion may even outpace mechanical abrasion.

5. Materials loosened and removed are carried by traction (rolling and sliding), saltation, suspension, and solution. This transport is due to both the backwash and the *undertow*—a diffuse seaward return of water along the bottom. In some situations, too, longshore currents and *rip currents*—concentrated seaward flow lines—transport sediment away from the watermark.

6. Effective wave cutting is generally restricted to times of stormy and high seas, when wave energy is at a maximum.

The materials eroded at the watermark are deposited in several possible locations. The major part of the coarse sediment is usually dragged out by the backwash, undertow and rip currents, to be deposited at shallow depths just offshore. However, suspended clays and solubles may only be deposited at great depths on the sea floor.

Refracting waves, particularly in combination with longshore currents, favor lateral transport of sand and pebbles. As a result of waves striking the beach obliquely, particles move laterally. The swash rushes up at an oblique angle. The sand particles swept along subsequently ride back down—but perpendicular to the shoreline—with the backwash. Each transport consequently involves a net lateral shift in position, and, over long periods of time, great masses of material are transported *along* the coast rather than away from it. This process of *beach drifting* is complemented by *longshore currents* that move sands through shallow, offshore waters (Figure 11–3). The overall effects of refraction are to create local areas of deposition along complex shorelines: Erosion is concentrated on rocky promontories, while sand accumulates in bays. On straight coasts, refraction initially favors a rough balance of erosion and deposition, eventually tending to form long sandy or gravelly beaches.

As in the case of wave erosion, deposition is not a continual process. Deposition offshore is most prominent as storms begin to subside. Deposition inshore along the beach is most prominent during periods or seasons of low wave energy, and sand banks may build up under rocky headlands, only to be destroyed completely during the next storm. Swash deposition above the average watermark as well as beach drifting are most effective at times of relatively high wave energy, although major storms have an erosional impact.

Tidal processes in coastal proximity have geomorphic impact. But few, if any, characteristic landforms are primarily due to tidal influences. The role of the tides in coastal evolution involves a number of processes:

1. Wave erosion is most effective with large tidal amplitudes. As the watermark moves up and down the shore, waves attack a broader

zone and can sculpture more prominent forms (Figure 11–1).

2. Tides may keep inlets between lagoons and the sea open. Whatever the origin of such *tidal inlets,* they are enlarged, deepened, and preserved by the scour of ebbing tides that sweep sediment out to sea.

3. Coastal lagoons in tidal zones are periodically flooded and drained. The resulting salt-water environment is colonized by low vegetation known as *salt marsh* in mid-latitudes, by swamp forest or *mangroves* in tropical regions (Figure 11–12). The tide penetrates along inlets and then surges up dendritic channels that crisscross low-lying mudflats. Ebbing waters retreat by the same *tidal creeks,* kept free of sediment even as the lagoon gradually fills up with suspended matter and organic residues.

4. Broad, shallow coastal streams may be conspicuously affected by tidal activity. Thus, for example, the Hudson River rises and falls 4.5 feet (1.4 m) at its mouth, and the same damming effect produces a river tide of 3.0 feet (0.9 m) near Troy, New York—131 miles (209 km) upstream. Such *tidal rivers* can behave like an arm of the sea, and in some instances *tidal bores* travel upstream at rates of up to 10 miles per hour (16 km/hr) and with amplitudes of 15 feet (4.5 m) or more. The mouth of a tidal river is called an *estuary,* characterized by intertidal sand bars and mud shoals, often forming a salt marsh (see Figure 11–14).

5. Longshore currents along some complicated shorelines are accelerated by tidal flows, and, in a general way, tidal currents can affect extensive tracts of the continental shelves (see Chapter 13).

11-4. FEATURES OF ROCKY COASTS

There are two basic types of coast: First, there are shorelines with prominent erosion at the watermark, as waves break at the shore. Depending on rock type and local terrain, such coasts have either cliffs with sandy coves or straight, sandy beaches; bedrock is usually visible at or near the shore. Second, there are shorelines with wave erosion below the watermark, as waves break offshore in shallow water. Here there are sandy barrier islands, lagoons, mudflats, and deltas or estuaries; bedrock is not seen. The western coastline of North America is a *rocky coast* of the first type, as is the eastern coast north of Cape Cod. The remaining Atlantic and Gulf of Mexico littoral is a *coastal plain* of the second type.

Cliff Coasts. Many rocky coasts terminate in cliffs that rise abruptly from the watermark. In fact, the basic characteristics of a cliff coast in profile are: (1) the *cliff* itself, either steep or vertical; (2) the knickpoint or undercut *notch,* at or just above the watermark; and (3) the

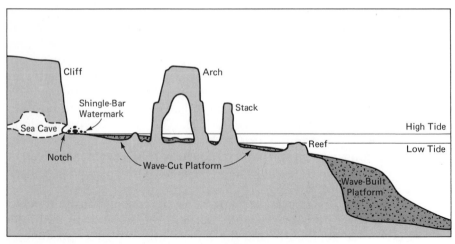

Figure 11–4. Composite Section of a Cliff Coast.

sandy and pebbly *platform* sloping gently seaward below the notch (Figure 11–4).

Wave erosion undercuts the cliff and so widens the platform. Overhangs may be created above the notch, or sea caves excavated at the watermark by either mechanical abrasion or chemical corrosion. Cliff faces retreat by talus and rockfalls or debris slides, steepening or maintaining a constant angle through time. The height also gradually increases, with local relief in the order of 10 to 200 feet (3 to 60 m) or more.

The platform commonly has a gentle gradient (less than 5 degrees or so), and may attain a width of several hundred yards or meters. Materials eroded inshore are eventually swept out by the undertow, to be deposited a little offshore. Consequently, the platform grades over from a wave-cut or *abrasional* platform (Figure 11–5) to a wave-built or *depositional* platform. The latter may be reworked by longshore currents into a longshore bar. Due to the constant agitation of the water, clays or silts are seldom deposited in the shore zone, and sands or well-rounded gravels form the characteristic deposits. On sandy beaches, pebbles are flattened by repeated sliding, while pebbles on rocky platforms can be churned around in potholes, producing spheroidal pebble shapes.

In seas or lakes having little or no tidal influence, the mean watermark commonly coincides with the notch or knickpoint at the foot of the cliff. Where tides are prominent or where heavy surf develops during storms, the high tide or storm shoreline may be at the foot of the cliff, but the mean watermark is found somewhere down on the

Figure 11–5. A Rocky, Wave-Cut Platform During Low Tide, Atlantic Coast of Northwestern Spain. (Karl W. Butzer)

platform. Another kind of cliff coast *plunges* to well below sea level, so that the platform, if any, is completely submerged, with the notch developed somewhere along the cliff face (Figure 11–6,A).

In horizontal arrangement, cliff coasts are commonly irregular, with rocky headlands and sandy inlets (Figure 11–7). Cliffs are steepest on promontories, and erosional remnants may stud the platform as rocky islands or rock spires (*stacks*), rock *arches* linked to the cliff, and

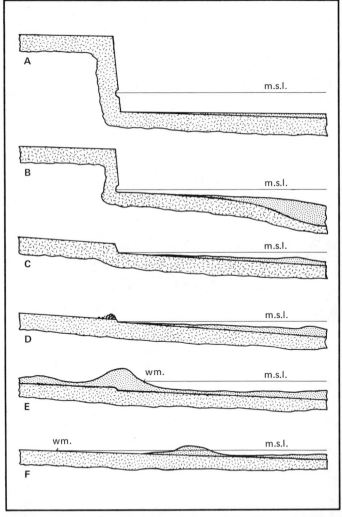

Figure 11–6. A Classification of Coastal Profiles: (A) Plunging Cliff Coast, (B) Cliff Coast with Notch, Wave-Cut, and Depositional Platforms, (C) Nip Coast, (D) Shingle-Beach Coast, (E) Beach-Ridge Coast, (F) Bar-and-Lagoon Coast (wm = watermark; m.s.l. = mean sea level). (After Butzer, 1962.)

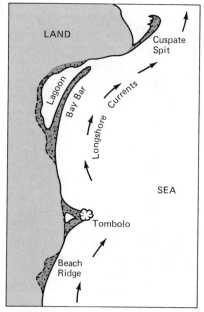

Figure 11–7. Complex Erosional and Depositional Coastline, with Bay Bar, Beach Ridge, Cuspate Spit, and Tombolo.

partially submerged *reefs*. Sandy beaches are common in bays and other indentations, while sand bars (*tombolos*) may also connect rocky islets to the coast (Figure 11–8).

Influence of Rocks and Structure. The configuration and the profile of cliff coasts are modified by lithology and structure. Lithology affects rates of wave erosion and cliff recession, leads to differential erosion, and determines the significance and type of weathering, as well as the nature of the beach deposits.

Cliff recession is slow and almost imperceptible in resistant bedrock. However, unconsolidated or weak materials behave differently. Glacial till and outwash, for example, are rapidly undercut at rates of as much as several feet per year. In fact, rapid coastal recession along the shores of Lake Erie and on the eastern coast of England is made possible by loose, Pleistocene sediments or more recent coastal sands. Volcanic ash also offers little resistance. Steep cliffs or projections cannot be supported; instead, mass movements such as debris slides and slumping attempt to counteract wave undercutting by establishing new equilibrium slopes. Consequently, storm after storm, the base of the cliffs is undermined, leading to compensating collapse features on the slopes above. In this way, large amounts of fresh sand and

Figure 11–8. Reef Linked to Coast by Tombolo, Northwestern Spain. (Karl W. Butzer)

gravel are repeatedly supplied to the beach zone, building up beach ridges or attached bars.

Differential erosion is most conspicuous in consolidated rocks of variable lithology. The location of promontories, stacks, marine arches, and sea caves is often closely related to differential resistance, as is the degree of irregularity or *articulation* of the shoreline.

Differences of rock resistance are frequently linked to weathering properties. As in the case of other geomorphic forces, weathering is an important prerequisite to wave erosion of most rock types. Surface weathering of cliff faces as well as weathering at the watermark aid in cliff recession and facilitate wave erosion by enlarging joints, fissures and bedding planes. Thus frost weathering may be important in colder zones, chemical attack in warmer climates. Limestones and, to a lesser extent, dolomites are particularly weak. The majority of sea caves are developed in limestones and in lime-cemented sandstones, as a result of corrosion and wave erosion at the water-mark. Salt spray attacks cliff faces, enlarging cracks by crystal growth and creating solution cavities that can be rapidly enlarged to brine-filled potholes. Corrosion features are also found on crystalline rocks of tropical coasts.

The abundance of sediment on the platform as well as the relative proportions of sand and gravel vary according to rock type—much in the same way as the load of a stream. The degree of development of beach ridges, attached bars, and coastal dunes varies according to how much sediment is available: durable rocks provide little, whereas sandstones and glacial drift provide generous amounts. The paucity of quartz sand along limestone coasts is partly compensated for by ground up mollusk shells, which provide lime sand.

Structural factors can exert considerable influence on coastal morphology, and the primary location of shorelines is more often than not determined by large-scale structural alignments such as fault zones and rock inclinations (Figure 11–9). Intermediate-scale features such as straight coastal segments, crenulations, inlets, and headlands may be circumscribed by faults, minor fractures, joint systems, and the like. At an even smaller scale, wave attack is aided by fractures and jointing, and sea caves may develop along major joints, small faults, or in weaker dikes. Finally, the overall gradient and ultimate profile of a coast may be profoundly influenced by rock inclination.

Beaches and Attached Bars. Sandy beaches, with or without gravel, develop in the gently sloping, shallow inlets that interrupt most cliffed coasts. Frequently, too, sandy beaches fringe straight coasts with moderate gradients. In each case spilling waves are typical, with longshore currents and beach drifting. Below the watermark, the smooth, sandy platform slopes at angles of about 3 to 8 degrees. Above the watermark

Figure 11–9. Plunging Cliff Coast Formed by Inclined Strata of Variable Resistance, Northern Mallorca. (Karl W. Butzer)

the ground rises sharply to a ridge of sand or gravel that curves to a crest at about the swash level during stormy seas. The height of such *beach ridges* reflects on wave energy and varies from 1 or 2 feet (30 or 60 cm) to 10 or even 25 feet (3 or 8 m) (Figure 11–6,E). Frequently, several such ridges are backed up against each other, and a new ridge can form during each major storm. The tops of most beach ridges are susceptible to wind erosion, and eolian features are often well developed where sand is abundant. Sand dunes are common in the rear of beach ridges on windward shores.

Bedrock is not unusual in the shore zone of a beach-ridge coast. Sometimes, rocky outcrops are so extensive as to interrupt the ridges. Such erosional segments may form incipient cliffs or *nips* (Figure 11–6,C) with a relief of from 2 to 10 feet (0.6 to 3.0 m); they are intermediate between the sandy beach and cliff coastal types. In other instances, the beach ridge is replaced by gravel accumulations known as *shingle beaches* (Figure 11–6,D).

Attached bars develop along irregular rocky coasts with abundant sediment and strong longshore currents (Figure 11–7). *Bay bars* close off large coastal indentations, while *spits* develop downcurrent of headlands that mark major changes in coastal geometry. Converging longshore currents can build up *cuspate bars* in front of convex coastal sectors, and a variety of bars may be found attached to islands and stacks, including tombolos tied to the coast. These bars rise from the coastal platform and extend above sea level, creating zones of shallow water. If not breached periodically by storm waves or kept open by tidal action, attached bars can eventually close off embayments, forming lagoons or mudflats.

Coral Reefs. In clear salt waters of the tropical oceans, coral reefs are characteristic of rocky coasts. Coral consists of the calcareous skeletons of primitive organisms that accumulate layer by layer on long stalklike projections (solitary coral) or irregular, rocklike masses (colonial coral). Biologically, coral polyps are animals, although they look like plants. The fleshy part resembles a jelly fish, with small tentacles protruding from a base fixed to dead coral or rock. Typical dimensions are 0.4 to 0.5 inches (about 1 cm). The polyp feeds from the sea water, and lime particles build up in the cells at the base of the organism. As the polyp is replaced, the calcareous base remains behind as a layer of bizarrely patterned lime rock.

Corals are exclusively found in salt water, with a salinity of 2.7 to 4.0 percent, free of suspended sediment, and preferably agitated by wave action. They also require a solid base of rock to grow out from, oriented to the open sea, but at shallow depths—preferably less than 135 feet (45 m). Due to the lack of adequate light, coral growth is normally impossible in turbid, silt-laden waters or below a depth of about 200 feet (60 m). Some corals, mainly of the solitary kind, can grow in cool or temperate waters. Most of the colonial forms, however, thrive only at temperatures of 77 to 86°F (25 to 30°C) and do not tolerate winter temperatures of less than 64°F (18°C). As a result, reef-growing coral is not found outside of the tropical oceans, and it is absent near the mouths of rivers and muddy streams.

Several shore forms are provided by coral reefs:

1. *Fringing reefs,* attached directly to the shore platform
2. *Barrier reefs*, extending up from a rocky base at depths of more than 300 feet (90 m) and separated from the shoreline by broad, deep lagoons
3. *Atolls*, coral islands consisting of circular or arcuate reefs, interrupted by inlets or boat passes, surrounding a lagoon with water depths of 100 to 300 feet (30 to 90 m)

Whereas fringing reefs can be adequately explained in relation to modern shore platforms, barriers and atolls are more problematical, since their basal reef coral is now found at depths far too great for coral growth. Such cases are interpreted by relative movements of land and sea—both by continued subsidence of submerged platforms, volcanoes, and sea mounts or by fluctuations of ocean level.

Some of the more important geomorphologic features of a coral reef (Figure 11–10) include:

1. The steep external slope, formed of coral, coralline sand, and talus, dipping steeply (at angles of 25 to 45°) down to considerable depths
2. The reef crest above the external slope, formed of growing colo-

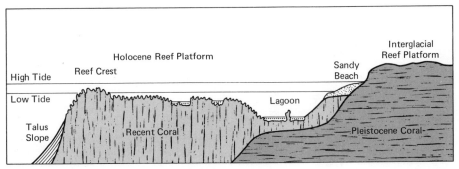

Figure 11–10. A Fringing Coral Reef.

nies of lime-secreting algae, active coral, ground-up coralline sand, and loose coral rock; the crest surface is at about mean sea level
3. The reef platform, to the rear of the crest, often several hundred yards or meters wide, composed of dead and living coral below low-tide level
4. The lagoon (if present) with a sandy floor, interrupted by sporadic coral growths below the low-tide mark

These constituent parts are developed in varying degrees on most coral coasts, although they are best represented in fringing reefs.

11-5. FEATURES OF COASTAL PLAINS

When gradients on either side of the watermark are extremely gentle (less than 1 degree), a distinctive coastal morphology may develop. Under such conditions the coast builds seaward, and there is little attrition of the land area. This is usually the case along the shores of flat coastal plains that have minimal relief and sluggish drainage.

Offshore Bars. In very shallow waters, waves strike bottom to break well offshore, thus expending their energy on the sea floor. Eroded materials are thrown up and then forward, to be deposited at a little distance from the shoreline. The resulting shoals and longshore bars develop parallel to the coast. As their relief increases and erosion deepens the broad, shallow trough on their seaward side, waves are eventually able to break against the edge of the submerged bar. The form is accentuated until an *offshore bar* (or *barrier island*) appears above low tide. Subsequent development by swash and beach drifting is analogous to that of beach ridges and attached bars, and dimensions are similar (Figure 11–11). However, the primary source of sand remains

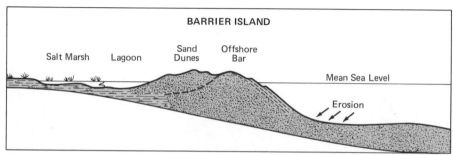

Figure 11–11. Offshore Bar-and-Lagoon Coast.

erosion of the sea floor, augmented by ground-up molluskan shell. Longshore currents and beach drifting ensure lateral transport and growth, with development of cuspate forms and spits.

When offshore bars rise to the high-tide mark and beyond, the surface is commonly reworked by wind action. Dunes may build up in the rear of the bar, widening it and creating an irregular relief.

Lagoons and Mudflats. The shallow waters cut off by an offshore or attached bar are often converted into *lagoons,* connected to the open sea by narrow channels or tidal inlets. Wave activity is at a minimum here. However, fresh water influx from the land introduces suspended and dissolved sediment, part of which is trapped in lagoons. A specialized vegetation, adapted to salt or brackish waters, may colonize the margins and shoals, helping to bind more sediment. Eventually, the lagoonal area can be reduced to a marsh or tidal mudflat with salt-tolerant plants (Figure 11–12,A). In addition, dunes may encroach from the offshore bar, sometimes sweeping right across the flats.

Corresponding to the salt marsh of tidal mudflats in mid-latitudes are the *mangrove swamps* of the tropics (Figures 11–12,B, 11–13). In both environments there is a gradation from fresh to salt water, with submergence sporadic or periodic (once or twice daily) or even permanent. Mangroves include several genera of swamp trees that have varying degrees of salinity tolerance, require loose and permanently wet subsoils, but that do not tolerate temperatures below 32°F (0°C). In

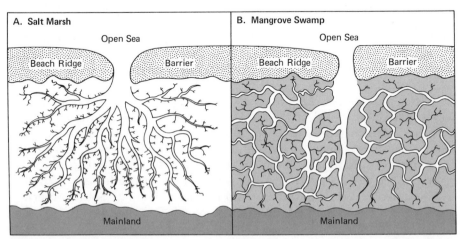

Figure 11–12. Salt Marsh (A) and Mangrove Swamp (B). (After Allen, 1965. With permission.)

Figure 11–13. Younger Mangrove Colonization in Reef-Protected Lagoon, Grand Cayman Island. The original stands had been destroyed six years earlier by Hurricane Hattie. (Courtesy of J. D. Sauer)

the ideal case, pioneer genera first colonize shoals and banks below low-tide level, binding sediment and building up organic residues. After 20 or 30 years the pioneer shrubs and trees give way to a mature mangrove forest, now at the elevation of the intertidal zone. As these formations colonize seaward, salinity initially increases as the tides are no longer able to flush out the root zone. Finally, after gradual desalinization, other forest species begin to colonize the new land surface. Mangroves are widespread along tropical coasts, near river mouths and wherever coastal lagoons or mudflats are present. They may also grow along shorelines sheltered by fringing reefs.

Estuaries and Deltas. *Estuaries* are semienclosed bays receiving fresh-water influx and subject to tidal influences. Heavier salt water underlies silt-laden fresh water, permitting great concentrations of suspended sediment. Locally, there may be tidal mudflats with salt marsh or mangroves. Estuaries are best developed at the mouths of major rivers where tidal amplitudes are great. Good examples are provided by Chesapeake Bay, the Amazon and La Plata rivers in South America, and by several rivers draining into the North Sea (Figure 11–14).

Deltas have been discussed earlier (Chapter 8). They are best developed where tidal amplitudes are small, although there are some notable exceptions. Strong coastal currents also inhibit delta develop-

Figure 11–14. The Estuary of the Minho River, with Spain to the Left, Portugal to the Right. (Karl W. Butzer)

ment. The basic delta geometry of birdfoot, arcuate, and estuarine types reflects on the balance of forces between (1) wave energy, (2) tidal influences, and (3) sediment load. Birdfoot deltas form where sediment influx is high, wave energy and tidal amplitude small. Estuarine deltas form where tidal influences are paramount, arcuate deltas under more intermediate conditions. Spits, cuspate bars, and tidal mudflats are often conspicuous at the margins of a delta, as a result of wave action, longshore movements, and tidal influences (Figure 11–15).

Although estuaries, deltas, and tidal mudflats are not restricted to the low-gradient shores of coastal plains, they are especially characteristic of such mixed marine-terrestial environments.

Coastal Dunes. Eolian activity of geomorphic impact is possible on all coastal sectors where sands are abundant at or above the high-tide mark. If there is a constant supply of fresh sediment, disturbance by storm waves can impede colonization by plants while the salt splash and spray inhibits most forms of vegetation. Consequently, the plant cover and root mat may be inadequate to bind loose sand.

Most common are the lines of nondescript beach foredunes that sometimes run parallel to beach ridges. These form irregular mounds,

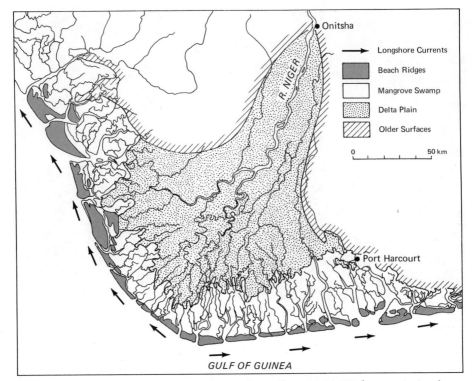

Figure 11-15. The Niger River Delta. (After Allen, 1965. With permission.)

typically 10 to 20 feet (3 to 6 m) high, and locally undercut by wave action or deflated to form blow-outs. Farther afield there may be more extensive fields of dunes, of types common to the continental interiors (see Chapter 19).

Chapter 12

COASTAL LANDFORMS (II): MODELED BY TERRESTRIAL PROCESSES

Along some coasts, streams or glaciers form characteristic shorelines of mixed type. These include deltas and estuaries, as well as valley glaciers and icecaps that calve into the open sea or feed extensive coastal ice shelves. Equally common are shorelines modeled primarily by terrestrial forces that are legacies of the past. In particular, the Pleistocene and Holocene have seen continued and rapid, real and apparent changes of sea level, which, for individual coastal sectors, varied as much as 500 feet (150 m) below to 350 feet (115 m) and more above the present relative position of land and sea. These changes reflect on tectonic deformation, isostatic readjustment, relative relief of ocean basins and continental margins, and actual fluctua-

tions of oceanic water volume due to the growth and wastage of continental glaciers. During the Pleistocene glaciations, world sea level was in the order of 300 feet (90 m) or more lower, at a time when stream or glacial action sculptured valleys and plains in relation to these lower base levels. Even though sea level has since recovered, drowned glacial landscapes remain unmistakable despite remodeling by waves, and along some of the Earth's coastlines stream-carved landscapes remain almost equally prominent. These submerged form-complexes of preeminently terrestrial origin are matched by extensive sweeps of "raised" beaches and once associated coastal dune fields. Whatever their origins, most of the Earth's coastlines are today being modified by man, who, often unwittingly, has now become another major force in coastal development.

12-1. THE CHANGING LEVEL OF THE SEA

Worldwide Sea-Level Changes? Coastal landforms are complicated by the fact that sea level does not remain constant through time. On a local scale, nonoutlet desert lakes rise and fall rapidly in response to climatic changes. Even in moist climates, where lakes are drained by rivers to the sea, small but perceptible changes of level can be observed. The world oceans also fluctuate in level, although in response to different mechanisms and at a very slow rate. It is important to distinguish between changes in level of the interconnected oceans and seas on the one hand and of interior lakes on the other. The fluctuations do *not* correspond, and the background factors are usually quite different.

Interior lakes that are beyond the reach of tidal influences respond to short-term climatic changes—over a period of a few years, decades, or even centuries. The two major variables are precipitation and evaporation over the drainage basin. Evaporation reflects temperature, cloud cover, and wind speed and may be as important as rainfall in modifying lake levels. If a lake has no outlet, the level responds rapidly to changes in water influx, as well as to the precipitation-evaporation balance over the lake surface itself. The Great Salt Lake, the Dead Sea, and Lake Rudolf in East Africa are examples of this type. If the lake has an overflow, changes of level are very subdued in amplitude. Lake Michigan and Lake Victoria are examples of this second type. Lake

Michigan has fluctuated with an amplitude of only 5 feet (1.6 m) since the beginning of records in 1860, Lake Victoria with an amplitude of 10 feet (3 m) since 1898. By contrast, the Dead Sea has fluctuated in the range of 11.5 feet (3.5 m) between 1930 and 1960, Lake Rudolf about 64 feet (19.5 m) since 1896.

The level of the oceans and connected seas, however, is barely affected by changes of regional climate. Instead the world oceans respond to four major mechanisms:

1. *Tectonic Deformation.* On a local or regional scale tectonic deformation may raise or lower coastal strata. Thus, for example, uplift continues intermittently among the horst blocks along the Gulf of California, or in the Coast Range of the Californian coast. Individual earthquakes may even be sufficiently severe to displace local shorelines. Vulcanism can change the elevation of the land, as in the case of the growth of the island of Hawaii by successive lava flows. Explosive eruptions, such as the destruction of the islands of Tamboro and Krakatoa, can also reduce a land mass. However, tectonic deformation is restricted in terms of area. It is neither worldwide nor synchronous, although whole tectonic provinces may be affected on an appreciable scale. Most of the mountainous coasts of the Mediterranean Sea have apparently been upraised some 300 to 500 feet (90 to 150 m) during the past 2 or 3 million years, and uplift along the California coast has been greater still.

2. *Isostatic Readjustment.* A different effect has been registered in areas once under the Pleistocene glaciers. The great weight of the ice masses served to compress the Earth's crust. This temporary isostatic depression at any one time may have amounted to one-third of the thickness of the ice sheet. As the ice melted, the crust rebounded with a corresponding degree of uplift. The result has been that the coasts of Canada and Scandinavia have been rising throughout the Holocene, at rates of as much as 10 feet (3 m) per century and often more rapidly than worldwide changes of sea level. Accordingly, there are many well-developed but abandoned shorelines of postglacial age along these particular coasts at elevations of 600 feet (180 m) and more.

3. *Capacity of the Oceans.* Through geological time the shape of the great ocean basins has changed very slowly—but nonetheless appreciably—in response to large-scale movements of the Earth mantle and crust. In response, it is probable that the capacity of the ocean basins to hold water has also changed as the basins changed their configuration and became deeper or shallower. Furthermore, the continental shelves are also liable to change and, depending on their slope, can also readjust to other changes by isostatic readjustment. Finally, whole segments of the continental margins appear to rise or sink, relative to the sea, at slow rates but over long periods of time. Whatever

the explanation, on some occasions the oceans appear to have spilled far up over the edges of the continents for long periods of geological time; on other occasions they have receded. There is at least some reason to believe that relative sea level, on a worldwide scale, was as much as 330 feet (100 m) higher during the Pliocene period, some 3 to 5 million years ago. Since then the overall trend has been downward.

4. *Amount of Ocean Water.* The growth and wastage of large glaciers on the continents provides a proved mechanism to lower or raise ocean level. A hydrological balance exists between the water in the oceans, in the atmosphere, in lakes and streams, in the groundwater, and in ice sheets or glaciers. If the icecaps grow, water must eventually be withdrawn from the oceans, which provide the only significant reserve. If the present glaciers of Antarctica, Greenland and other polar or mountain regions would melt today, the meltwaters would all be returned to the oceans, so that world sea level would theoretically rise by about 220 feet (66 m). A sudden thaw of this magnitude is, of course, quite improbable. But a slight but widespread recession of glaciers during the twentieth century is slowly adding more water to the oceans, and over the past 50 years sea level has been rising at a rate of 4.5 inches (11 cm) per century.

Estimates have been made of the volume of ice held in the Pleistocene glaciers. At the maximum of the Pleistocene glacial stages, world sea level was, according to various estimates (see Mörner, 1971, and Donn et al., 1962, for the full range of opinion), 280 to 525 feet (85 to 159 m) lower than it is today. During the waning stages of the last glacial stage, meltwaters were returned so rapidly to the oceans that sea level may at times have risen by as much as 5 feet (1.5 m) per century.

Such fluctuations in the volume of ocean water are known as *glacial-eustatic* changes. They provide a mechanism that can theoretically raise sea level to about 220 feet (66 m) *above* the present or lower it to perhaps 480 feet (145 m) *below* that value. This is a surprising range of 700 feet (210 m).

Pleistocene Sea Levels. The rate at which coastal landforms evolve today is in most instances slow. If world sea level can and does fluctuate rapidly, it becomes pertinent to know whether sea level is sufficiently stable to permit full-fledged, "mature" coastal forms to develop. In fact, do the visible coastal features reflect present or past stands of the sea? For reasons such as these it is necessary to know about the sea-level changes of the Pleistocene and the more recent past. Only then can coastal landforms be seen in a proper time dimension and evaluated in a meaningful perspective.

Three trends can be recognized during the 2-million-year course of the Pleistocene:

1. There has probably been a gradual deepening of the ocean basins or an upraising of the continental margins to simulate a downward trend or regression of world sea level in the order of perhaps 330 feet (100 m).
2. Superimposed upon this slow downward trend were a series of rapid glacial-eustatic fluctuations, corresponding to at least eight major glacial spasms. During each glacial, sea level dropped perhaps 300 feet (90 m) or more below "normal." Slight positive fluctuations or *transgressions*, in the range of 10 to 25 feet (3 to 8 m), may have been caused by slightly greater melting of residual ice masses during several interglacial phases. The resultant curve is fairly complex (Figure 12–1).
3. In addition to these worldwide changes, varying degrees of local deformation can be detected in all coastal areas next to mountains affected by the Alpine orogeny. Here intermittent uplift has been the rule, and a few areas such as the California coast have even experienced appreciable changes during the past 100,000 years or so.

The western and arctic coasts of the New World and the seashores of the Baltic have also been particularly mobile during the past 10,000 years, as Earth's crust rebounded after the weight of continental glaciers was removed. In more stable tectonic regions, positive or negative movements of the land have resulted from isostatic readjustment to the weight of water added or withdrawn from the continental shelves in response to glacial-eustatic transgressions or regressions.

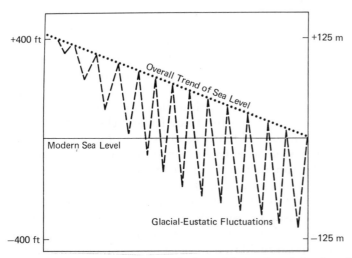

Figure 12–1. Schematic Trend of World Sea Level During the Pleistocene.

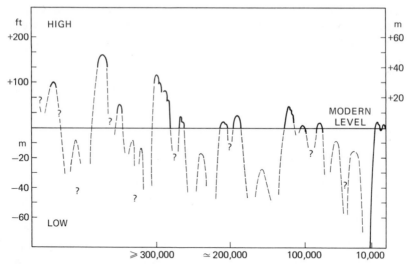

Figure 12–2. Relative Sea Levels Recorded on Mallorca. The record prior to 300,000 B.P. is temporally distorted. (After Butzer, 1975. Copyright © 1975 by Mouton Publishers, The Hague.)

For the purposes of coastal geomorphology, two significant applications emerge from an understanding of Pleistocene sea levels (Figure 12–2). First, during successive interglacials, world sea level was higher than it is today. At such times, particularly during the Last Interglacial (from about 125,000 to 75,000 years ago), shore features were cut or built that remain conspicuous on many coasts. They now lie anywhere from 5 to 100 feet (1.5 to 30 m) or more above the watermark. But even during the Last Interglacial, sea level was not stable, and multiple beaches frequently formed at individual locations. As a result, in many places a series of high or "raised" shoreline features can be observed rising above the modern beach. Although fragmentary, these relics of the Pleistocene are definitely a part of the constellation of coastal forms.

Second, for almost 60,000 years, during the course of the Last Glacial, world sea level was much lower than today. This is partly reflected by now submerged coastal features at depths of almost 500 feet (150 m) on the continental shelves (see Chapter 13). Far more important, however, is what happened on land surfaces now exposed at the coast. Rivers cut down and readjusted their profiles to a new equilibrium, sometimes creating fluvial landforms that have been difficult for subsequent coastal processes to camouflage. Glaciers bore down over what is now the shoreline, gouging out a rock-modeled sculpture of glacial forms such as those of Maine and Norway, or depositing the thick blankets of drift in Boston Bay or Long Island, where glacial origins remain evident millennia after the event. The fact is that wave action has remained incidental along many coastlines, so impressive is the legacy of fluvial or glacial landforms created during the Last Glacial, prior to 8500 B.C.

Holocene Sea Levels. With exception of a few tectonically unstable areas, relative movements of land and sea during the past 10,000 years have been primarily of the glacial-eustatic kind. The maximum cold of the Last Glacial was achieved about 25,000 or 20,000 years ago, and the ice sheets began to wane rapidly by about 14,000 years ago. In about 10,000 years over 90 percent of the volume of ice sheets was returned to the oceans, an amount that, if not compensated by coastal readjustments, would cause a rise of sea level in the order of 330 feet (100 m). This comparatively rapid transgression flooded out broad sectors of the continental shelves, drowning whole constellations of fluvial and glacial landforms. In fact, both fluvial and coastal processes have barely had sufficient time to adjust to a new equilibrium during the past 6,000 years. The result is that many coasts still appear to be "drowned" as an aftermath of the postglacial rise of sea level.

Radiocarbon dating of Last Glacial deposits in northern Canada as well as of shoreline features along a number of coasts implies that present sea level was only attained about 4000 B.C.—shortly before the beginning of the historical era in the Near East. Fairbridge (1960) has suggested that the biblical record of the deluge may refer to this postglacial transgression and its effects in drowning the Persian Gulf and other coastal plains. However, even a rise of the sea by 6 inches (15 cm) in 10 years would hardly be noticeable to the average person! Instead, if they are to be considered as historical documents, the Gilgamesh and the biblical records of a great Mesopotamian flood suggest a catastrophic but local event (see section 1-1).

During the historical era sea level has continued to oscillate about its present level in response to minor advances and retreats of the existing glaciers. Lind (1969) has studied these changes on the Bahamas, tectonically a stable area. On the basis of radiocarbon dating he suggests a complex sequence of sea-level oscillations that can be approximately dated (corrected to calendar years) as follows:

Fourth and third millennia B.C.	Near present level
3000 (?)–2000 B.C.	As much as 10 feet (3 m) *above*
2000–1200 B.C.	As much as 10 feet (3 m) *below*
1200–200 B.C.	6.5 feet (2 m) *above*
200 B.C.–400 A.D.	As much as 6.5 feet (2 m) *below*
400–1200 A.D.	3.5 feet (1 m) *above*
1200–1400 A.D.	As much as 3.5 feet (1 m) *below*
1400–1600 A.D.	1.5 feet (50 cm) *above*
1600 A.D. to present	Present level or slightly *below*

Archeological evidence from the Mediterranean Basin agrees with the hypothesis that sea level was lower during much of the classical Greek and Roman period, although many Northern Hemisphere coasts lack any evidence of relatively "high" Holocene shorelines. Mörner (1971) has attempted to reconcile the increasingly contradictory evidence on this controversial issue by invoking differential readjustments of crustal components of the continental margins to the increased volume of water resulting from ablation of the Pleistocene glaciers.

Whatever their true details, the oscillations of sea level since 4000 B.C. have been too brief to produce distinctive landforms in most parts of the world. But rapid oscillations are a fact in areas with narrow continental shelves, and they do help explain why contemporary coastal platforms are poorly developed along sheltered coasts where tidal amplitudes are small. In most instances where valid comparisons can be made in the Mediterranean region, wave-cut platforms of interglacial age are much broader than those of postglacial times. The instability of sea level also helps explain the multiple beach ridges and barrier islands seen on many coasts. Repeated oscillations of sea level favor renewed adjustments of the erosional-depositional balance, and the resulting landforms are better developed. So, for example, the width and the height of the offshore and attached bars of the Gulf-Atlantic coasts of the United States are far greater than can be explained by a simple, one-phase sequence of development.

12-2. LANDFORMS REFLECTING CHANGES OF SEA LEVEL

High Shorelines. Coastal forms that reflect relatively high sea levels fit the same categories as contemporary coastal types, although two types are best preserved and most diagnostic:

1. *Cliff Coasts.* Relict cliffs with fragments of platforms provide good evidence for former shorelines, and the watermark can often be identified with accuracy from notches, abandoned sea caves, and other phenomena (see Figure 11–1). The dating of such features is frequently possible on the basis of mollusks cemented into former beach sands or gravels. Ages are obtained by the radiocarbon method, using the rate of decay of radioactive atmospheric carbon, or by one of several uranium methods that utilize the breakdown of minute quantities of primary uranium into daughter elements. Such techniques show that most high shorelines outside of the Arctic date back to the Last Interglacial. The related beach deposits and sea-cave fillings are frequently linked with Early Man sites in the Old World. The species assemblages and isotopic chemistry of the mollusks give further information on prevailing water temperatures, showing that many but not all Pleistocene interglacials were as warm as today. Older "raised" shorelines can also be identified from earlier parts of the Pleistocene.

2. *Offshore Bar-and-Lagoon Coasts.* Great parallel ridges of sands, interrupted by flat surfaces, can be traced along parts of the U.S. Atlantic coast or along the northern shores of Egypt. These extend to as much as 350 feet (110 m) above modern sea level and represent offshore bars formed in relation to different interglacial sea levels.

High shorelines are widespread along most world coasts, but they are most conspicuous where tectonics have uplifted a coast, sometimes producing steplike tiers of abandoned beaches. This is so on the unstable coasts of California and Peru as well as in areas rebounding from crustal compression under the Pleistocene ice sheets.

Drowned Coasts of Fluvial Origin. On some coasts stream dissection was strongly accentuated during the glacial-eustatic regressions of the Pleistocene. Rivers cut down to 300 feet (90 m) or more beneath modern sea level, often creating deep and prominent valleys. During the postglacial rise of sea level, some of these valleys were drowned; they have only been partly filled with modern sediments and remain partially submerged. Such drowned river or stream valleys are collectively known as *rias,* named after prominent estuaries along the northwest coast of Spain. Coastlines dominated by rias are designated as *ria coasts* (Figure 12–3,A). Rias exhibit dendritic or geometric drainage patterns. Occasionally, alluvial terraces are preserved along their exposed slopes, and entrenched meanders are not uncommon. A classic example of drowned meanders can be seen along the river mouths that empty into Chesapeake Bay.

The localization of ria coasts is still poorly understood. One prerequisite is that Pleistocene fluvial processes were more active than those of today. In lower mid-latitudes drowned stream inlets can be attributed to greater cold and accelerated runoff (see Chapter 18). In some low-latitude deserts the explanation also involves increased precipitation. In all cases, rias imply that postglacial sedimentation could not match the rise of sea level. At present these are areas of limited

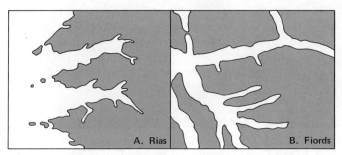

Figure 12–3. Trace of (A) Ria and (B) Fiord Coastlines.

stream competence or erosion. In some instances, tectonic deformation has also been at work, favoring coastal subsidence, as in the case of Chesapeake Bay.

Special forms of rias, with regional names, are provided by semi-arid or arid environments, where valley cross-sections are steep walled and flat floored, while modern streams are intermittent or almost defunct (Figure 12–4).

Drowned Coasts of Glacial Origin. Wherever Pleistocene mountain glaciers carved out valleys down to the coast, great troughs were created that are difficult to fill with sediment. The resulting *fiords* are often 2,000 feet (600 m) or more deep, with vertical walls and flat, rock-cut floors. The diagnostic U-shape is further complemented by hanging valleys. Due to their regularity they are navigable by ocean-going vessels for many miles inland. Similar, deeply indented *fiord coasts* (Figure 12–3,B) with characteristic plunging cliffs are developed in Norway, southern Alaska, and British Columbia, as well as in southern Chile and in New Zealand.

In the case of continental glaciers, the ice planed off broad surfaces down to bedrock—with the undulating or lineated configuration of ice-scoured plains. A fine example of this type is provided by the irreg-

Figure 12–4. Drowned Inlet Cut by Greater Pleistocene Stream Discharge and Modified by Cutting of a 7-Foot (2-m) Interglacial Beach, Bay of Palma, Mallorca. (Karl W. Butzer)

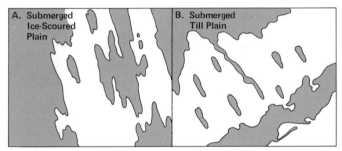

Figure 12–5. Trace of Drowned Plains of Glacial Erosion (A) and Deposition (B).

ular, rocky coast of Maine (Figure 12–5,A). Drowning of low-level drift plains is also possible, classic examples being Long Island (a partly submerged terminal moraine with adjacent outwash plain) and the Massachusetts coast (with terminal moraines, till plain, and drumlin fields) (Figure 12–5,B). Bunker Hill, in Boston Bay, is a partly submerged drumlin. Wave-remodeled drift plains are also found along the shores of the Baltic and North seas.

Inactive Coastal Dunes. Other relicts from the past, found along many coasts of the tropics and subtropics, include cemented or immobile dune sand, occasionally aggregated into dune fields. Such sands are mainly composed of lime, blown up from sandy beaches with ground-up molluskan shell. In some instances, eolian activity is minimal today and there may not even be a ready source of sand. The majority but not all of such relict coastal dunes accumulated during the course of Pleistocene regressions (Butzer, 1975), when sea level was dropping and exposing the fresh sands of beach platforms to deflation. Some inactive coastal dunes have been partly submerged by the postglacial rise of sea level.

12-3. A CLASSIFICATION OF COASTAL LANDFORMS

This discussion of coastal types and their many origins has probably given glimpses as to why the Earth's coasts are one of the most fascinating themes open to geomorphologists. Landforms at the coast can be studied as the pages of a book can be read, revealing insights into the history of the planet. Variations from one coast to another provide a challenge to search for explanations, both in terms of structural backdrop and coastal evolution through time.

As it is, there is no foolproof rationale for coastal types—why they happen to occur in the patterns they do and how they are supposed to evolve from incipient to well-developed examples of an individual

Table 12-1. A CLASSIFICATION OF COASTAL LANDFORMS

I. COASTS MODELED PRIMARILY BY WAVE ACTION

		FEATURES DUE TO WAVE ACTION	FEATURES DUE TO TIDAL OR TERRESTRIAL PROCESSES
A. *Rocky Coasts*	1. Cliff sectors	a) Notched or plunging cliffs b) Abrasional, depositional platforms c) Stacks, reefs, arches, caves d) Longshore bars	a) Chemical corrosion or solution b) Frost weathering c) Talus and rockfalls (consolidated rock) d) Slumping and debris slides (unconsolidated rock)
	2. Sand and shingle sectors	a) Beach ridges b) Shingle beaches c) Nips d) Longshore bars	Coastal dunes
	3. Attached bars	a) Bay bar b) Spit c) Cuspate bar d) Tombolo bars	a) Tidal lagoons or mudflats, including saltmarsh or mangrove swamp, tidal creeks, and inlets b) Nontidal lagoons and marsh
	4. Coral reefs	a) Fringing reef b) Barrier reef c) Atoll	Lagoons
	5.	High Pleistocene shorelines, including cliffs, platforms, ridges, and coral platforms	Fossil dunes

I. COASTS MODELED PRIMARILY BY WAVE ACTION

	FEATURES DUE TO WAVE ACTION	FEATURES DUE TO TIDAL OR TERRESTRIAL PROCESSES
B. *Coastal Plains*	Offshore bars, including sectors with cuspate bars, spits, or longshore bars	a) Lagoons and mudflats, with salt marsh or mangrove swamp, tidal creeks, and inlets b) Estuaries and deltas c) Coastal dunes

II. COASTS MODELED PRIMARILY BY TERRESTRIAL PROCESSES (WAVE ACTION SUBORDINATE)

A. *Fluvial*	1. Contemporary stream action	a) Estuaries b) Deltas (including birdfoot, arcuate, and estuarine types)
	2. Pleistocene stream action: drowned landscapes shaped primarily by fluvial dissection (including rias)	
B. *Glacial*	1. Contemporary glacial action: valley glaciers or icecaps calving into open sea or feeding ice shelves	
	2. Pleistocene glacial action: drowned glacial landscapes	a) Drowned glacial troughs (fiords) b) Drowned ice-scoured plains c) Drowned drift plains

class. Sea level is far too unstable to provide opportunity for land-forms to evolve unidirectionally without many complicating factors. It is simpler to order the recognizable types according to their diagnostic features. They are summarized in Table 12–1, which lists those land-forms commonly found adjacent to each other, distinguishing forms due primarily to wave action from those modified or controlled by other processes.

12-4. MAN AND THE COAST

A great deal can be said about the relevance of coastal landforms to man and for his use of the land. Coasts have been vital as a source of marine food and as a means of communications since prehistoric times. More recently, they have attracted people as vacationlands and homesites. To do justice to these topics would require a chapter by itself, and many geographers have indeed concerned themselves with the applications of coastal geomorphology to land use, settlement, or industrialization (Mitchell, 1974). There are, however, two central themes that equally concern geomorphologists, engineers, and people who have developed farmlands, preserved parks, built homes, or op-erated harbor installations by the sea:

Coastal Recession. In coastal regions where unconsolidated sediments are readily undermined by wave attack, cliffs retreat rapidly at times of stormy seas, and sandy beaches, bars, and spits change in shape. In part, this is a matter of platform widening, a process that will go on indefinitely. In part, it is due to changes in the coastal profile, similar to adjustments in the longitudinal profile of a river. Some adjustments are natural, resulting from gradual changes of coastal configuration and articulation through time (Figure 12–6). Others are a result of human interference, as coastal equilibrium attempts to compensate for the dredging of harbor channels, construction of piers, jetties, quays, and other disturbances of the balance of erosion and deposition. In many instances, too, people have been naïve enough to build roads or other structures much too close to a zone of intensive geomorphic activity.

Costly and often unsuccessful attempts have been made to com-pensate for coastal recession. Sea walls, wood pilings, and concrete re-vetments are constructed along cliffed coasts, or great masses of crude rock are built up around the watermark to absorb wave impact. Along sandy beaches long piers (groynes) are built out to catch the sand drifting by: Material eroded in one area is carried by longshore cur-rents until it enters the sheltered waters to the lee of the groyne. Locally, this will favor deposition over erosion, and so halt further at-trition wherever refracting waves are the culprit. The material ac-

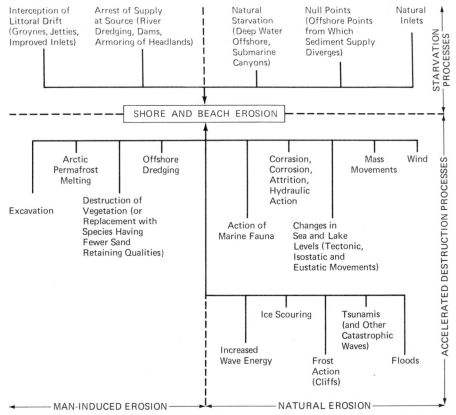

Figure 12–6. Processes of Natural and Man-Induced Coastal Erosion. (After Mitchell, 1974. With permission.)

cumulating between individual groynes must, however, be obtained elsewhere, so that adjacent or intervening areas suffer from "sediment starvation" and accelerated coastal recession. Perhaps the most urgent conservation measure is to save coasts from "development," not from destruction. At a time when industrial development and gross national products need to spiral exponentially in order to keep the labor force in work, the very existence of "undisturbed" coasts is threatened—by suburban sprawl, industrial blight, and pollution by shipping or industrial wastes. Perhaps it would be more sensible to begin to restrict coastal development and to keep existing beaches open to people who would like to enjoy them.

Storm Catastrophes. Every now and then hurricanes and flood surges smash into coastal lowlands, causing severe damage and taking many lives. Severe gales combine with exceptionally high tides to cause equal destruction, as may the tsunamis generated by earthquake shocks in the ocean floor. These natural disasters have plagued man from prehistoric times, not only taking their toll of ships at sea but also destroying fishing villages, harbors, and agricultural lands. Little can be done to eliminate the impact of such catastrophes. They will continue at intervals wherever man tries to wring his livelihood from the seacoast. And, as coastal development continues without abatement, their frequency increases in terms of human damage. The only real solution is land planning that considers the magnitude and frequency of natural coastal hazards. Without such planning—and enforcement—the best one can hope for is competent prediction and effective advance-warning systems, so that rudimentary precautions can be taken.

12-5. SELECTED REFERENCES

Allen, J. R. L. Coastal geomorphology of eastern Nigeria: beach-ridge barrier islands and vegetated tidal flats. *Geologie en Mijnbouw,* Vol. 44, 1965, pp. 1–21.

Bascom, W. L. Beaches. *Scientific American Offprint,* No. 845, 1960, 12 pp.

Bird, E. C. F. *Coasts.* Cambridge, Mass., M.I.T. Press, 1968, 246 pp.

Butzer, K. W. Coastal geomorphology of Majorca. *Annals, Association of American Geographers,* Vol. 52, 1962, pp. 191–212.

————. Pleistocene littoral-sedimentary cycles of the Mediterranean Basin. In K. W. Butzer and G. L. Isaac, eds., *After the Australopithecines.* The Hague, Mouton, and Chicago, Aldine, 1975, pp. 25–71.

————. *Recent History of an Ethiopian Delta: The Omo River and the level of Lake Rudolf. Research Papers, University of Chicago, Department of Geography,* No. 136, 1971, pp. 1–184.

Chappell, J. Late Quaternary glacio- and hydro-isostasy on a layered Earth. *Quaternary Research,* Vol. 4, 1974, pp. 405–428.

Coleman, P. J. Tsunamis as geological agents. *Journal, Australian Geological Society,* Vol. 15, 1968, pp. 268–273.

Davies, J. L. *Geographical Variation in Coastal Development.* Edinburgh, Oliver and Boyd, 1972, 204 pp.

Donn, W. L., W. R. Farrand, and M. E. Ewing. Pleistocene ice volumes and sea-level lowering. *Journal of Geology,* Vol. 70, 1962, pp. 206–214.

Fairbridge, Rhodes W. The changing level of the sea. *Scientific American,* Vol. 202, 1960, pp. 70–79.

Guilcher, André. *Coastal and Submarine Morphology.* Translated from the French by B. W. Sparks and R. H. W. Kneese. London, Methuen, 1958, 274 pp.

King, C. A. M. *Beaches and Coasts,* 2nd ed. London, Edward Arnold, and New York, St. Martin, 1972, 600 pp.

Larsen, C. E. Variation in bluff recession in relation to lake level fluctuations along the High Bluff Illinois Shore. Chicago, Illinois Institute for Environmental Quality, Document No. 73-14, 1973, 73 pp.

Lauff, G. H., ed., *Estuaries*. Washington, D.C., American Association for the Advancement of Science, Publication 83, 1967, 757 pp.

Lind, O. *Coastal landforms of Cat Island, Bahamas. Research Papers, University of Chicago, Department of Geography*, No. 122, 1969, pp. 1–156.

McGill, J. T. Map of coastal landforms of the world (1:25 million). *Geographical Review*, Vol. 48, 1958, pp. 402–405.

McIntyre, W. G., and H. J. Walker. Tropical cyclones and coastal morphology in Mauritius. *Annals, Association of American Geographers*, Vol. 54, 1964, pp. 582–596.

Mitchell, J. K. *Community response to coastal erosion. Research Papers, University of Chicago, Department of Geography*, No. 156, 1974, pp. 1–209.

Mörner, N. A. Late Quaternary isostatic, eustatic and climatic changes. *Quaternaria*, Vol. 14, 1971, pp. 65–83.

Psuty, N. P. Beach-ridge development in Tabasco, Mexico. *Annals, Association of American Geographers*, Vol. 55, 1965, pp. 112–124.

Schwartz, M. L., ed. *Barrier Islands*. Stroudsburg, Pa., Dowden, Hutchinson and Ross, 1973, 464 pp.

Shepard, F. P. *Submarine Geology*, 2nd ed. New York, Harper & Row, 1963, 557 pp.

———, and H. R. Wanless. *Our Changing Coastlines*. New York, McGraw-Hill, 1971, 580 pp.

Steers, J. A. *The Coast of England and Wales in Pictures*. London, Cambridge University Press, 1960, 146 pp.

———, ed. *Applied Coastal Geomorphology*. Cambridge, Mass., M.I.T. Press, 1971, 227 pp.

Stoddart, D. R. Ecology and morphology of Recent coral reefs. *Biological Reviews*, Vol. 44, 1969, pp. 433–498.

Thom, B. G. Mangrove ecology and deltaic geomorphology: Tabasco, Mexico. *Journal of Ecology*, Vol. 55, 1967, pp. 301–343.

Zenkovich, V. P. *Processes of Coastal Development*. Translated from the Russian by O. G. Fry. Edinburgh, Oliver and Boyd, 1967, 738 pp.

Chapter 13

SUBMARINE TOPOGRAPHY

The oceans comprise 71 percent of Earth's surface. They include the shelves and slopes of the continents, with their combination of marine and terrestrial sediments. The continental slopes in particular are very steep and dissected by deep, submarine canyons. The deep-sea floors have proven to be complex, including flat sea-floor plains, local clusters of seamounts, extensive systems of mid-ocean ridges (comparable, in terms of their dimensions, to mountain chains on land), and long chains of island arcs with parallel, deep trenches. The sea floors have been growing as a result of spreading of lithospheric blocks or plates. This process of plate tectonics underlies the now proven hypothesis of continental drift. Zones of plate compression are favored areas of mountain building today, and the history of continental drift since Paleozoic times explains the distribution of the major world mountain chains.

*The more important topographic features of the seas
will be discussed in the course of this chapter, together
with current hypotheses as to their origins.*

13-1. THE REALM OF THE SEAS

Our planet is unique in the solar system, not only in that it has an atmosphere containing oxygen but in that it has a partial covering of water, or H_2O. It remains a mystery how Earth retained its water at the time the planet came into existence. Volatile gases such as neon were almost completely lost, so that Earth's water, insofar as it was retained, had to be bound up in more stable chemical and mineral compounds (Cloud, 1968). Eventually, during the first billion years of Earth's history, this water was released as a liquid. Since then an equilibrium of sorts has been maintained. Unlike the moon or Mercury, which were too small to retain their liquid water, or Mars, which only held on to a trace, Earth had sufficient mass and gravity to keep its mantle of water and air.

Today the oceans account for 71 percent of Earth's surface—roughly 60 percent of the Northern Hemisphere and 80 percent of the Southern Hemisphere. The average depth of the seas is almost 12,500 feet (3,790 m), compared with an average elevation of the land of just 2,900 feet (875 m) (Figure 13–1). The highest peak on the continents, Mount Everest, has an elevation of 29,028 feet (8,848 m), compared with a depth of almost 40,000 feet (12,000 m) for the greatest

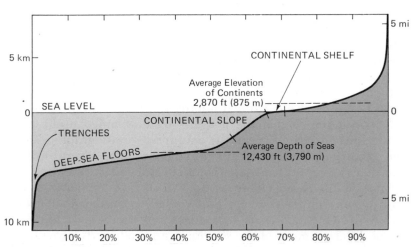

Figure 13–1. Relative Distribution of Land and Water for Earth. (Modified after Dietrich and Ulrich, 1968.)

ocean deep, the Marianas Trench of the Pacific Ocean. In all, the depth of the seas is so much greater than the elevation of the continents that the average elevation of the entire Earth surface is almost 8,000 feet (2,430 m) *below* sea level.

As far as the dated history of the planet can be read from the rock record, the seas have been with us for some 3,600 million years. The continents, always changing in their configuration, were originally bare of vegetation and other forms of life. The earliest organic substances—algae and bacteria—first came into existence in the seas, some 3,100 million years ago. Even during the Cambrian era (600 to 400 million years ago), there still were no plants on the continents and the oxygen concentration of the atmosphere was as little as 1 percent that of today. The ocean holds the clue to the origins of life as well as to the mechanics of tectonic deformation. In fact, the many mysteries of the deep have been the focus of intensive geophysical research during the past 20 years, and the discoveries have been as momentous as those in the fields of human evolution and molecular biology.

The study of the oceans involves many disciplines and comprises the topography of the sea floors, the physics of the waters, and the biology of the living creatures. Oceanographic work dealing with submarine topography and geology is highly relevant for geomorphology. Partly, it is a matter of perspective, since some 71 percent of the Earth's surface is a matter of submarine "landforms." More important perhaps is that oceanic phenomena often provide evidence to explain land features.

The recent breakthroughs in submarine research have only been possible as a result of sophisticated technical equipment:

1. Echo-sounding apparatus (sonar) can provide a continuous trace of ocean depths, recording topographic irregularities in full detail. More complex devices of this type can also bombard the ocean floor with acoustical impulses that are reflected by the floor and by materials or structures of different density or consistency—as much as 2 miles (3.2 km) below the sea bottom.
2. Various means of mechanical photography and sampling on the sea floors have been made possible by unmanned probing devices, lowered from specially equipped oceanographic ships. Long "cores" of sediment, spanning as much as several million years, are removed from depths of several miles, revealing surprising details about changes of ocean surface temperature and salinity through time. Lavas that welled up on the sea floor are removed and dated by isotopic decay methods, indicating the age of different parts of the sea floors and indirectly providing a key to a new understanding of how tectonics operate there. In the same way, oil drilling—from floating rigs near the edges of

the continents or on isolated islands—continues to provide information on rates of sedimentation and sea-floor subsidence.

3. Diving operations, all the way from the simple scuba equipment necessary to explore coral reefs to the deep-sea diving saucers developed by Jacques Cousteau, are opening up the seas to direct human observation.

13-2. THE CONTINENTAL MARGINS

The edges of the continents are submerged by the world ocean to a variable depth and width, depending on their topography. These margins include two basic units: (1) the gently sloping platform known as the *continental shelf* and (2) the rapid descent, formed by the *continental slope,* down to the deep-sea floor. The continental shelves account for 7.5 percent of the area of the oceans, the continental slopes for some 8.5 percent.

The Continental Shelf. For the whole world the continental shelf has an average width of 46 miles (78 km) and an average maximum depth of 435 feet (133 m) at its margin (see Figure 13–1). The range of variation is great, with depths of 60 to 1,800 feet (20 to 550 m) at the edge of the continents and widths ranging from zero to almost 950 miles (1,500 km). The average gradient of the shelf is only 0.1 degrees, much smoother than even the flattest surfaces exposed on land.

In detail, the continental shelves do exhibit some relief. On about 60 percent of their surface there are low hills with a relief of 60 feet (18 m) or more. Furthermore, about 35 percent of the shelf surfaces are marked by shallow basins or valleys with a relief of more than 60 feet (18 m). Only about 30 percent of the present shelf area is covered with recent marine beds. These include muds deposited in coastal lagoons and shelf basins, silts swept out to sea beyond the mouths of large rivers, chemical precipitates, and biological sediments such as shell debris. Sand deposition is largely confined to the shore zone, to depths of less than 35 to 70 feet (10 to 20 m).

The greater part of the continental shelves were dry land during the glacial-eustatic regression of sea level marking the Last Glacial. As a result, approximately 70 percent of the area is mantled by terrestrial deposits laid down about 25,000 to 10,000 years ago on river and coastal plains. At that time, broad expanses of lowland favored the development of floodplains, delta plains, and marshland. Forests grew here, with a mosaic of marsh vegetation, and the organic mucks have yielded a record of pollen, indicating the changes of vegetation through time. Along the Atlantic seaboard of the United States (Emery, 1967), cold-climate forests of pine and spruce were only replaced by warmth-loving woodlands of oak and hickory some 12,500 years ago. Animal bones include mammoth and mastodon. At the former shore—now at depths of 380 to 450 feet (115 to 145 m)—wave action produced coastal landforms similar to those at the modern shoreline. On many other continental shelves there also are traces of submerged barrier islands, sea cliffs, and shore platforms (Emery, 1968). They are normally veneered with modern sediments that obscure relief and fill the stream-cut valleys that once emptied onto these terraces. However, most glacial troughs now drowned as fiords are so deep that their traces can be easily followed far out onto the shelf.

Geomorphologic Processes of the Continental Shelf. Contemporary processes operating on the floor of the continental shelves include the following:

1. *Wave Agitation of Sediments in the Shore Zone.* Since waves are not effective at depths exceeding wavelength, their impact is restricted to the coastal perimeter at depths of less than 30 feet (9 m). However, wave action is complemented by longshore and rip currents that may be effective to depths of 100 feet (30 m).

2. *Tidal Influences and Bottom Currents.* Where tidal amplitudes are great, the in and out of the tides accelerates slow bottom currents, deflecting them shoreward during the flow, seaward during the ebb stage. The overall effect is to redistribute sediments across the shelf, where they form a ripple-marked mantle of silt and fine sand. At the same time, shallow, discontinuous channels are kept open where tidal currents are prominent. In other cases, strong currents may develop submarine bars with conspicuous topographic expression.

3. *Erosion by Suspended Sediments.* Wherever large quantities of silt and clay enter the sea, at the mouths of large streams, differences of density are produced, sometimes with geomorphic impact. In the mouth of an estuary, for example, silt-laden fresh waters are relatively light and therefore flow over saline and cooler waters that penetrate the estuary as a bottom current. But these temperature and salinity contrasts disappear offshore, so that the suspended sediment often forms a current of high-density, turbid water that sinks, gaining momentum as it sweeps seaward along the floor of the shelf. Erosional channels can be cut by such *turbidity currents* wherever large streams debouch into the sea and particularly near the break of gradient to the continental slope. Valleys of this kind are typical at the submarine terminus of deltas.

4. *Mass Movements.* Headward erosion of existing submarine valleys is primarily a result of debris slides, slumping, siltfalls, and sandfalls in rock or unconsolidated sediment. As steep "walls" of silt or sand collapse, they may provide avalanches of detrital material that glide rapidly at 15 to 50 miles per hour (24 to 80 km/hr) as turbidity

currents, eroding the bottom farther downslope. Earthquake shocks commonly trigger mass movements and turbidity currents.

5. *Glacial Deposition from Drifting Ice.* Glaciers calving at the coast consist of dirty ice, with variable proportions of sediment of all particle sizes. Icebergs that break off like this can transport tons and tons of glacial debris into warmer waters, where they eventually melt. This glacial drift accumulates at the bottom of polar seas or wherever currents transport floating ice into warmer latitudes.

6. *Tectonic Deformation.* Movements along fracture zones, block faults, and synclines are fundamental in determining the broad lines of shelf topography. These internal forces continue to create or accentuate irregularities such as basinlike depressions and tectonic valleys.

The Continental Slope. The continental slopes are delimited from the shelves by a marked break of gradient. The slope itself may be straight or smoothly curved, regionally interrupted by one or more platforms. Generalized slopes near the continental break range from 1 to 10 degrees, averaging about 4.3 degrees; in other words, they are by no means precipitous. The general gradient decreases as the slope merges with the deep-sea floor at depths of about 10,000 to 15,000 feet (3,000 to 4,500 m). The continental slope is steepest where mountain ranges border the coast and particularly where faults demarcate the shoreline and allow little or no development of a continental shelf. On the other hand, the slope is very gentle near the terminus of river deltas.

The most common material exposed on the continental slope is mud, accounting for perhaps 60 percent of the surface; sands comprise about 25 percent, rock and gravel 10 percent, shell beds and organic oozes 5 percent. None of these sediments is directly of terrestrial origin, although the muds were ultimately derived from the discharge of large rivers. Wave agitation, tidal influences, or bottom currents play little or no role on the continental slope. Instead, mass movements and turbidity currents are two of the most important agencies, next only to diastrophism.

Submarine Canyons. Unlike the continental shelf, the slope is anything but regular in its configuration. Beyond doubt, the most conspicuous features are *submarine canyons* (Figure 13–2). By definition, these are rock-walled, V-shaped, winding valleys that extend down the continental slope. Tributary valleys, often dendritic, are common. The canyons begin abruptly on the lower shelf, occasionally extending to coastal proximity near the mouth of some, but not all, large rivers. Such rock-cut canyons terminate on the lower continental slope, but may continue in the form of shallow, levee-bordered channels that run across broad fans of sediment debouching onto deep-sea plains. The floor of a submarine canyon is normally filled with alternating lay-

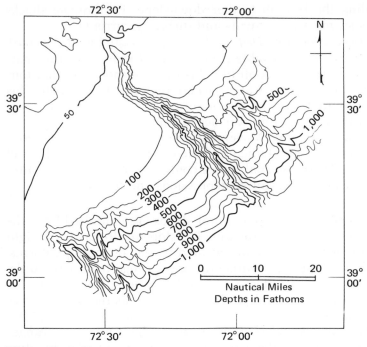

Figure 13–2. The Hudson Canyon on the Continental Slope off New York. (After Shepard, 1963. Copyright © 1963 by Harper & Row, New York.)

ers of sands and clays, and similar beds build up the associated *submarine fan.*

Dimensions of submarine canyons vary. The length of the rock-cut segments may exceed 75 miles (120 km), and longitudinal gradients decrease downvalley from initial averages of as much as 3 to 8 degrees. Valley walls commonly have a relief of 1,000 feet (300 m) or more, and some submarine canyons have dimensions and slopes comparable to those of the Grand Canyon. The only basic difference in form with deeply incised stream valleys is that the submarine canyons are comparatively straight, without entrenched meanders of small radius.

Contemporary research indicates that submarine canyons have multiple origins. Many are primarily a result of faulting. Others are due to turbidity currents and mass movements. Some may also be the result of stream cutting during periods of low sea level; this type would subsequently have been modified by marine processes or maintained by turbidity currents. Most of the large submarine canyons are found near the mouths of large rivers of great age and that have long provided masses of suspended sediment suitable for turbidity currents. Finally, "built-in" patterns of up-and-down-canyon currents, only in part related to tidal alternations, have been recorded. Available evidence suggests that many if not most submarine canyons require millions of years to form.

13-3. UNDER THE DEEP SEAS

General Features. The deep seas beyond the continental slope account for about 84 percent of ocean area. It was once thought that most of the deep seas formed flat, featureless surfaces, known as sea-floor plains. As knowledge of submarine topography and geology increased, it became evident that sea-floor plains were of limited distribution. Instead, there is a broad range of forms, of different origins, lending to the ocean floor a complexity and diversity almost comparable to that of the land surface (Figure 13–3).

Three major topographic constellations can be recognized:

1. *Sea-floor plains*, locally studded with groups of seamounts
2. *Mid-ocean ridges* and *rises*, a geometrically organized topography of great submarine relief and irregularity
3. *Trenches and island arcs*, that is, parallel chains of volcanic islands accompanied by linear deeps

The salient features of these categories will be discussed later. Complex tectonics and vulcanism are responsible for most of this topographic relief.

The sediment mantle of the deep seas varies in thickness and composition. It is thickest on the sea-floor plains, near the foot of the mid-ocean ridges and the continental slopes. On the other hand, at the surface of the ridges and rises there may be little or no sediment. These deep-sea sediments have several different sources.

First, much of the material was originally of terrestrial origin, as, for example, the silts and sands introduced from the continental shelf by turbidity currents and mass movements. Another instance is that of the glacial drift deposited by melting icebergs off the New England coast or into the Antarctic seas. Eolian dust, volcanic ash, as well as meteorite particles further contribute to bottom sediments.

The second major component of deep-sea sediments is organic, provided by the calcareous and siliceous skeletons of the algae or *plankton* that float in the surface waters. The skeletons of these organisms accumulate in a clayey matrix providing *organic oozes*, variously named after the characteristic forms (calcareous foraminifera and pteropods, siliceous radiolaria and diatoms). Furthermore, slumping around coral reefs can introduce coralline sand and white, lime mud.

A third factor in deep-sea sedimentation is inorganic precipitation and alteration of existing deposits. The minerals crystallized in sea

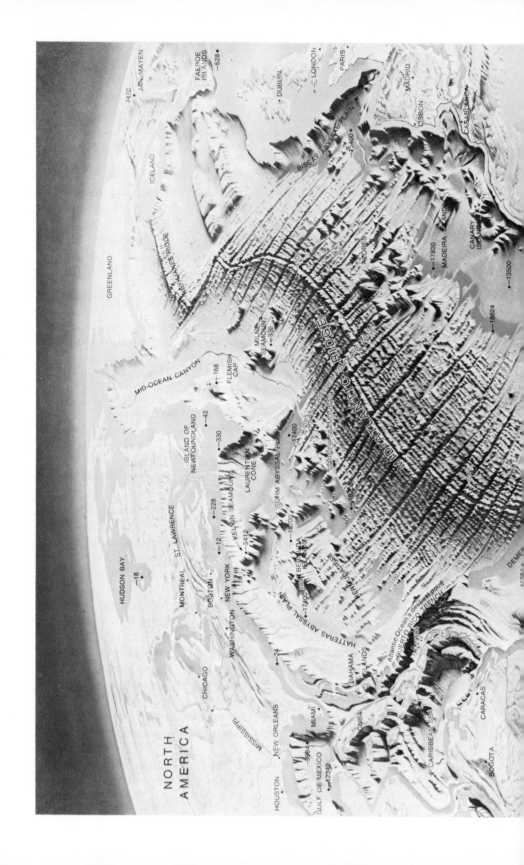

The Atlantic Ocean Floor

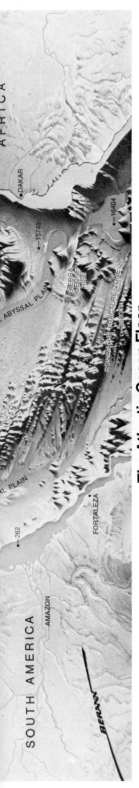

Condensed cross-section from Puerto Rico to England showing comparative depths.

Figure 13–3. Submarine Topography of the North Atlantic Ocean. (Painting by Heinrich Berann, courtesy of Alcoa)

water include manganese, iron, nickel, calcium, or magnesium in their make-up, while alteration can produce clay minerals.

Sea-Floor Plains. Sea-floor plains are best developed between the continental slopes and the mid-ocean ridges. Typical depths are 10,000 to 20,000 feet (3,000 to 6,000 m). Many such plains are truly flat and featureless, mantled by organic oozes or brown clays of mixed origin. Others form shallow submarine fans, built up of silt, sand, and fine siliceous plant debris originally derived from river discharge. Such fans are best developed downslope of deep submarine canyons and adjacent to large rivers with much suspended matter. In fact, the greatest fans are found beyond the Ganges and the Indus deltas. Typically, the surface of such turbidity deposits is runneled with fan valleys that are bordered by leveelike ridges and that ultimately bifurcate in delta fashion. Less apparent in its origin is the enigmatic mid-ocean canyon of the North Atlantic, stretching for almost 4,000 mi (6,400 km) (see Figure 13–3) southward from the Labrador Basin. It may be a chute repeatedly followed by large-scale turbidity currents that derive their material from glacial debris.

The major relief of the sea-floor plains is provided by groups of submarine volcanoes called *seamounts*. They rise abruptly 9,000 to 12,000 feet (2,700 to 3,700 m) above the sea floor, with steep sides and conical forms. If volcanic activity persists over periods of 10 million years or more, seamounts may show up above the ocean surface as volcanic islands, such as the Canaries or the Hawaiian groups. As activity subsides, the peaks of these islands are liable to be planed off by wave erosion; eventually, the sinking of the underlying ocean floor brings the flattened tops down to 5,000 feet (1,500 m) or more below sea level. Such wave-truncated, drowned seamounts are called *tablemounts* (or guyots). They commonly provide the foundation for coral atolls in tropical waters.

Mid-Ocean Ridges. The most fascinating field of deep-sea study has been provided by the mid-ocean ridges. Little was known about them, beyond their basic existence, before about 1960. Today they are a major focus of attention, holding a key to the understanding of crustal deformation.

The mid-ocean ridges extend down the axes of the major oceans, often following a rectangular, zigzag pattern and branching off into less prominent "plateaus" or rises. The major ridges are approximately 400 to 1,000 miles (1,200 to 3,000 km) wide, with an average relief of 5,000 to 10,000 feet (1,500 to 3,000 m). Yet the crests typically lie 5,000 to 8,000 feet (1,500 to 2,400 m) *below* sea level. Only a few peaks, with a relief of 12,000 to 20,000 feet (3,600 to 6,000 m) rise above the sea surface, as, for example, those of the Azores Islands. Consequently, in their dimensions the mid-ocean ridges rival or exceed the Rocky Mountains or the Andes.

These great underwater mountain ranges do not, however, resemble the ranges of the continents in anything but size.

1. Seen in profile, a mid-ocean ridge resembles two slabs of lithosphere upraised at their contact (Figure 13–4). A grabenlike rift runs between them, tapping magma reservoirs of batholith dimensions. The slabs of earth crust consist largely of successive lava flows, extruded along riftlike fissures and cooled at the bottom of the sea with strange pillowlike surfaces. Rising above these great slabs of lava there are endless rows of gigantic volcanic cones, the great majority of which never reach the ocean surface. In the absence of glaciers and streams, these underwater ranges are not subject to erosional forces other than mass movements and turbidity currents. Slopes are therefore steeper and probably smoother than those of most mountains on the continents.

2. Seen in plan, the mid-ocean ridges resemble a jigsaw puzzle of rock rectangles separated by great faults. The rift running down the center is a zone of fissure flows, with lavas being added to either side of the rift as the segments move *apart*. The fractures that cross transverse to the ridge axis are called transform faults, and individual segments move sideways along them while they spread outward along the zigzagging central rift (see Figure 13–4).

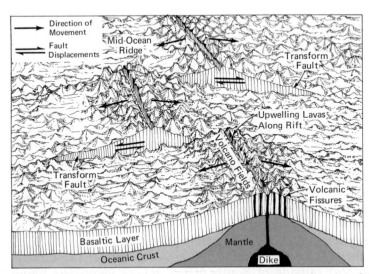

Figure 13–4. Submarine Features of a Mid-Ocean Ridge. (Modified after H. W. Menard, The deep-ocean floor. Copyright © 1969 by Scientific American, Inc. All rights reserved.)

The mid-ocean ridges are centers of continuous crustal deformation and vulcanism on an impressive scale. The frequency of seaquakes along these belts of ocean (see Figure 2–7) gives but a small inkling of what goes on in the deep. The dynamism of the ridges is fundamental to the principle of continental drift.

Deep Trenches and Island Arcs. At various places the ocean floors are marked by deep gashes with below sea-level depths of 25,000 to 35,000 feet (7,500 to 10,500 m) or more. These long narrow depressions are steep sided and V-shaped, with rocky bottoms relatively free of sediment. Known as deep-sea trenches, the floors of these deeps lie 10,000 to 20,000 feet (3,000 to 6,000 m) beneath the sea-floor plains. Frequently, they run parallel to narrow oceanic ridges or to subcontinuous island arcs, mainly volcanic in origin. A striking example of these is provided by the Aleutian Island Arc and the associated Aleutian Trench (see Figure 13–5). Trenches are also found offshore the Circum-Pacific cordilleran belts created during the Alpine orogeny. A case to point is the Peru-Chile Trench off the Andean shores of South America (Figure 13–5).

As in the case of the mid-ocean ridges, trenches and island arcs form belts of intensive earthquake activity. Their origin is intimately linked to continental drift. (The basic submarine forms and processes outlined previously are summarized in Table 13–1.)

13-4. CONTINENTAL DRIFT

Wegener's Theory. In 1910 Alfred Wegener, of the University of Hamburg, proposed a revolutionary theory. First, he gathered a number of geological and paleontological facts to argue for an earlier view that South America, Africa, India, and Antarctica had once all formed part of a single supercontinent, which he called *Gondwana*. He then proposed that Gondwana had broken up during the Mesozoic era (beginning about 230 million years ago). The individual continental blocks, consisting of lighter sial rock, had simply drifted apart, presumably by gliding slowly over the denser sima layer of the Earth's crust. Wegener explained the Alpine-age mountain belts of Eurasia as a consequence of India's and Africa's colliding with Eurasia, thereby buckling up the crust. The Americas were thought to be drifting westward, and the western cordilleras were explained very much as bow waves thrown up by a ship plowing through the sea: The leading edges of the New World continents were buckling as a result of frictional drag.

Wegener's theory of *continental drift* soon came under fire and received little support outside of South Africa, where geologists found the empirical evidence too compelling to overlook. During the 1960s, after several decades of abusive criticism, Wegener's basic ideas were

Table 13-1. A SYNOPSIS OF SUBMARINE TOPOGRAPHY

CATEGORY	FORMS AND FEATURES	PROCESSES
A. CONTINENTAL SHELF	a) Drowned depositional forms of terrestrial origin (floodplains, delta and coastal plains) b) Drowned and partly filled erosional forms of terrestrial origin (river valleys, glacial troughs) c) Drowned coastal landforms (barrier islands, sea cliffs, shore platforms, coastal dunes) d) Basins and flats of marine deposition e) Submarine valleys and canyons	1. Terrestrial agencies (running water, ice) operating during glacial-eustatic regressions 2. Coastal agencies (waves, wind) operating during glacial-eustatic regressions 3. Mass movements (debris slides, slumping, turbidity currents) 4. Glacial drift (iceberg debris) 5. Tidal and bottom currents 6. Marine deposition or precipitation of suspended or dissolved materials 7. Diastrophism and vulcanism
B. CONTINENTAL SLOPE	a) Inclines and platforms of marine deposition b) Submarine canyons	1. Marine deposition or precipitation of suspended or dissolved materials 2. Mass movements and turbidity currents 3. Glacial drift 4. Diastrophism and vulcanism
C. SEA-FLOOR PLAINS	a) Submarine fans and fan valleys b) Flat plains of marine deposition (brown clays and organic oozes) c) Seamounts and tablemounts	
D. MID-OCEAN RIDGES	a) Fault-block plateaus, ridges, and rises b) Volcanic seamounts c) Spreading rift fissures d) Transform fault scarps	1. Diastrophism (including plate tectonics) 2. Vulcanism 3. Mass movements and turbidity currents 4. Marine deposition or precipitation of sediments
E. TRENCHES AND ISLAND ARCS	a) Deep trenches b) Volcanic seamounts and islands	1. Diastrophism and vulcanism 2. Mass movements and turbidity currents

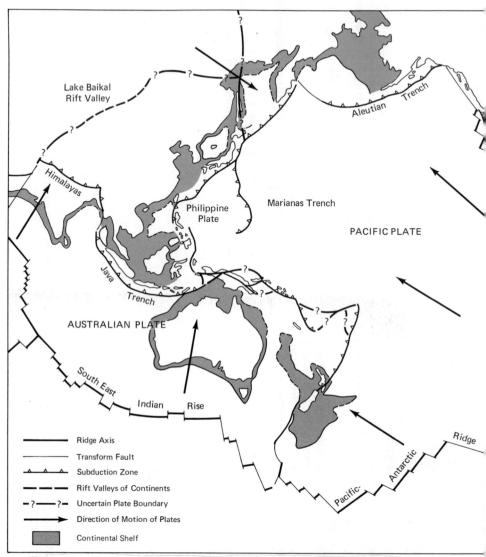

Figure 13–5. Tectonic Plates of Earth's Lithosphere. (Modified after John F. Dewey, Plate tectonics. Copyright © 1972 by Scientific American, Inc. All rights reserved.)

vindicated by deep-sea research. Today, a modified form of the theory of continental drift, known as *plate tectonics*, provides some rather fundamental, if not revolutionary, explanations for the patterns of structure and deformation apparent on the continents (Hallam, 1973).

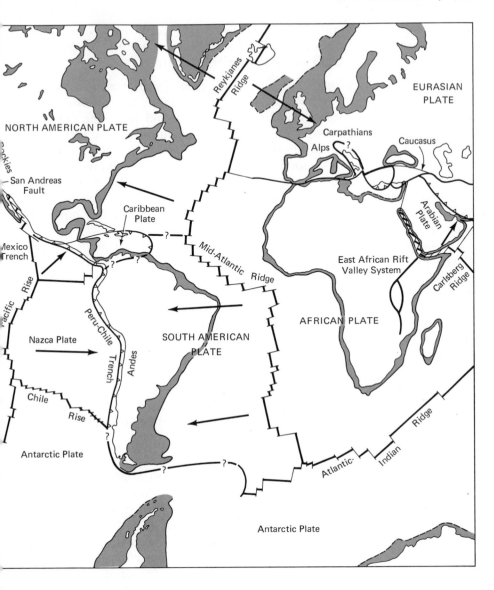

Plate Tectonics. The critical amendment necessary to Wegener's hypothesis is that the sial blocks do not move across the sima layer. Instead, geophysicists now hold that the entire lithosphere is comparmentalized into 6 to 8 major plates, each about 45 to 60 miles (70 to 100 km) thick, including segments of both the ocean floor and the continental masses. These rigid plates appear to move across the plastic zone of the mantle, the asthenosphere. The major plates and their present direction of movement is shown by Figure 13–5. From this premise two active processes are inferred, corresponding to zones of spreading or compression along the different plate contacts.

1. *Zones of spreading* occur wherever the major plates move apart. Fissures open here and hot, plastic material from the mantle rises between them. This upwelling basalt solidifies and joins the trailing edge of each plate (Figure 13–4). In existing ocean basins the intrusion of large bodies of magma helps to produce a symmetrical mid-ocean ridge axis, offset by transform faults and with other features as described previously. A spreading center may also open when a continent splits, forming a linear deep such as the Red Sea or the Gulf of California. Rates of spread amount to as much as 4 inches (10 cm) per year.

2. *Zones of compression* develop where the leading edge of a plate probes against another plate. They may collide rapidly (at rates of 2.4 to 4 in, or 6 to 10 cm, per year), in which case one crustal plate plunges under the other, sinking into the plastic asthenosphere (*subduction*), where plate material is absorbed and destroyed at the same rate at which it is created in the zone of spreading. The impact of one plate diving 60 miles (100 km) or more beneath the other is to produce a deep trench, with volcanoes and island arcs (if the overlying plate edge is a sea floor, see Figure 13–6,A) or folded mountains (if the overriding edge is a continental margin, see Figure 13–6,B). When plates meet at slower speeds (less than 2.4 in or 6 cm, per year), both plates buckle, raising a fresh range of folded mountains between them (Figure 13–6,C).

In other words, plate tectonics can be summed up in the notion that the lithosphere consists of plates that are continuously being enlarged at one edge and destroyed at the other. Where plates float apart, the continents, which are embedded in them, also drift away from one another. New material from the asthenosphere is added along the zones of spreading, forming mid-ocean ridges under the seas, rift valleys on the land. Adjustments of the rigid crustal plates along the centers of spreading leads to further compartmentalization, into smaller blocks that move laterally along transform faults. On the other hand, in zones of plate convergence and collision, compression may lead to gentle buckling and folding of the adjacent plates or to a catastrophic plunging of the leading plate down and under, into the plastic mantle. Deep trenches and adjacent island arcs or mountain belts are best explained in this way.

Continental Drift During the Geological Past. There are a number of reasons why continental drift, by the mechanism of plate tectonics, must be postulated for the geological past (Table 13–2):

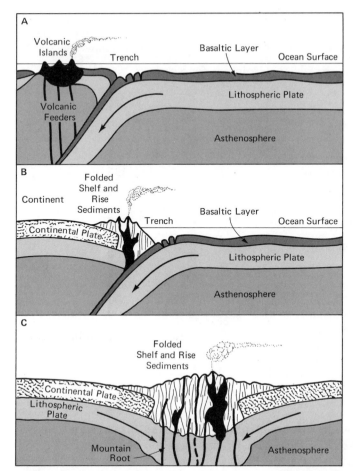

Figure 13–6. Mountain Building as a Result of Plate Collision: (A) Island Arc and Trench, as Two Oceanic Plates Meet; (B) Folded Mountains and Trench, as Oceanic and Continental Plates Meet; and (C) Folded Mountains, as Two Continental Plates Meet. (Based in part on Dewey and Bird, 1970, and Dietz, 1972.)

1. The present deep-sea floors are relatively young: There are few sediments or lavas older than 80 million years, and none is older than 150 million years. In other words, the present floors have been formed afresh by deep-sea spreading during that same span of time. This is corroborated by the age of successive basalt increments that have welled up along the mid-ocean ridges.
2. The continental margins, delimited by the continental slope of various continents, match surprisingly well (Figure 13–7). In fact, with the aid of a computer, Bullard and others (1965) were

Table 13-2. THE GEOLOGICAL TIME SCALE

ERA	PERIOD	EPOCH	BEGINNING DATES (MILLIONS OF YEARS)
Cenozoic	Quaternary	Holocene (or Recent)	0.01
		Pleistocene	2
	Tertiary	Pliocene	5
		Miocene	23
		Oligocene	36
		Eocene	58
		Paleocene	65
Mesozoic	Cretaceous		135
	Jurassic		190
	Triassic		225
Paleozoic	Permian		280
	Carboniferous	Pennsylvanian	320
		Mississippian	360
	Devonian		400
	Silurian		440
	Ordovician		500
	Cambrian		600
Cryptozoic (Precambrian)			4.6 billion years

Based primarily on recent potassium-argon dates.

able to show that the continents along the Atlantic can be fitted together like the pieces of a jigsaw puzzle. Individual rock provinces and even structural details can be traced point-by-point from Africa to South America or from Europe to North America. Similar contacts can be surmised for the other segments of Gondwana.

3. The occurrence of continental glaciation simultaneously in Brazil, southern Africa, India, and Australia is verified for a part of the late Paleozoic era, some 290 to 270 million years ago. This can best be explained by a unified Gondwana supercontinent, then located in Antarctic latitudes (Figure 13-7). This surprising idea is supported by the record of magnetization of lavas and iron-bearing rocks on these continents, a permanent record that helps locate their former relative position with respect to the magnetic poles.

4. Finally, the continents once part of Gondwana had near-identical floras in late Paleozoic times.

In assessing the history of the oceans, it now appears that the Atlantic only began to open up in mid-Triassic times (about 200 million

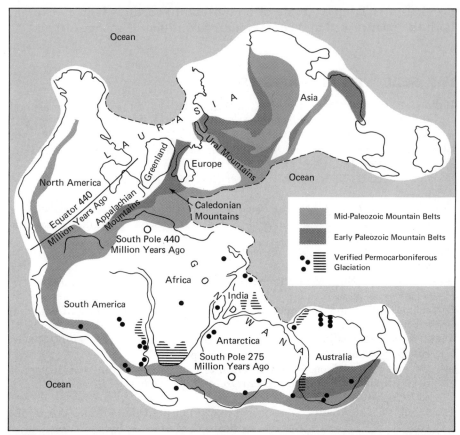

Figure 13–7. The Paleozoic World: Pangaea Amid the Universal Ocean, Panthalassa. (Modified after John F. Dewey, Plate tectonics. Copyright © 1972 by Scientific American, Inc. All rights reserved. Data for the Permocarboniferous glaciation, compiled from various sources, are incomplete for Antarctica.)

years ago), the Indian Ocean in Permian times (after about 280 million years ago). Prior to then, during much of the Paleozoic era, the Earth's continents were gathered in a universal land mass, *Pangaea,* composed of two, partly contiguous supercontinents, with *Laurasia*—composed of North America and Eurasia—forming a counterpart to Gondwana (Dietz and Holden, 1970). These supercontinents formed during the early Paleozoic by the inward motion of several separate blocks. And before that? Continents must have split up several times, forming new oceans, while on other occasions they collided and were welded together into new supercontinents.

Thus continental drift not only explains the origins and forms of the ocean floors but offers a rationale for mountain building. It also provides a striking example of how present processes help to explain both the past record and contemporary continental configurations.

13-5. SELECTED REFERENCES

Bullard, E. C., J. E. Everett, and A. G. Smith. The fit of the continents around the Atlantic. *Philosophical Transactions, Royal Society of London,* Vol. A-258, 1965, pp. 41–51.

Cloud, P. E. Atmospheric and hydrospheric evolution on the primitive Earth. *Science,* Vol. 160, 1968, pp. 729–736.

Cousteau, Jacques-Yves. *The Living Sea.* New York, Harper & Row, 1963, 325 pp.

Cox, Allan, ed. *Plate Tectonics: Readings in Sea-Floor Spreading, Global Tectonics, and Magnetic Reversals.* San Francisco, Scientific American, Freeman, 1973, 702 pp.

Dewey, J. F. Plate tectonics. *Scientific American Offprint 900* (May 1972), Vol. 226, pp. 56–68.

———, and J. M. Bird. Mountain belts and the new global tectonics. *Journal of Geophysical Research,* Vol. 75, 1970, pp. 2625–2647.

Dietrich, Günter, and Johannes Ulrich. *Atlas zur Ozeanographie.* Mannheim, Bibliographisches Institut, 1968, 76 pp. (in German with English key).

Dietz, R. S. Geosynclines, mountains and continent-building. *Scientific American Offprint 899* (March 1972), Vol. 226, pp. 30–38.

———, and J. C. Holden. Reconstruction of Pangaea: breaking and dispersion of continents, Permian to present. *Journal of Geophysical Research,* Vol. 75, 1970, pp. 4939–4956.

Emery, K. O. The Atlantic continental margin of the United States during the past 70 million years. *Geology Association of Canada, Special Paper,* No. 4, 1967, pp. 53–70.

———. Relict sediments on continental shelves of the world. *Bulletin, American Association of Petroleum Geologists,* Vol. 52, 1968, pp. 445–464.

Guilcher, André. *Coastal and Submarine Morphology.* Translated from the French by B. W. Sparks and R. H. W. Kneese. London, Methuen, 1958, 274 pp.

Hallam, Anthony. *A Revolution in the Earth Sciences: From Continental Drift to Plate Tectonics.* New York, Oxford University Press, 1973, 128 pp.

Heezen, B. C., and C. D. Hollister. *The Face of the Deep.* New York, Oxford University Press, 1970, 700 pp.

Hood, D. W., ed. *Impingement of Man on the Oceans.* New York, Wiley-Interscience, 1971, 738 pp.

Horsfield, Brenda, and P. B. Stone. *The Great Ocean Business.* New York, Coward-McCann, 1972, 360 pp.

Hughes, N. F., ed. *Organisms and Continents Through Time.* London, The Palaeontological Association, Special Paper, No. 3, 1973, 334 pp.

Menard, H. W. The deep-ocean floor. *Scientific American Offprint 883.* Vol. 221, No. 3, 1969, pp. 126–145.

Morgan, W. J. Rises, trenches, great faults and crustal blocks. *Journal of Geophysical Research,* Vol. 73, 1968, pp. 1959–1982.

Pickard, G. L. *Descriptive Physical Oceanography*, 2nd ed. New York, Pergamon, 1968, 200 pp.

Scientific American. *The Ocean*. San Francisco, Freeman, 1969, 140 pp.

———. *Continents Adrift*. San Francisco, Freeman, 1972, 172 pp.

Shepard, P. *Submarine Geology*, 2nd ed. New York, Harper & Row, 1963, 557 pp.

Shepard, F. P., and H. R. Wanless. *Our Changing Coastlines*. New York, McGraw-Hill, 1971, 579 pp.

Sullivan, W. *Continents in Motion*. New York, McGraw-Hill, 1974, 399 pp.

Wyllie, P. J. *The Dynamic Earth*. New York, Wiley, 1971, 416 pp.

Highly informative are the graphic representations of world ocean topographies in:

Goode's World Atlas, Skokie, Ill., Rand McNally, 14th ed. 1974, pp. 221–228.

National Geographic Magazine. Washington, D.C., 1967–1971. Maps of the Indian, Atlantic, Pacific, and Arctic ocean floors.

Chapter 14

STRUCTURE AND LAND FORM (I): MOUNTAIN BELTS AND SHIELDS

14-1. Rocks, structure, and relief

14-2. Hills and mountains created primarily by diastrophism

—Fold mountains and valleys
—Fault-block ranges and valleys

14-3. Landscapes modeled by vulcanism

—Volcanic hills and mountains
—The role of intrusive vulcanism
—Volcanic plains and plateaus

14-4. Ancient crystalline shields

The gross land form of the continents includes a full spectrum of plains, plains with hills, tablelands, hill country, and mountainous terrain. These form classes are of many possible origins and directly or indirectly result from tectonic forces emanating from Earth's interior. Diastrophism and vulcanism create positive relief by folding anticlines, raising fault-block mountains, general upwarping, and building volcanic cones or ash and lava uplands. Diastrophism also produces negative relief in synclines, in downfaulted grabens, and in general by downwarping or subsidence. Consequently, although the gradational agents may proceed to irregularize the surface by deep dissection, at times increasing relief, the requisite potential energy is commonly provided—indirectly—by the tectonic setting. In all of this, differential weathering and erosion that reflect on rock resistance may lead to systematic patterning of relief forms and drainage lines at both the large and small-scale levels. These same rock contrasts, particularly when combined with gentle or almost im-

perceptible deformations, favor several possible combinations of structural or lithological landforms found well beyond the domain of mountain building in either the distant or recent past. At the generalized level, the land form of the continents includes several major component classes that are all to some degree modified or even controlled by structure and lithology: (1) depositional plains, built of alluvium, till, loess, dunes, or coastal sediments; (2) erosional plains and plateaus cut into horizontal, gently warped, or inclined sedimentary strata; (3) plains and plateaus eroded in lavas and volcanic ash; (4) young mountain belts of the recent geological past, including fold mountain systems, fault-block ranges, and volcanic peaks; (5) old mountain belts in which repeated erosion and tectonic revitalization have created a more complex relief strongly influenced by lithological contrasts; and (6) ancient shields of metamorphic and intrusive igneous rocks that constitute the eroded roots of mountain systems dating to the dawn of Earth history. These variable continental settings are, in detail, modified by differing combinations of gradational agents that reflect on the ecological balance of the major world environments (see Chapters 16 to 20). The present chapter considers young and old mountain belts, as well as ancient shields. Landscapes on horizontal or warped sedimentaries, including karst, are analyzed in Chapter 15, together with a synopsis of continental land form.

14-1. ROCKS, STRUCTURE, AND RELIEF

Writing in the years 1021–1023 A.D. the Persian doctor and philosopher Avicenna (ibn-Sina) was the first to formulate a surprisingly modern hypothesis of mountain building. He was familiar with the Greek and Roman authors and began to write a commentary on Aristotle's earth sciences, but he seems to have found them less informative than his own observations and deductions.[1] In the resulting discourse Avicenna

---

[1] See F. D. Adams, *Birth and Development of the Geological Sciences*, New York, Dover Reprint, 1954, pp. 333–335.

attributed mountains to either a direct agent—violent earthquakes that raise the crust and create elevations—or an indirect process—selective erosion by wind and water, whereby deep valleys are cut and irregular, high relief imparted as a byproduct. His writings further show that he understood how sedimentary rocks were deposited in the sea, that they were solidified by compaction and sedimentation, and that they were subsequently exposed to the slow but cumulative and relentless impacts of erosion.

Most of the surface relief of Earth's crust does owe its origins directly or indirectly to forces within Earth's interior. Diastrophism creates mountain ranges or hill country by folding or block faulting (see section 2-10). Vulcanism builds up craters and cones of lava or ash (see section 2-5). Landscapes uplifted by epeirogenic movements obtain sufficient potential energy for streams to begin vigorous dissection, thus permitting tablelands or irregular hills to evolve. Even in areas where uplift is unimportant, warping of rocks of differential resistance leads in the end to landscapes with striking patterns of erosion.

The continental distribution of topographic forms—such as plains, tablelands, hills, or mountains—is either directly determined or indirectly preconditioned by geological history. This geological "past" is recorded in successive stages of sediment accumulation or volcanic accretion and in repeated phases of tectonic deformation. It is directly reflected at the microscale in the nature and variability of rock types, at the mesoscale in the inclination and other structural attributes of these rocks, and at the largest scale in the available relief of the land. At the regional or continental level, the effects of lithology and structure are conspicuous if not paramount in many landscapes, no matter how actively the agents of erosion and deposition have intervened. It is these broad categories of land form that have traditionally been the focus of *structural geomorphology* and that are the subject of this chapter.

In discussing the broad terrain characteristics and form types implied by the term "land form," it is useful to define first some basic categories of surface form. Such a system has been devised by Hammond (1964) and can be applied to topographic maps or to arbitrary subdivisions as small as about 5-by-5 miles (8-by-8 km). The major types of terrain are as follows:

Plains. Over 50 percent of the surface is gently sloping, with a local relief of less than 300 feet (90 m).

Tablelands. Over 50 percent of the surface is gently sloping, with a local relief of over 300 feet (90 m), while most of the gentle slopes are found in the upper half of the elevation range.

Plains with Hills or Mountains. Over 50 percent of the surface is gently sloping, with a local relief of over 300 feet (90 m), while most of the gentle slopes are found in the lower half of the elevation range.

Open Hills or Mountains. Between 20 and 50 percent of the surface is gently sloping, with a local relief of 300 to 1,000 feet (90 to 300 m) (open hills) or over 1,000 feet (open mountains).

Hills or Mountains. Less than 20 percent of the surface is gently sloping, with a local relief of 300 to 1,000 feet (90 to 300 m) (hills) or over 1,000 feet (300 m) (mountains).

Obviously these are commonplace terms, except perhaps for the designation "tableland," which is similar in meaning to the more common word "plateau." But attaching a specific meaning to each term is convenient for purposes of succinct description and to avoid ambiguity. The origins and features of these basic terrain types can be readily outlined and discussed from a genetic viewpoint.

14-2. HILLS AND MOUNTAINS CREATED PRIMARILY BY DIASTROPHISM

Fold Mountains and Valleys. The most common type of mountain range is related to parallel or subparallel folds, with the anticlines corresponding to ranges, the synclines to valleys (Figure 14–1,A). Complex rock folding (see section 2-10) is the primary type of deformation, with ancillary faulting and vulcanism on a large or small scale. Sets of linear ranges and great, complex chains of mountains—known as cordilleran belts—are formed in this way.

Fold mountain systems generally have a common sequence of development. They are situated in former geosynclines, shallow and deep seas, on the continental shelf and slope, where marine sediments accumulated over millions of years over sinking portions of the earth's crust (Figure 13–6,B). Lateral compression by crustal plates establishes a general axis of sinking. Eventually, the geosyncline is elimi-

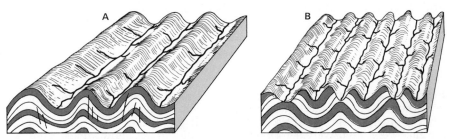

Figure 14–1. Block Diagrams of (A) Young Parallel Fold Mountains and (B) Old Parallel Fold Mountains, Once Eroded and Then Uplifted to Allow Renewed Dissection into the Weaker Rock Strata (white).

nated as crustal plates converge, collide, and buckle up. Rocks of the continental margins, as well as sedimentary and volcanic strata from the sea floor, are folded into mountains, while new volcanic material is intruded from below the lithosphere. In detail, the old geosyncline has been converted to a contorted mass of folded and faulted strata, injected with masses of magma. Deformation, vulcanism, and uplift commonly continue long afterward, particularly if continued plate convergence eliminates an existing ocean and brings two geosynclines into apposition (Figure 13–6,C). The Appalachian fold belt of the eastern United States appears to have been created in this fashion (Figure 14–2), over 350 million years ago, prior to reopening of the North Atlantic Ocean by renewed crustal spreading that began 225 million years ago (Dietz, 1972). In this way classical theory of the internal mechanics of mountain building can apparently be reconciled with the broader framework of plate tectonics (Dickinson, 1971).

From the moment they are exposed on the continents, fold mountains and valleys are continuously sculptured by gradational agents. These processes are exceptionally active in high country: Running water is potent, since available relief is great; mass movements are accelerated by the relief and irregularity of the terrain, as well as by relatively cold and wet climates in response to altitude; glaciers are not uncommon as a result of abundant snow and cool summers at very high elevations. The combined effect of water, gravity, and ice serves to change the shape and relief of fold mountains. At first, slope and irregularity are accentuated; ultimately, they may be reduced until the range is eroded to a plain studded with low hills. However, mountain building takes place much more rapidly than does denudation. An orogenic phase may take a few million years to create a mountain range; contemporary and subsequent denudation require several tens of million years before the range can be reduced to a plain.

Earth has seen an almost interminable succession of orogenies during the course of its history, reflecting the consolidation and fragmentation of continents and the train of events ensuing as the lithospheric plates move. These orogenies have been of variable intensity and complexity and have affected only limited areas—along certain plate contacts—in each instance. The major orogenies have recurred

Figure 14–2. Evolution of the Appalachian Fold Belt During the Paleozoic (Stages A–D). The former sediments of the continental shelf were folded into what is now the Ridge and Valley Belt, while the thicker, slope sediments to the east were intensively deformed to a cordilleran belt, now almost completely eroded. (From Robert S. Dietz, Geosynclines, mountains and continent-building. Copyright © 1972 by Scientific American, Inc. All rights reserved.)

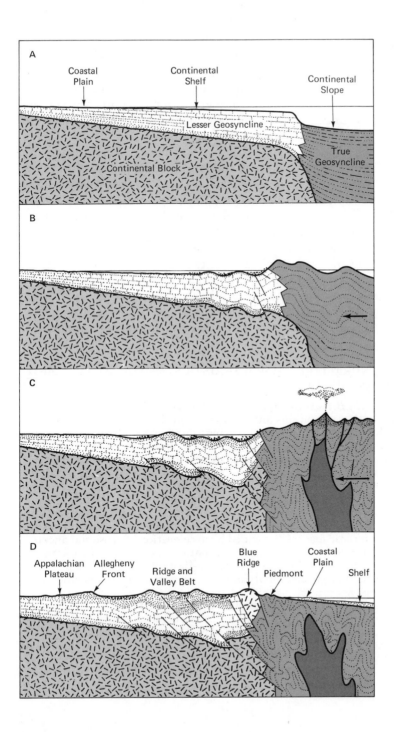

at apparent intervals of 20 to 100 million years (see Table 14–2). But for all practical purposes in geomorphology, these periods of mountain building fall into three groups (Table 14–1):

1. Relatively young orogenies which, during the past 75 million years, have created cordilleran belts that maintain a fresh aspect and high relief. These are responsible for what are called *young fold mountains* (see Figure 14–1,A). Uplift, deformation, earthquakes, and, possibly, vulcanism continue today, while gradational agents are very active. The coastal ranges of California and parts of the Rocky Mountain system provide examples of young fold mountains.

2. Orogenies of 100 to 500 million years in age were responsible for a number of cordilleran belts of intermediate age. These have been eroded to plains on one or more occasions, and their present relief is due to renewed uplift, tilting, deformation, and vulcanism. Revitalization of tectonic activity is commonly a side effect of mountain building elsewhere, and the greatest realm of old mountain belts—in Asia—owes much of its existence and most of its prominence to repeated uplift and block faulting during the Alpine orogeny (see Table 15–1). In the process of exhumation or rejuvenation, structural and lithological details and contrasts are subject to renewed attack and accentuation by the gradational agents. Folded geosynclines consequently have repeated leases on life. The resulting *old fold mountains*

Table 14–1. GENERALIZED CHARACTERISTICS OF YOUNG, OLD, AND ANCIENT MOUNTAIN BELTS

A. *Young Mountain Belts* (Less than 100 Million Years)
Belts of high, fold mountains and valleys, major fault-block ranges, rift valleys, volcanic mountain complexes and island arcs—all arranged in subparallel chains of mountains (cordilleran belts). Provide structural land-form of moderate to great relief, slope, and irregularity, commonly in linear arrangements.

B. *Old Mountain Belts* (100 to 500 Million Years Old)
Former cordilleran belts now upraised or exhumed by block faulting, epeirogenic uplift, or tilting and accentuated by fresh vulcanism. Further relief and many characteristic features provided by acceleration of erosional agents— strongly controlled by fracture and joint patterns, differential erosion, and intrusive volcanics. Provide structural land-form of low to moderate relief, slope, and irregularity, commonly of linear or geometric arrangement.

C. *Ancient Crystalline Shields* (greater than 500 Million Years)
Complex metamorphic rocks with multiple intrusions, intensively fractured by intersecting fault networks. Repeated and thorough denudation has destroyed relief, although block faulting may accentuate forms, through intensified erosion in relation to differential resistance and patterns of fractures or intrusions. Provide surfaces of variable but generally low relief with conspicuous geometric arrangement of forms.

are different in all but the essentials of the structural geometry. Ranges commonly match with units of resistant rock and other erosional remnants that may or may not correspond with the original synclines, that is, an inversion of relief is possible (see Figure 14–1,B). Ongoing tectonic activity is generally limited, and gradational agents are less active than in the case of young fold mountains, reflecting lower relief and a lesser degree of irregularity. The Appalachian ridge-and-valley country is a good example of old folded mountains.

3. Ancient orogenies, dating back from beyond 500 million years to the early years of the planet, also created cordilleran belts. These have been denuded so often and the rocks so intensively metamorphosed in the wake of repeated deformation that they have expended their lease on life. Remnants of this kind play no role in contemporary fold mountains. Instead they commonly coincide with extremely stable portions of the continents, favoring the development of a subdued *shield* topography (section 14-4).

Fault-Block Ranges and Valleys. Mountain building inevitably involves some faulting when crustal blocks are rigid and subject to sudden and intensive stresses. Block faulting may occur hand in hand with folding and accounts for the orogenic deformation of some former geosynclines. Frequently, too, fold mountains are accentuated by repeated block faulting once denudation has begun to reduce the weight of the crust, favoring isostatic rebound. Finally, block faulting is the characteristic mode of deformation in intrusive igneous metamorphics, as well as rocks that are far more rigid than young sedimentary strata.

Block faulting operates on a scale fully comparable with mountain folding. Subcontinental land masses can be chopped up and differentially raised or tilted. Great batholiths can be upthrust into a mountain range the size and grandeur of the Sierra Nevada. Fault scarps of all dimensions can be created where tilted blocks continue to rise, or where horst-and-graben structures are in process of formation. Finally, in addition to these many types of fault-block mountains or valleys, there are the great rift fractures. Rift faulting involves horst-and-graben structures developing on a continental scale along zones of crustal spreading. The classical case is that of the rift valleys of eastern Africa (Figure 14–3) and their extension along the Red Sea into the Dead Sea-Jordan graben and beyond (see Figure 13–5).

The drainage of both fold and fault-block mountains and valleys is commonly although by no means always aligned in peculiar geometric arrangements. Primary tributaries may join mainstreams at right angles, while secondary tributaries are elongated and run parallel to the mainstreams. Such a drainage pattern resembles an old-fashioned garden trellis and is so described as *trellis* drainage (Figure 14–4,B).

Table 14-2. MAJOR EPISODES OF MOUNTAIN BUILDING AND VULCANISM

MAJOR ERAS OF MOUNTAIN BUILDING	PHASES OF MOUNTAIN BUILDING AND MEDIAN AGES	MAJOR OLD WORLD EVENTS	MAJOR NEW WORLD EVENTS
Mesozoic to Recent ("Alpine") Orogenies	Pleistocene (1.5 million years)	Vulcanism, uplift, and continued deformation in Mediterranean area, eastern Africa, central Asia, Indonesia, and the Pacific Island Belt	Vulcanism, uplift, and continued deformation in the Western Cordilleran Belt, associated island arcs, and the mid-ocean ridges
	Late Tertiary (6 million years)	Folding, uplift, and local extrusive vulcanism in the Apennines, Carpathians, Atlas, Himalayan Ranges, and the Pacific archipelagos; block faulting and vulcanism in eastern Australia; major rift faulting in East Africa; uplift of shields and old mountain belts elsewhere	Block faulting and uplift of Cascades, Sierra Nevada, Rockies, Great Basin; folding of Coast Ranges; vulcanic activity in Central America and central Andes; uplift of shields and old mountain belt
	Mid-Tertiary (40 and 25 million years)	Folding and local vulcanism in southern Europe (Alps and Pyrenees), western Asia, the Atlas, Indonesia, and the Pacific archipelagos; extrusive vulcanism in eastern Africa and the Deccan Plateau; block faulting in central Asia and eastern Africa	Folding in Pacific Mountain System, Middle America, and northern Andes; flood basalts of Columbia Plateau extruded.
	Basal Tertiary (65 million years)	Folding in parts of southern Europe and North Africa, western Asia, and Indonesia	Major folding of Rocky Mountain System, Caribbean Island Arc, and southern Andes
Mesozoic Orogenies	Early Cretaceous (125 million years)	Folding in the Caucasus and New Zealand	Folding of Pacific Mountain System and Andes, with widespread vulcanism

	Late Triassic to early Jurassic (180 million years)	Folding and vulcanism in eastern, northeastern, and southeastern Asia and in South Africa	Block faulting in eastern United States; extrusive vulcanism in western United States and southern Brazil
Late Paleozoic ("Hercynian" or "Appalachian") Orogenies	Mid-Carboniferous to mid-Permian (300 and 260 million years)	Folding and vulcanism in western and central Europe, northwestern Africa, the Urals, central and northeastern Asia, eastern Australia	Folding of Appalachian geosyncline and of rocks in Arkansas-Oklahoma, eastern Canada, and Bolivia and Argentina; extrusive vulcanism in western United States
Mid-Paleozoic ("Caledonian" or "Acadian") Orogenies	Late Ordovician to late Devonian (450, 420, 390, and 360 million years)	Folding and vulcanism in British Isles, Norway, central Asia, southeastern Australia and Tasmania; intrusive vulcanism in Gondwana shields	Folding and intrusive vulcanism in northeastern United States, eastern Canada, and the Yukon Plateau
Precambrian to Early Paleozoic Orogenies	Several protracted and complex orogenies recorded within the crystalline shields of all continents. These eras are dated approximately 490–680, 1000–1200, 1800–2000, and 2800–3000 million years.		

After Umbgrove (1947), Machatschek (1955), Bederke and Wunderlich (1968), Dott and Batten (1971), and others. Approximate median ages of orogenies based on recent potassium-argon dates.

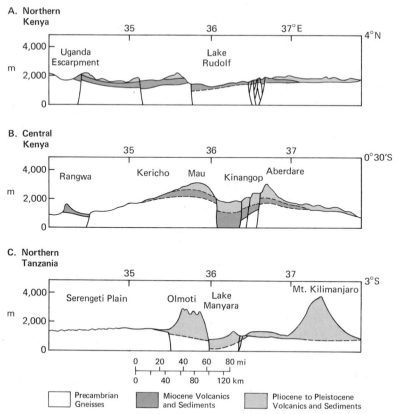

Figure 14–3. Sections Across the Eastern Rift Valley in Kenya and Tanzania. (From Baker, Mohr, and Williams, Geology of the eastern rift system of Africa. *Geological Society of America, Special Paper* No. 136, Boulder, Colorado, 1972. With permission.)

When the major stream and its tributaries make right-angled bends and intersections, without perfect parallelism of the side streams, drainage is *rectangular* in pattern (Figure 14–4,C). However, most of the low-order streams and many of the high-order, through-streams may show the random branching or *dendritic* pattern (Figure 14–4,A) that is the norm with undeformed rocks of uniform resistance. Drainage organization can also be *antecedent*, that is, preserved from a period prior to folding or uplift. This is the case when major streams are able to cut into bedrock at a rate equal to or greater than that of uplift. In this way antecedent streams may cut across a mountain range, regardless of the grain of the land.

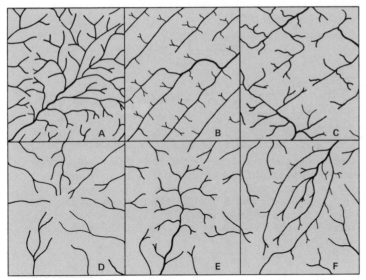

Figure 14–4. Drainage Patterns: (A) Dendritic, (B) Trellis, (C) Rectangular, (D) Radial, (E) Centripetal, and (F) Annular.

14-3. LANDSCAPES MODELED BY VULCANISM

Volcanic Hills and Mountains. Many of the world's highest mountain ranges are capped by volcanic cones that frequently cluster in chains or groups. Isolated volcanoes or groups of volcanoes are also found outside of the cordilleran belts, although most of these are associated with rift valleys, mid-ocean ridges, or island arcs—all zones of crustal spreading or collision (Figure 14–5). This relationship underscores the role of vulcanism as a major component of crustal deformation and mountain building. Volcanoes contribute relief to the continental land masses as well as the sea floors by constructing rock edifices that vary considerably in shape, relief, and slope. They are generally associated with younger mountain belts.

Shield volcanoes consist of massive accumulations of lava, mainly basalt, that erupt quietly from a number of craters or fissures. These volcanoes resemble great domes, with gentle slopes of 2 to 4 degrees on their lower flanks, steepening a little to 5 degrees or more near their summits. Although not impressive in terms of abrupt relief, shield volcanoes are of enormous size. The Island of Hawaii consists of a group of overlapping shield volcanoes that rise from 15,000 feet (5,600 m) *below* sea level to 13,800 feet (4,200 m) *above*. This relief of 29,000 feet (8,800 m) is matched by a base 100 miles (160 km) in diam-

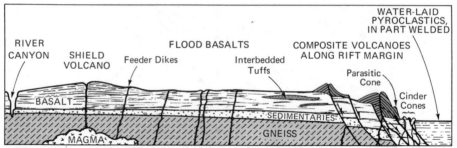

Figure 14–5. Complex Volcanic Landscape. In the case of the Ethiopian example, shown here in highly simplified form, flood basalts and shield volcanoes built up over an eroded continental surface during the Eocene and Oligocene; Miocene downwarping and Pliocene to Pleistocene rift faulting of the Ethiopian Rift led to acidic vulcanism along its margins, creating many, large composite volcanoes and filling depressions with thick accumulations of ash.

eter. As in the case of all large volcanoes, streams radiate outward and downslope to form *radial* drainage patterns (Figure 14–4,D).

Cinder cones reflect on explosive vulcanism and contrast strongly with shield volcanoes. Instead of lava, they consist almost entirely of volcanic cinders and ash that have been blasted out of crater vents. These pryoclastics are normally of a more acid variety of igneous rock, initially rich in gases and liable to violent eruption. Cinder cones are comparatively small, varying from 50 to 1,500 feet (15 to 450 m) in height, but with steep slopes of 25 to 30 degrees. Unlike shield volcanoes, which take millions of years to form, cinder cones are created in a few months or years. Sunset Crater (see section 2-5) is a good example.

Composite volcanoes result from protracted eruptions of cinders and lavas that maintain the general shape of a cinder cone, with slopes of 20 to 30 degrees. However, the relief is much greater and Mount Etna (10,740 ft, or 3,250 m) in Sicily or Fuji-san (12,400 ft, or 3,750 m) in Japan are mighty mountains indeed. Most of the world's active volcanoes are of this type. Their violent and seemingly unpredictable character has been responsible for most of the high toll in lives and destroyed property through the length of recorded history. The burial of Pompeii and Herculaneum by cinders from Vesuvius in 79 A.D. and the fiery destruction of St. Pierre, Martinique, by white-hot ash from Mount Pelée in 1902 are two dramatic examples. But similar eruptions have destroyed countless other settlements in Java, Central America, and other lands.

Large volcanoes sometimes develop deep, saucer-shaped depressions within their cones. Known as *calderas*, dimensions may range to as much as 9 miles (14 km) across and 2,000 feet (650 m) deep. Some calderas have resulted from gigantic eruptions that blasted the tops off composite volcanoes—such as Tamboro in Indonesia. The majority, however, are attributed to internal collapse, by downfaulting as the magma reservoir is drained off underground or through lower-situated vents and fissures. Many calderas harbor lakes and will normally develop *centripetal* drainage, radially inward along the inner slopes of the depression (Figure 14-4,E).

The different types of volcanoes may occur singly or in multiple groups. Thus, for example, cinder cones are commonly found along the flanks of shield or composite volcanoes but can also occur in swarms (see Figure 2-3).

The Role of Intrusive Vulcanism. Intrusive volcanics (see Chapter 2) contribute to surface irregularity when exhumed by erosion. Fields of cinder cones, for example, only have limited longevity once activity ceases. The surface ash, cinders, and incidental lavas are eventually removed—exposing the neck and a network of dikes and sills of relatively resistant igneous rock. Many such necks form striking landmarks, and Shiprock in northwestern New Mexico has a relief of 800 feet (245 m). Dikes may be sufficiently large and durable to form ridges several miles in length, while sills can emerge as hard caps of rock preserving small plateaus. Even more massive are laccoliths and batholiths, particularly where large complexes of intrusive igneous rocks have been block-faulted into prominence during the course of orogenic spasms (see Figure 14-6). In fact there often is a temporal sequence in mountain building:

1. The initial phases of mountain building are marked by extrusion of basic lavas.
2. Subsequently, the major mountain lineaments are created by folding, accompanied by intrusive vulcanism deep within the crust.
3. As erosion continues to level these fold mountains, block faulting—accompanied by further extrusive vulcanism—begins to upraise and expose the acidic batholith masses at the mountain roots.

In this way intrusive rocks normally contribute to the mass and relief of the mountain belts.

Volcanic Plains and Plateaus. Not all volcanic activity creates mountains. In some areas and at certain times surfaces have been mantled by deep spreads of volcanic ash or buried under great floods of horizontal lavas (see Figure 14-5).

Ash plains consist largely of pyroclastics in the dust size, that is,

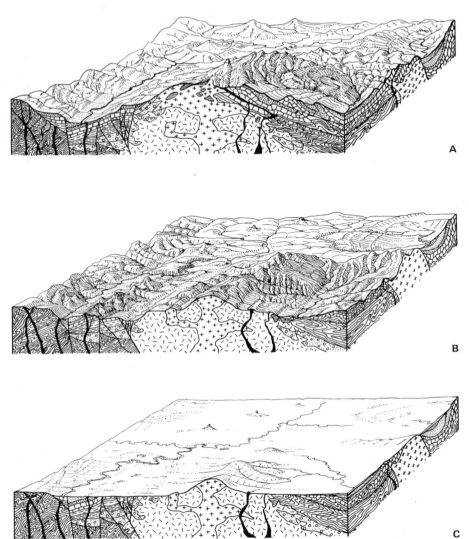

Figure 14–6. A Possible Sequence of Erosion in a Region of Complex Rocks and Structure. The central batholith is intruded by a laccolith and overlaid by recumbent folds to the right. Igneous veins also intrude the metamorphics and block-faulted sedimentary strata at the left. Erosion rapidly removes the extrusive volcanics and the weathering, exposed sedimentaries. The end result approaches a crystalline shield. (Drawn by Erwin Raisz, from Strahler, *Physical Geography*, 3rd ed. Copyright © 1969 by John Wiley & Sons, Inc., New York.)

tuff grade. Tuffs may be distributed by wind over a distance of 100 or 200 miles (150 or 300 km) in the wake of a violent eruption, mantling the land surface or collecting on stream floodplains and lake bottoms. Thicknesses can build up rapidly to tens or even hundreds of feet (3 to 30 m). When the ash falls in the incandescent state, it may weld extensive layers of ash into lavalike rock of great resistance. Thus when streams begin to carve into ash deposits, they commonly leave broad, flat uplands preserved by hard, welded ash strata or by highly permeable ash beds. Such surfaces are widespread in the great rift valleys of East Africa and in other, smaller volcanic regions.

Flood basalts are highly fluid and erupt from large fissures rather than from shield volcanoes, and they spread out over vast areas. During the geological past, flood basalts have periodically erupted to create fresh lava surfaces. During the Mesozoic and the Tertiary, fissure flows—over millions of years—built up great, horizontal accumulations of basalt in several areas, including southern Brazil, the Deccan Plateau of India, the Central Siberian Plateau, the Columbia Plateau, and the Ethiopian Plateau (Figure 14–5). The area of each of these spreads varies from 75,000 to 200,000 square miles (200,000 to 500,000 sq km), and their average thickness is close to 2,500 feet (800 m). In detail such lavas consist of horizontal strata of variable thickness, sometimes with columnar structures, and locally interbedded with stream and lake deposits. The successive layers of resistant rock maintain flat interfluves, even after protracted and intensive dissection. As a result, upland plains, tablelands, or fretted plateau edges are the characteristic landform. In fact, some authors simply refer to these flood lavas as "plateau basalts." Collectively, volcanic plains and plateaus account for less than 2.5 percent of the continental surfaces. However, their size and regional prominence give them more than incidental importance.

14-4. ANCIENT CRYSTALLINE SHIELDS

Large parts of all of the continents consist of stable masses of very ancient and complex metamorphic rocks. These *shields* have undergone repeated faulting, volcanic intrusion, and denudation. The original rocks generally exceed 500 million years in age, and they have been deformed by several orogenies. Successive fracturing and uplift, followed repeatedly by erosion down to about sea level (Figure 14–6), have removed all but the very roots of these ancient mountain belts (see Table 14–1).

Characteristically, the shields are areas of low relief with gentle slopes and rolling topography. Wherever epeirogenic uplift or upwarping has provided potential energy, running water or ice has effected differential erosion. As a result, valleys typically occur in geometric patterns related to complex networks of old fracture lines and multiple intrusions (Fig. 14–4,C). Topography is a matter of rolling plains, locally grading over into hill country.

This description applies to most of the crystalline shields of the Americas, Eurasia, and Australia. Yet recent block faulting or uplift has not been excluded from these stable land masses. Segments of the

Laurasian shields have been caught up in the Alpine cordilleran belts, while small outliers such as the Adirondacks of New York have been upraised as isolated blocks. Above all, however, the Gondwana shield of Africa has been shattered along the great rift valleys, as well as up-warped into the various highland swells (see Appendix B). Some of the resulting hills or mountains rival the dimensions of fault-block ranges in the cordilleran belts. As a result, the surfaces of the crystalline shields are variable in their relief and irregularity, even though they may be structurally similar, with uniform mesolandforms over wide areas. Crystalline shields account for 23 percent of the basic land form configuration of the continents (see Table 15–1). They are prominent on all of the continents except Eurasia.

Chapter 15

STRUCTURE AND LAND FORM (II): SEDIMENTARY COVERS

Almost half of the continental surfaces are formed by depositional plains or by erosional plains and plateaus cut across sedimentary rocks. Of particular interest in this chapter are the structural landforms created by differential erosion in warped or subhorizontal rock strata of variable resistance. These include extensive escarpments or prominent ridges, with inclined back slopes, which are commonly arranged around large-scale domes or basins. Equally significant are solution forms that provide detail and sometimes determine the general appearance of landscapes developed in soluble limestones or dolomites. Such karst features include sinkholes of various sizes that may coalesce to form valley systems with underground drainage. The overall distribution of gross land-form categories on the continents is represented by a series of maps and summarized in Table 15–1.

15-1. STRUCTURAL INFLUENCES BEYOND THE MOUNTAIN BELTS

The effects of structure upon both young and old mountain belts are readily apparent, since gross land form and a great many smaller-scale features are a direct or indirect consequence of the patterns of rock deformation and lithology. Yet mountain belts of whatever age and type account for only 52 percent of the continental surfaces (see Table 15–1). The remaining areas fall into one of three possible categories of genetic land form:

1. *Depositional Plains.* About 25 percent of the continental surfaces are mantled with Pleistocene to Recent deposits that create very young landforms, only indirectly affected by bedrock properties. These include flat alluvial plains, dissected alluvial surfaces or loess plains, rolling or dissected till country, and undulating fields of sand dunes. Since these landform types primarily reflect on gradational processes, they are considered in Chapters 16 to 20. However, the basic location of many depositional plains is dictated by areas of tectonic subsidence—either synclinal downfolding or epeirogenic downwarping—so that the structure of the Earth's crust is never irrelevant.

2. *Erosional Plains and Plateaus on Sedimentary Rocks.* About 21 percent of the land surface has been modeled directly in sedimentary rocks—horizontal or gently warped—of pre-Pleistocene age. The resulting erosional landscapes range from dissected plains to hill country and plateaus. They are modified to some degree or other by differential erosion in different lithologies and by the attitude of the rock strata. In most cases, erosion gradually reveals and accentuates the structural-lithological features. These peculiarities begin once more to lose their prominence only when dissection and denudation have thoroughly reduced the landscape.

Erosional plains normally originated as marine, lake, or alluvial surfaces that were exposed to erosion at some point. Available relief was usually provided by a tectonic impulse, and this potential energy helped to determine the intensity of dissection. In the case of horizontal or subhorizontal rocks, resistant caprock or hard, mid-slope strata favor the preservation of flat interfluves or stepped surfaces due to differential erosion. In the case of gently or moderately warped rocks, a number of structural-lithological landforms are possible. Consequently, plains, hills, and plateaus developed on horizontal or warped sedimentaries commonly display landforms of "structural" type. Despite any differences of slope forms from one environment to another, the large-scale landforms reflect primarily on available relief, rock inclination, and differential lithology. These features will be discussed further.

3. *Volcanic Plains and Plateaus.* These landscapes have been intro-

Table 15-1. DISTRIBUTION OF LAND FORM CLASSES ON THE CONTINENTS (EXCLUDING GREENLAND AND ANTARCTICA)

CLASS	NORTH AMERICA	SOUTH AMERICA	EUROPE	ASIA	AFRICA	AUSTRALASIA	WORLD
Depositional Plains	18%	21%	31%	22%	26%	28%	25%
Erosional Plains and Plateaus in Sedimentaries	19	25	30	12	30	24	21
Volcanic Plains and Plateaus	4	2	1	4	4	0	2
Young Mountain Belts	31	21	11	19	2	7	14
Old Mountain Belts	5	1	14	33	1	10	15
Ancient Crystalline Shields	25	30	13	10	37	31	23

duced already (section 14-3 and Figure 14–5). They are similar in form to erosional plains or plateaus developed in horizontal sedimentaries.

15-2. STRUCTURAL LANDSCAPES IN HORIZONTAL AND WARPED SEDIMENTARIES

When rock strata of differential resistance are essentially horizontal, tabular landscapes are liable to develop. Flat interfluves with steep-sided, V-shaped valleys form where resistant caprock mantles the uplands. Prominent facets or structural benches mark the valley sides. Intensive dissection can favor tableland development as stream activity detaches tabular *mesas* or smaller *buttes* (Figure 15–1) from the unconsumed uplands, producing an open hill country and, often, plains studded with residual hills. At the same time, resistant caprocks and other differences in lithology favor backwearing processes, generally preserving the tabular arrangement of the langscape. Dendritic drainage patterns are characteristic.

When horizontal sedimentary strata are essentially uniform in terms of lithology, their structural aspects are usually only apparent during the initial stages of dissection. A major exception is provided by landscapes in highly soluble limestones (section 15-3).

When rocks of differing resistance are inclined over broad areas,

Figure 15–1. Small Mesas and Buttes Stud This Erosional Surface in Egyptian Nubia. (Karl W. Butzer)

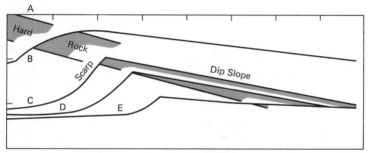

Figure 15–2. Computer-Simulated Development of a Cuesta in Gently Dipping Rocks of Variable Resistance. (Modified after Ahnert, 1973. With permission.)

one of several landform patterns may emerge. Perhaps the most common is the *cuestaform* plain. When rocks of differential resistance are gently tilted at 5 to 10 degrees, erosion tends to undermine softer strata. Escarpments of variable elevation are created in this way, forming steep cliffs (25 to 90 degrees) rimmed by resistant rocks (Figure 15–2). Beyond this *escarpment* the *dip slope* falls away gently along the inclination of the rock strata. In plan view, cuesta escarpments frequently occur in parallel or concentric lines, their relief and spacing varying according to the thickness and tilt of the different rock units. All degrees of dissection are possible, from a straight, sheer cliff to a fretted upland edge with fine drainage texture and numerous detached outliers.

When sedimentary rocks show steeper regional inclinations, of 10 to 25 degrees, erosion again wears back the inclined strata by undermining, forming more steeply tilted cuestas known as *hogbacks* (Figure 15–3). These hogbacks are arranged in concentric, elliptical rings that rise inward in *domes* (Figure 15–4, A), outward in *basins* (Figure 15–4, B). Since basins and domes vary greatly in their dimensions, there is a continuum of angles between cuestas and hogbacks. Thus, for example, cuestaform plains fringe a great basin structure in the Great Lakes area, centered over Lower Michigan. In general, cuestas and hogbacks disrupt existing drainage patterns by their reversed slopes and subparallel escarpments. In the case of basin or dome structures, streams may develop in an *annular* drainage pattern (see Figure 14–4, F).

15-3. LIMESTONE LANDSCAPES DUE TO UNDERGROUND SOLUTION

In many parts of the world, there are extensive exposures of soluble limestones and dolomites that may develop peculiar landforms. Often these peculiarities are only a matter of detail, at the level of microlandforms (Figure 3–6). Sometimes, however, distinctive landscapes

Figure 15–3. Echo Cliffs, North-Central Arizona. Strong monoclinal folding and subsequent erosion have created a striking hogback ridge. Highway 89 runs along the foot of the escarpment. (Karl W. Butzer)

Dome

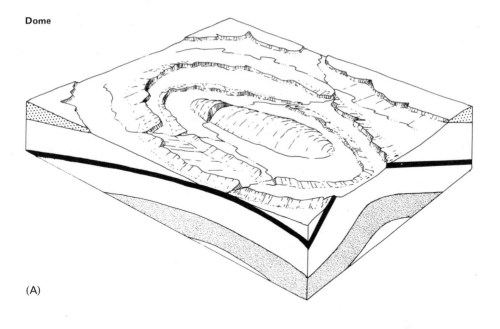

(A)

Basin

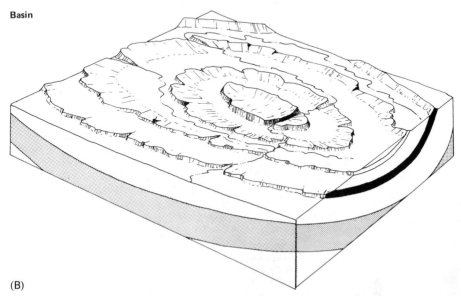

(B)

Figure 15–4. (A) Dome Structure with Hogback Ridges. (B) Basin Structure with Hogback Ridges. (After Hamblin and Howard, *Physical Geology Laboratory Manual.* Copyright © 1967 by Burgess Publishing Co., Minneapolis.)

develop that differ on a much larger scale. These reflect lithological conditions and are primarily modeled by underground waters. Underground waters can work so effectively in some limestones and dolomites that surface discharge and occasional streams are swallowed up in expanded, funnel- or saucer-shaped sinkholes (Figure 15–5). Most commonly, the water simply filters through the soil and rubble at the bottom of the sinkhole. Sinkholes vary considerably in size from a deep, vertical shaft, or *ponor,* to a simple but well-defined depression, or *doline,* and a multiple doline called an *uvala.* Small dolines may be as little as 25 feet (8 m) in diameter and 5 feet (1.5 m) deep. Well-developed dolines or uvalas commonly range from 200 to 1,000 feet (60 to 300 m) wide and 30 to 150 feet (9 to 45 m) deep. They are roughly circular in plan, with steep or near-vertical rock faces. Several sinkholes may coalesce to form open depressions of irregular outline, occasionally as much as a mile or more in length. Such complex collapse hollows, or *polja* (singular, *polje*), are flatfloored and may or may not have external drainage.

Streams that drain into sinkholes commonly flow through a series of interlocking underground caverns. Mammoth Cave, Kentucky, is an exceptional example of such a cavern system, with a total length of 197 miles (315 km) of subterranean hollows and passages. These are arranged along five levels, representing major horizontal rock units, with a vertical depth of up to 500 feet (150 m). Sooner or later underground streams reemerge as springs along some distant valley-side, or, occasionally, a large stream will issue from an external cave. If subterranean drainage is sufficiently prominent, there can be a number of

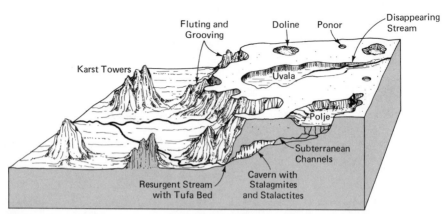

Figure 15–5. Degrees of Dissection and Denudation of a Karst Landscape. These different features are not necessarily found in direct association, and the tower karst (*left*) is restricted to humid tropical environments.

"abandoned" valleys that lack streams. *Interrupted* drainage patterns may therefore be characteristic.

Landscapes with disappearing streams, "dry" valleys, and numerous sinkholes and polja are designated as *karst* topography (Figure 15–5), named after hill country of this type in Yugoslavia. Karst is best developed in horizontal limestones of high solubility and less than 10 percent nonsoluble impurities. In particular the presence of montmorillonitic clays, which swell and seal off joint cavities, will inhibit karst development. This means that the more bizarre, well-developed karst landscapes with interrupted drainage are not too common. However, some karstic features due to corrosion or underground solution can be found in almost all compact limestones. Important karst landscapes are developed in parts of France, central Kentucky, western Texas, the lake district of central Florida, and the Yucatan Peninsula. Even more striking is the karst topography of certain Caribbean islands and parts of southern China (see section 20-7). Perhaps the best-known sinkholes are the great cenotes near some of the Mayan ruins of northern Yucatan. The so-called Well of Death, at the Temple of Chichen Itza, is a vertical-walled sinkhole measuring 200 feet (60 m) across and 130 feet (39 m) deep, about half of it filled with water.

As long as karst caverns remain immersed in water, they are prone to continued erosion and enlargement. Eventually, however, the groundwater level may fall and water movement in the caverns is reduced to a trickle. Water, rich in dissolved carbonates, oozes through cracks in the rock of the roof and walls of a cave. As this water drips down onto the floor, some of it evaporates. If it evaporates on the ceiling, precipitation of the soluble lime forms icicles known as *stalactites.* The corresponding, blunt knobs of calcium carbonate deposited on the cave floor are called *stalagmites.* If a large stalactite links up with a knob of stalagmite, a column is formed. Stalactites, stalagmites, and columns are collectively known as *dripstone.* Films or sheets of water also serve to precipitate solubles, forming thin, wavy layers of calcium carbonate known as flowstone or *travertine* (Figure 15–6). The various cave deposits normally have a whitish color but may be tinted by oxides of iron (brown or red), manganese (black), or copper (greenish blue). Fresh-water carbonates can also be deposited where springs or seeps emerge at the surface. In addition to normal travertines and limey oozes, plants are encrusted with lime and ultimately leave a spongy rock (tufa) with organic impressions.

Accessible caves and caverns of karstic origin can provide underground "landscapes" of almost unsurpassed beauty and form major tourist attractions in widely dispersed places such as the Majorcan caves of the Balearic Islands: Carlsbad Caverns, New Mexico; and the Jenolan Caves, New South Wales, Australia. In addition to the great size of some of these caverns, attractive displays of dripstone and mul-

Figure 15–6. Cave Limestones of Las Monedas, Santander Province, Spain. Stalagmite boss (below) is matched by hanging stalactite. Flowstone cascades are well developed in background. (Karl W. Butzer)

ticolored, cascading flowstones have long drawn attention. At an earlier date, prehistoric man and animals inhabited many caves, leaving their bones or artifacts to be sealed on cave floors by lime deposits, rockfalls, ash, or animal dung. And the famous cave art of northern Spain and southwestern France was painted on the walls of limestone caverns some 12,000 to 25,000 years ago.

15-4. LAND FORM OF THE CONTINENTS

The overall history of the continents in the wake of continental drift has already been touched on in section 13-4. It is now appropriate to summarize the component parts of the continents. This can be done by outlining their land form and topographic expression in Appendix B and by illustrating the salient features by land-form maps of the six continental landmasses, excluding Antarctica (Figures 15–7 to 15–12). From these continental maps and from Table 15–1 it can be seen that each continent consists of five essential parts:

1. An ancient crystalline shield
2. One or more segments of old mountain belts
3. A young cordilleran belt or part thereof
4. A partial mantle of subhorizontal sedimentaries resting on the shields or fringing the mountain belts
5. Scattered areas of relatively young sediments (Pleistocene to Recent), many of them in broad synclines and other regions of subsidence

For details the reader may consult Appendix B.

15-5. SELECTED REFERENCES

Ahnert, Frank. Coslop 2—a comprehensive model program for simulating slope profile development. *Geocom Programs*, Vol. 8, 1973, pp. 99–122.

Baker, B. H., P. A. Mohr, and L. A. J. Williams. Geology of the eastern rift system of Africa. *Geological Society of America, Special Paper, No. 136*, 1972, pp. 1–67.

Bederke, Erich, and H. G. Wunderlich. *Atlas zur Geologie*. Mannheim, Bibliographisches Institut, 1968, 75 pp. (in German with English key).

Cloud, P. E. Rubey conference on crustal evolution. *Science*, Vol. 183, 1974, pp. 878–881.

De Jong, K. A., and R. Scholten, eds. *Gravity and Tectonics*. New York, Wiley-Interscience, 1973, 502 pp.

Dickinson, W. R. Plate tectonics in geologic history. *Science*, Vol. 174, 1971, pp. 107–113.

Dietz, R. S. Geosynclines, mountains and continent-building. *Scientific American Offprint*, No. 899, Vol. 226 (March 1972), pp. 30–38.

Dott, R. H., and R. L. Batten. *Evolution of the Earth*. New York, McGraw-Hill, 1971, 649 pp.

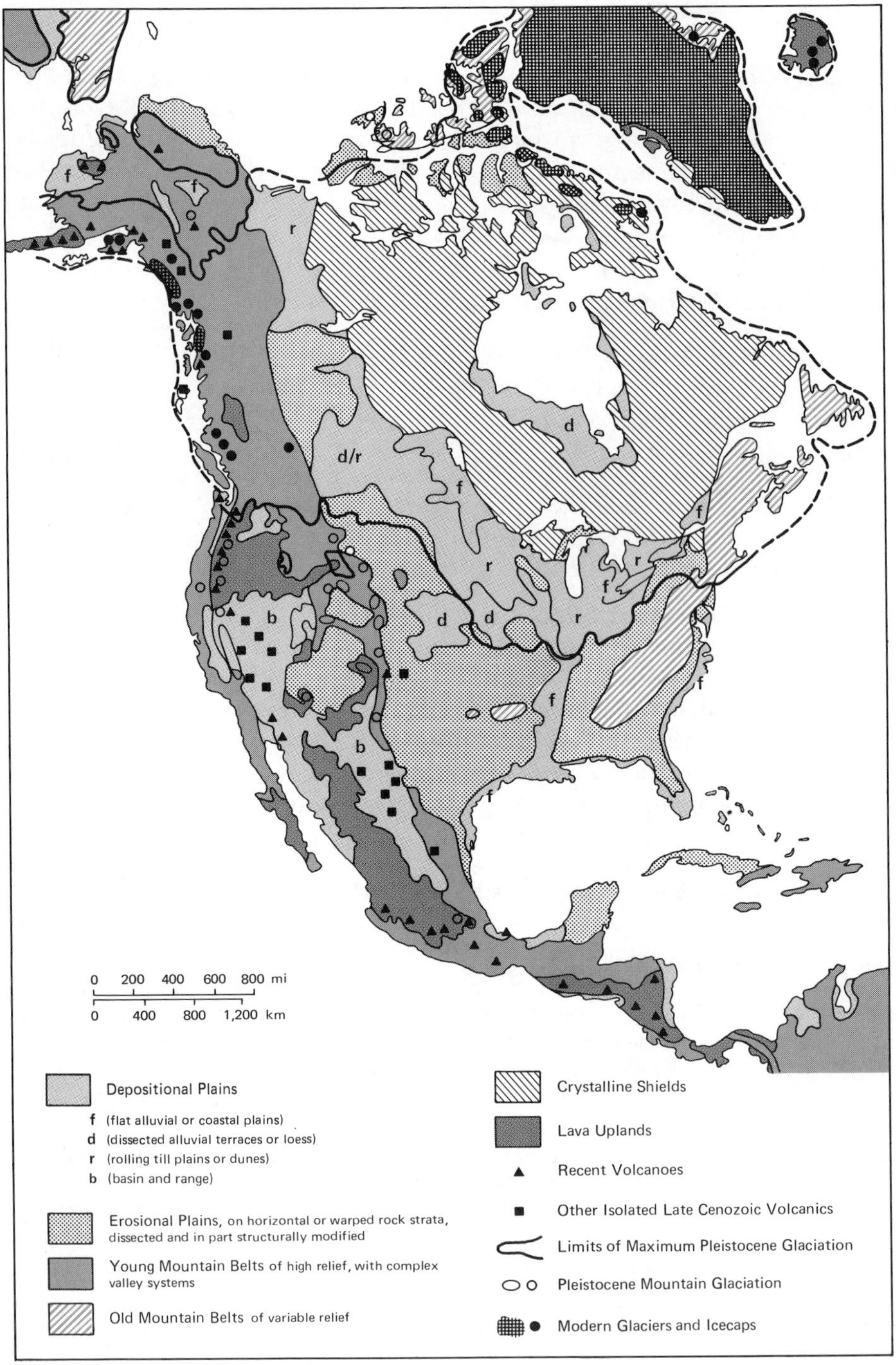

0 200 400 600 800 mi
0 400 800 1,200 km
Depositional Plains
f (flat alluvial or coastal plains)
d (dissected alluvial terraces or loess)
r (rolling till plains or dunes)
b (basin and range)
Erosional Plains, on horizontal or warped rock strata, dissected and in part structurally modified
Young Mountain Belts of high relief, with complex valley systems
Old Mountain Belts of variable relief
Crystalline Shields
Lava Uplands
Recent Volcanoes
Other Isolated Late Cenozoic Volcanics
Limits of Maximum Pleistocene Glaciation
Pleistocene Mountain Glaciation
Modern Glaciers and Icecaps

Figure 15–7. Land Form of North America. For identification and a description of the major units see Appendix B. (Base outlines for Figures 15–7 to 15–12 are from the Goode Base Map Series. Copyright by the University of Chicago Department of Geography.)

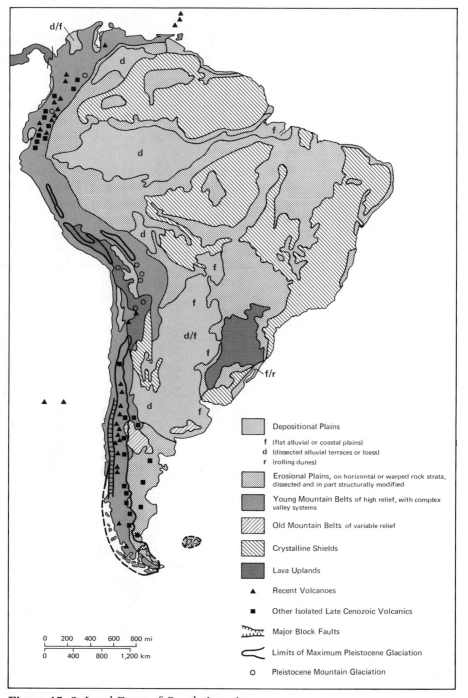

Figure 15–8. Land Form of South America.

Legend:

Depositional Plains
- f (flat alluvial or coastal plains)
- d (dissected alluvial terraces or loess)
- r (rolling dunes)

Erosional Plains, on horizontal or warped rock strata, dissected and in part structurally modified

Young Mountain Belts of high relief, with complex valley systems

Old Mountain Belts of variable relief

Crystalline Shields

Lava Uplands

▲ Recent Volcanoes

■ Other Isolated Late Cenozoic Volcanics

Major Block Faults

Limits of Maximum Pleistocene Glaciation

○ Pleistocene Mountain Glaciation

Scale:
0 200 400 600 800 mi
0 400 800 1,200 km

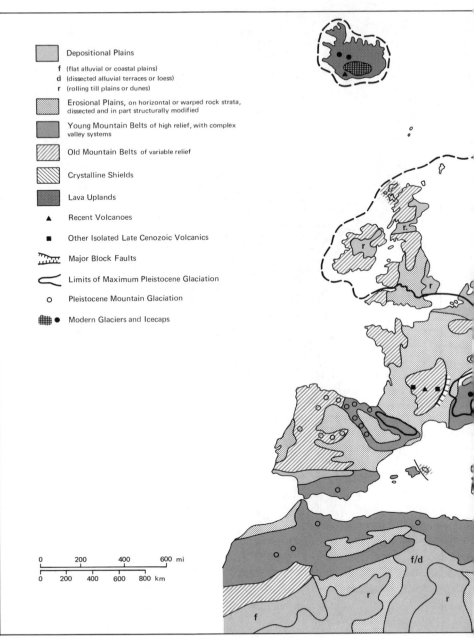

Figure 15–9. Land Form of Europe.

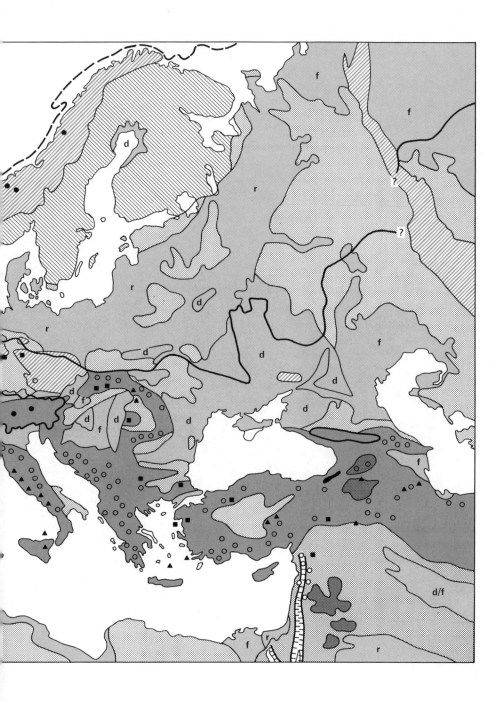

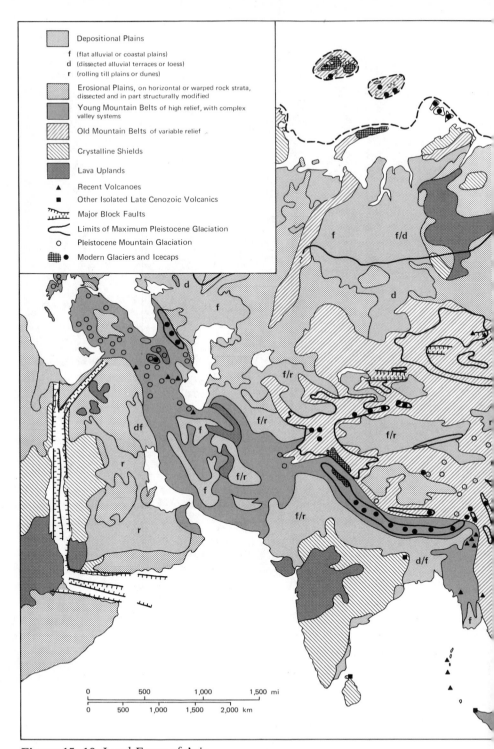

Figure 15–10. Land Form of Asia.

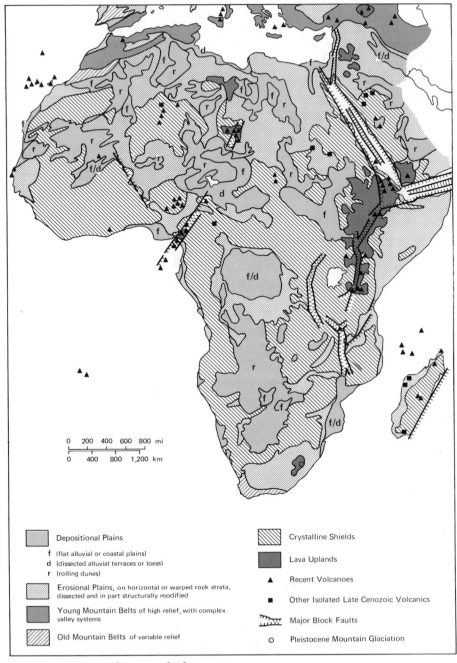

Figure 15–11. Land Form of Africa.

Depositional Plains

f (flat alluvial or coastal plains)
d (dissected alluvial terraces or loess)
r (rolling dunes)

Erosional Plains, on horizontal or warped rock strata, dissected and in part structurally modified

Young Mountain Belts of high relief, with complex valley systems

Old Mountain Belts of variable relief

Crystalline Shields

Lava Uplands

▲ Recent Volcanoes

■ Other Isolated Late Cenozoic Volcanics

Major Block Faults

o Pleistocene Mountain Glaciation

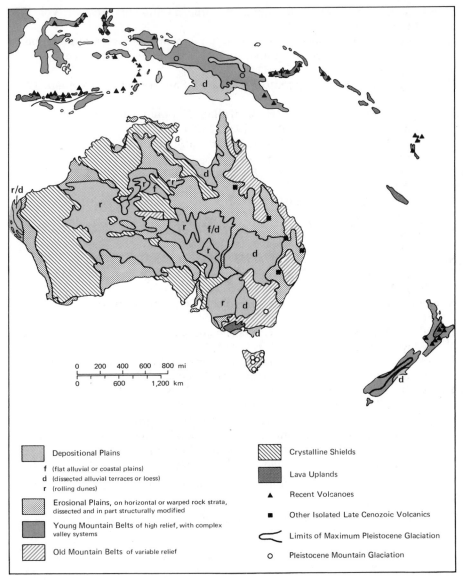

Figure 15–12. Land Form of Australasia.

Legend:

Depositional Plains
- **f** (flat alluvial or coastal plains)
- **d** (dissected alluvial terraces or loess)
- **r** (rolling dunes)

Erosional Plains, on horizontal or warped rock strata, dissected and in part structurally modified

Young Mountain Belts of high relief, with complex valley systems

Old Mountain Belts of variable relief

Crystalline Shields

Lava Uplands

▲ Recent Volcanoes

■ Other Isolated Late Cenozoic Volcanics

Limits of Maximum Pleistocene Glaciation

o Pleistocene Mountain Glaciation

Scale:
0 200 400 600 800 mi
0 600 1,200 km

Elders, W. A., et al. Crustal spreading in southern California. *Science,* Vol. 178, 1972, pp. 15–24.

Hammond, E. H. Analysis of properties in land form geography: application to broad-scale land form mapping. *Annals, Association of American Geographers,* Vol. 54, 1964, pp. 11–18.

Hunt, C. B. *Physiography of the United States.* San Francisco, Freeman, 1967, 480 pp.

Jennings, J. N. *Karst.* Canberra, Australian National University Press, and Cambridge, Mass., M.I.T. Press, 1971, 253 pp.

King, L. C. *Morphology of the Earth,* 2nd ed. Edinburgh, Oliver and Boyd, and New York, Hafner, 1967, 726 pp.

Machatschek, Fritz. *Das Relief der Erde,* 2 vols. Berlin, Borntraeger, 1955, 530 and 594 pp.

Ollier, Cliff. *Volcanoes.* Camberra, Australian National University Press, and Cambridge, Mass., M.I.T. Press, 1969, 177 pp.

Seyfert, C. K., and L. A. Sirkin. *Earth History and Plate Tectonics.* New York, Harper & Row, 1973, 504 pp.

Sparks, B. W. *Rocks and Relief.* London, Longmans, 1971, 404 pp.

Strahler, A. N. *Physical Geography,* 3rd ed. New York, Wiley, 1969, 733 pp. See Chap. 32–34.

Sweeting, M. M. *Karst Landforms.* New York, Columbia University Press, 1972, 362 pp.

Thornbury, William D. *Regional Geomorphology of the United States.* New York, Wiley, 1965, 609 pp.

Tricart, Jean. *Structural Geomorphology.* Translated from the French by S. H. Beaver and E. Derbyshire. London, Longmans, 1974, 305 pp.

Twidale, C. R. *Structural Landforms.* Canberra, Australian National University Press, and Cambridge, Mass., M.I.T. Press, 1971, 247 pp.

Umbgrove, J. H. F. *The Pulse of the Earth.* The Hague, M. Nijhoff, 1947, 358 pp.

Williams, P. W. The geomorphic effects of ground water. In R. J. Chorley, ed. *Water, Earth and Man.* London, Methuen, 1969, pp. 269–284.

Yatsu, Eiju. *Rock Control in Geomorphology.* Tokyo, Sozosha, 1966, 135 pp.

Structural landforms are best illustrated by topographic and geological maps. These are readily available in:

Hamblin, W. K., and J. D. Howard. *Physical Geology: Laboratory Manual,* 4th ed. Minneapolis, Burgess, 1975, 231 pp.

Chapter 16

CLIMATE, TIME, AND LANDFORMS

16-1. Structure versus environment
16-2. Time, potential energy, and climatic change
16-3. Climatic and historical geomorphology
16-4. Inherited soils
16-5. Landscape stability and geomorphic periodicity
16-6. Selected references

Gross structural patterns impart relief and determine the configuration of continental topography. On a smaller scale rock type and attitude further modify landforms. Yet given similar conditions of structure, slope, and available relief, landforms evolve distinctly where the balance of hillslope weathering and erosion differs in response to vegetation and climate. These climatic-environmental factors affect soil development and the constellation of gradational processes that operate in any one region. The result is that slope shape, slope inflections, lithological microforms, and drainage density combine to provide a distinctive scenic character to glacial, polar, humid, and arid landscapes. Such landscapes are, however, a composite product of repeated climatic shifts and a complex structural history that extends well back into the Pleistocene and even the Tertiary. In particular, geomorphologic processes have responded to climatic change by variable trends and rates and were profoundly modified by potential-energy flux in the wake of continuing orogenic deformation. The result has been a widespread periodicity between episodes of slow geomor-

phic change and ongoing soil development, on the one hand, and vigorous geomorphic erosion and sedimentation, on the other. These periodicities and the intervening discontinuities are largely responsible for the considerable legacy of inherited soils and landforms.

16-1. STRUCTURE VERSUS ENVIRONMENT

Large-scale land-form attributes and types are controlled or influenced by rock structure and lithology. So, for example, a high range may consist of parallel-fold or fault-block mountains with a trellis arrangement of drainage lines. In detail, the landforms will differ, depending on whether the peaks reach above the snowline and whether or not the range is either too dry or too cold to support forest vegetation and favor rapid chemical weathering. As a result, a cuestaform plain in western Europe will look drastically different from one in the Sahara. Or valleys in upland Siberia are somehow distinct from those developed in similar rocks and with comparable relief in a Brazilian rainforest. Also, when all other conditions are equal, the drainage texture in arid, as opposed to humid, environments is more coarse. There is, then, an array of both quantitative and qualitative landform properties that is influenced by the nature of weathering and by the peculiar interplay of gradational processes. These factors, in turn, are modified by the broad environmental controls of vegetation and climate.

It has already been emphasized (Chapter 5) that geomorphologists have a fundamental interest in hillslopes. Yet structure, slope, and available relief only provide the stage setting for hillslope evolution. Where the constellations of these parameters are similar, the ultimate ratio of downwearing to backwearing is influenced, if not determined, by the balance of hillslope weathering and erosion. In dry environments, soil development is slow and superficial, so that much bare rock is exposed on hillsides; as a consequence, soil does not build up on and bury the foot-slopes, with the result that mid-slope angles remain steep, with slope inflections sharp and angular. In humid environments, soils are deep, even on the average hillside where creep, slumping, and earthflows transport large quantities of regolith onto and across the foot-slope; as a result, mid-slope angles decline and slope inflections become less and less marked. In this fashion an environmental factor—the rate and intensity of soil development—influences the patterns of gradation and the details of land form.

The role of the environment in landform development is complex. Climate provides the potential for broad categories of vegetation, although the details of what a forest, grassland, or shrub community will look like are affected by local factors such as rock type, drainage conditions, slope processes (Table 16–1), and, most importantly, man. Climate and vegetation both influence the rate and type of weathering, although rock type, drainage, and relief remain important. Consequently, over broad areas soils tend to reflect weathering processes, vegetation, and, ultimately, climate (see section 4-8). Given a uniform period of time, the attributes most significantly affected would be the number of soil horizons, their depth and intensity of development; the kinds of clay minerals and other alteration products; the degree of leaching, eluviation, and illuviation; as well as the nature of organic or mineral concentrations in different soil horizons.

This ecological relationship can be taken further. Rate of weathering and depth of soil mantle are critical for the balance of weathering and erosion on the crest-slope, mid-slope, and foot-slope. As a result, the nature and velocity of gradational processes at work on a hillslope are affected, if not controlled, by the growing chain of interacting ecological variables. On slopes and interfluves the intensity of overland flow, the significance and types of mass movements, the rate of rill and sheet erosion, as well as the overall balance of weathering and denudation, can all be expected to respond. Similarly, within a river system the hydrological regime, the total sediment carried, the proportions of bed and suspended sediments, the parameters

Table 16–1. INTERRELATIONSHIPS OF REGIONAL AND LOCAL FACTORS IN LANDFORM DEVELOPMENT

REGIONAL FACTORS	PHENOMENA	LOCAL FACTORS
Temperature Radiation Moisture Soils	Vegetation	Mineral nutrients Rock porosity/permeability Surface drainage Slope and relief
Temperature Moisture Vegetation	Weathering and Soils	Rock type Surface drainage Slope and relief
Temperature Moisture Vegetation Weathering and soils	Gradational processes	Rock type Rock structure Slope inclination and length Available relief
Climate Vegetation Weathering and soils Gradational processes	Landforms	Rock type Rock structure Available relief

of channel shape and stream velocity, the patterns of cut and fill, and the balance of dissection and deposition can all be affected.

In the end, climate, vegetation, weathering, soil mantle, and gradational processes in their turn influence landform development (Table 16–1). It should not be surprising, therefore, that landforms generated in different environmental settings do vary. This variation from one environment to another would be one of degree, rather than kind, since rock type, structure, and available relief never lose their significance. It should most likely affect landform elements such as slope shape (straight-sided, concave, convex, or concavo-convex), slope inflections (angular or rounded), the expression of lithological variations (accentuated facets or subdued segments), and drainage density (coarse or fine). Although these are attributes that can be sensed subjectively, fortunately they can also be measured.

Soil depth and horizonation are features that can also be recorded and measured with relative ease. With the necessary back-up laboratory studies that agricultural use of soils has long made necessary, the mechanical and chemical properties of soils have been widely studied. Consequently, large-scale soil variations and continental patterning are fairly well understood. Many problems of soil genesis remain to be clarified, and there is much controversy concerning soil classification. Nonetheless, a considerable body of comparative material is available for an environmental analysis of ecological soil types and their distribution. This information is summarized by the map of the more common soil classes (Figure 16–1) to be discussed in the following chapters.

Macro- and mesolandforms are more difficult to express by objective parameters, and quantitative terrain studies on whatever scale are still unable to provide rigorous generalizations as to the significance of vegetation versus lithology at the local scale or of different regional climates at the subcontinental level. Similarly, quantitative studies of contemporary processes in various environments are generally limited to individual slopes or small watersheds and, above all, are too sparse to be considered fully representative of broad environments. Available quantitative data on regional denudation rates are similarly inadequate. In particular, various regional estimates of sediment yield and erosional rates based on river sampling have been offered by a number of authors. Yet all are based on short-term measurements, seldom standardized, fraught with assumptions (Meade, 1969), and in part mutually contradictory. Major difficulties lie in the paucity of usable data on bed-load sediments and in the fact that accelerated soil erosion primarily involves localized erosion and deposition that may show little output beyond small watersheds (Trimble, 1975).

Lacking such vital categories of background information, it is still impossible to provide a quantitative characterization of landforms specific to or representative of different environments. It remains equally difficult to assess the contribution of the lithosphere (rock type, structure, and available relief) as opposed to the influence of vegetation and climate in landform development. This should not, however, discourage study of the biospheric and atmospheric components of geomorphology. These represent yet another challenging perspective to the subject, one that assumes particular importance for the very reason that man affects geomorphic processes primarily through his "control" of vegetation. Finally, the reality of contrasting landform constellations in glacial, polar, humid, and dry environments has never been in serious doubt. It remains to be shown whether these scenic landscape differences can be articulated, quantified, and, as assemblages, shown to be significantly different.

16-2. TIME, POTENTIAL ENERGY, AND CLIMATIC CHANGE

Soils and landforms not only change continually, but they acquire their properties over long periods of time. Under optimal conditions of biochemical weathering, a moderately deep soil with distinct horizonation will take centuries or millennia to form. In arid or subpolar environments, with limited chemical weathering, soils of moderate depth and horizonation may require tens of millennia. Once established, soils are often quite durable and may undergo little apparent change through time. As a result, many deep soils still reflect weathering conditions prevalent during parts of the Pleistocene or the Tertiary; if these soils are removed, they do not form afresh but are replaced by other soil types. Such records of the past are called *relict soils*.

Depositional landforms, under optimal conditions of sedimentation, also require centuries or millennia to develop. Erosional landforms, on the other hand, develop over tens or hundreds of millennia, and some Tertiary erosional surfaces may have taken tens of millions of years to form. With such long time spans, erosional landforms seldom evolve under a single climatic regime; instead, rapid climatic changes during the Pleistocene may have favored now one, now another balance of weathering and erosion. Consequently, the landforms of today often reflect the composite effect of several constellations of processes, operating at different rates or even tending in different directions. Many landforms are therefore *inherited* and would not evolve under prevailing conditions.

The time dimension is inescapable in soil studies and geomorphology. Stated in its simplest terms, the time concept implies that soils and landforms change continuously but interdependently as new surfaces are created by deposition, emergence, diastrophism, or vulcanism. Even if the external climate and base level did remain constant, rates of weathering and denudation would change through time

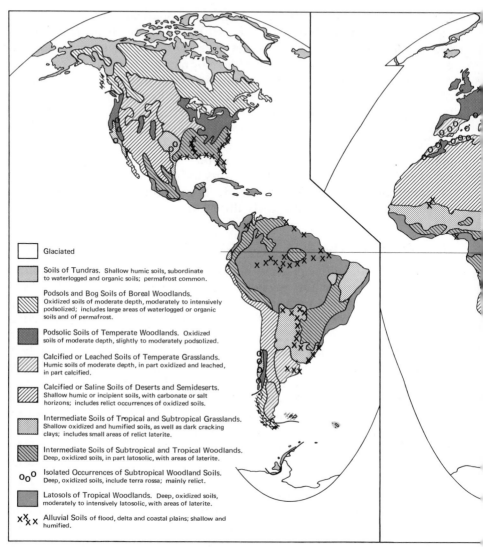

Figure 16–1. World Soil Distribution. (Compiled from various sources, including Ganssen and Hädrich, 1965. Base map copyright by The University of Chicago Department of Geography.)

as dissection or planation proceed, drainage conditions vary, and available relief is reduced. Furthermore, at any one time rates differ from place to place, reflecting patterns of regional climates, changing constellations of geomorphic processes, and differing potential energy. Consequently, a partly dissected, rugged landscape in one region may be much older than a smoother, well-eroded surface elsewhere. In fact, absolute ages are now available from many areas where surfaces and

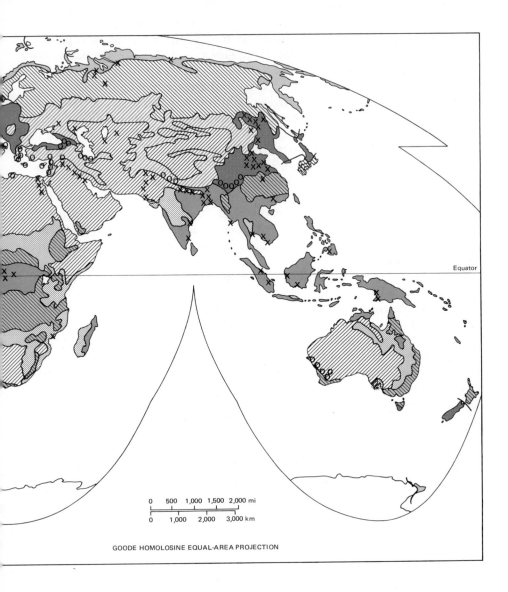

GOODE HOMOLOSINE EQUAL-AREA PROJECTION

sediments have been isotopically dated. The results are often unpredictable and caution strongly against the use of misleading and subjective terms such as "youth," "maturity," or "old age" to describe macrolandforms.

If relative base level varies through time, rates of dissection are decelerated and accelerated. Glacial-eustatic fluctuations of sea level have periodically affected stream systems in coastal proximity—in an elevation range of 500 feet (150 m) or more. Other global or regional changes of relative sea level during the course of the late Tertiary and the Pleistocene had similar effects over longer or shorter periods of time. Ups and downs of the land, during the course of epeirogenic and orogenic deformation, are a third possibility. Any of these controls can alternately increase or decrease potential energy. On a limited and temporary scale this may lead to minor floodplain adjustments. On a larger, long-term scale the particular constellation of geomorphic processes may be interrupted, as a new suite of functional landforms begins to develop. The best examples of such changes are the high planation surfaces that cut across parts of old mountain belts in the United States and western Europe; these were tectonically upraised during the late Tertiary or the Pleistocene, increasing their potential energy to the point that stream dissection is now actively cutting up the nonfunctional surfaces.

The significance of available relief for denudation rates can hardly be overemphasized. Ahnert (1970; see also Selby 1974) has standardized information on total denudation from large basins in North America and Europe to demonstrate a clear linear relationship between mean denudation rate and relief. In effect, for a wide range of arid to humid and temperate to alpine environments, these rates increased approximately twofold for every doubling of mean relief. This suggests that even in monsoon environments, where rainfall periodicity is at a maximum (see Jansen and Painter, 1974), potential energy will outweigh the impact of climate at the most general level, despite distinct differences in the proportions of dissolved and suspended sediment between arid and humid climates. Yet Ahnert's figures also imply that present, mid-latitude denudation rates are so slow that several tens of millions of years would be necessary to achieve the constant steady-state relief prerequisite to Davis' "stage" of peneplanation. Such extended episodes of tectonic quiescence simply did not occur during the Tertiary (see Table 14–2), so that other more efficient modes of areal denudation must be invoked to generate ancient planation surfaces.

The possible impact of climate during the Tertiary and the Pleistocene is no less complicated by change than was potential energy in the wake of repeated orogenic episodes and relative sea-level variations (see Figure 16–2). At their simplest, climatic changes can accelerate or decelerate existing geomorphic processes. They may also change the balance of weathering and erosion. So, for example, cold Pleistocene climates in what are now temperate woodlands favored backwearing instead of downwearing; at such times the rates of denudation increased dramatically. As a result, the past 10,000 years of subdued geomorphic development under temperate, Holocene conditions have more often than not hardly sufficed to change the inherited forms re-

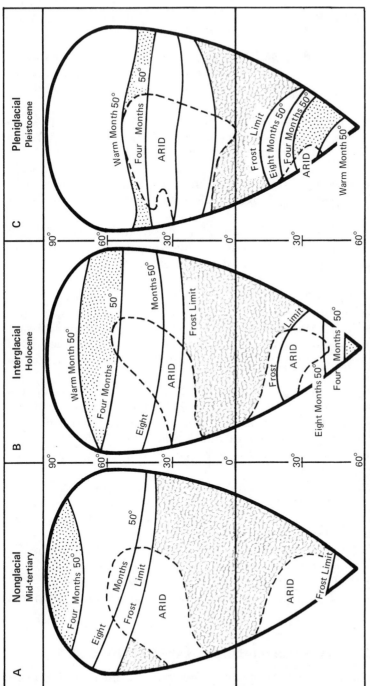

Figure 16-2. Descriptive Models of Climatic Zonation During the (A) Mid-Tertiary, (B) Holocene, and (C) Pleistocene Glacial. (The Holocene is modified after Trewartha, 1968. From Butzer, 1976.)

peatedly sculptured by 50,000 years and more of highly effective, polar processes. Even more drastic was the impact of Pleistocene glaciation.

The significance of the environment and time for soil and landform development can therefore be stated explicitly.

1. Rates of weathering, soil formation, dissection, and denudation vary from one environment to another, with local variations reflecting factors such as rock type, structure, slope, and available relief.
2. Soils and landforms require many centuries or millennia to achieve complex profiles or characteristic forms, with complex assemblages of land form requiring millions of years.
3. Soil mantles on inclined surfaces are being continuously renewed to compensate for the effects of denudation; they may also be completely destroyed during the course of stream dissection.
4. Deep soil profiles once established on flat, upland equilibrium surfaces may exhibit little net change through time; if they reflect past environments and different climates, they may be relict.
5. As dissection and denudation proceed, drainage density, terrain irregularity, and average slope increase at first but eventually decrease as uplands are reduced. If external factors did remain constant, potential energy would decrease steadily. When both average slope and available relief diminish, rates of dissection decelerate.
6. Potential energy may increase or decrease through time as a consequence of relative changes in sea level or on account of tectonic deformation or vulcanism. Long-term changes of this kind can terminate a sequence of development by drowning a subsiding lowland in alluvium or by converting an upraised lowland to an upland interfluve.
7. Pleistocene and Tertiary changes of climate (Figure 16–2) were of sufficient magnitude in most environments to affect rates or balances of weathering and erosion and, occasionally, to change the constellation of dominant geomorphic processes.
8. If the balance of weathering and erosion was different in the past, and geomorphic processes were more effective, contemporary landforms will include inherited features such as nonfunctional depositional surfaces and polygenetic erosional forms.

16-3. CLIMATIC AND HISTORICAL GEOMORPHOLOGY

Both systematic and regional geomorphology in continental Europe have long been affected by application of the environmental approach, the "climatic geomorphology" of the German and French schools. The 1926 meetings of the German geographers' association in Düsseldorf[1] were devoted to the theme of geomorphic processes and erosional landforms as seen in relation to regional variations of climate, vegetation, and soil mantle. The concept itself can be traced back at least as far as the first Russian soils geographers, at the close of the nineteenth century. The impact of this developing approach was a more systematic comparison of contemporary processes and forms with Pleistocene features. This in turn led to the recognition of paleoforms and of the distinct nature of Pleistocene environments and processes well beyond the margins of the glaciers, in mid-latitudes and the subtropics. Attention was turned to modern, unglaciated polar and high mountain environments. Ultimately, this regional and historical data, much of it still at an exploratory level, was integrated into several attempts at synthesis. The development of this approach can be traced in two anthologies (Rathjens, 1971; Derbyshire, 1973) and in the critical evaluation of Stoddart (1969). The single comprehensive statement is that of Tricart and Cailleux (1972), with several volumes on specific environments (see bibliographies to subsequent chapters).

Climatic geomorphology, as it emerges from the corpus of topical and regional literature, has been concerned with the possible recognition of unique assemblages of landforms related to sets of environmental parameters, primarily climate and vegetation. Almost all of the regional analyses have been part of historically oriented studies. Contrary to some opinions, the primary aim has not been to define morphogenetic regions on a world basis. Far from imposing a mental strait jacket on geomorphologic research, climatic geomorphology has provided a stimulating approach to some of the problems of "traditional" geomorphology. It has served to counteract the preoccupation with structural history (prevalent in Europe during the 1920s and 1930s) and now usefully complements the processual approach to hillslope and river development. The emphasis again is on *approach*. There are in fact very few geomorphologists even in Europe who would describe themselves as core practitioners of "climatic geomorphology."

In effect, the climatic-environmental framework is not intrinsically better or worse than other explanatory or predictive models that attempt to impose order arbitrarily, guided by some qualifying assumptions of the complex real world. Climatic geomorphology attempts to cope with the excessive complexity of natural parameters by holding variables such as structure, lithology, and man constant. In much the same way, implicit or explicit models for describing the evolution of

---

[1] Franz Thorbecke, ed., *Morphologie der Klimazonen, Düsseldorfer Geographische Vorträge und Erörterungen*, Breslau, F. Hirt, 1927, Vol. 3, 100 pp.

stream channels or drainage basins commonly are used to make simplifying assumptions that eliminate considerations of time, history, and sometimes even progressive change.

Another criticism that has been leveled at the climatic-environmental approach is that it is insufficiently quantitative. The fact is that quantitative data on form or process are readily available only on the localized scale of a river channel or a small watershed. On this microscale, most environmentally oriented geomorphologists also work with quantitative-analytical techniques such as sedimentology, soil analyses, and slope measurements. But no subfield of geomorphology has yet been successful in producing quantitative generalizations that adequately characterize either large drainage basins or regional landscapes. The challenge and the need to do so remain. In the meanwhile it is essential to keep the overarching problems of composite landscape analysis in constant view. These are the goals, and until they are indeed attainable by quantification, objective qualitative characterization is the next best thing.

The climatic-environmental approach has long ceased to be a European hobby. The basic perspectives, as opposed to premises, have been widely adopted in most other schools of geomorphology. This can, for example, be gauged by the flurry of such research activity in arid and tropical regions since about 1960. In many ways this approach has drawn attention to the fact that landforms are among the environmental assemblages that show planetary zonation. It is, therefore, a peculiarly geographical means of research. It lends itself equally well to concepts of three-dimensional ecozonation in the Earth's mountain belts. On account of its integration with ecological systems it is ideally suited for studies of man and the environment. Last but not least, the climatic-environmental approach has had a salutary effect on historical geomorphology, helping to redirect research to an analysis of features in their own right, thereby counteracting the mechanical, cyclic interpretation of polygenetic landscapes.

The important distinction between history and evolution in geomorphology must be explained. It is a matter of both time and approach. Historical geomorphology implies real time, as measured by isotopic or indirect semiquantitative techniques; such time is not to be confused with the Davisian equation of "stage of evolution" (read geometry) with time. Historical geomorphology seeks to record a stepwise evolution of past cause-and-effect systems, attempts to put these into a real temporal framework, and, ideally, stresses their role as points of departure for discussion of the contemporary system of process and form. By contrast, the evolutionary approach assumes systematic change through time, temporarily reversed in the wake of cyclical repetition, but with change otherwise considered as inevitable and progressive.

16-4. INHERITED SOILS

The close interdependency of weathering rates, soil formation, and geomorphic trends (downwearing versus backwearing) implies that soils and geomorphology are inextricably linked. This applies to both contemporary and historical situations. The concept of inherited soils has already been introduced. It is now appropriate to explain and amplify on the evolution of soils through time.

Many of the processes of soil formation are essentially irreversible, and the products of such processes persist long after a climatic change. In particular, deep, intensively weathered profiles with multiple horizons that escape erosion are seldom more than modified by subsequent weathering and pedogenetic processes, unless these happen to be of even greater intensity and duration. Unlike plant communities, which can be exterminated or replaced through time, soil profiles cannot be removed, except by erosion, and they cannot be totally replaced, except by pedogenesis of appreciably greater intensity and at least equal persistence.

The morphological changes of any given soil profile in response to ecological shifts will depend on the nature of the existing profile as well as on the direction of climatic change. Therefore, the record will favor more intensive kinds and rates of weathering. The soil profile cannot change instantly, and may never respond to new ecological conditions in order to become a new zonal soil. Consequently, individual morphological traits or dominant attributes, sometimes even the essential profile itself, may be unrelated to the prevailing environment. For this reason there are both *relict* soils (dominantly or entirely reflecting past conditions) and *polygenetic* soils (showing one or more distinctive traits reflecting past conditions).

Soil formation is seldom characterized by *uniform rates through time,* nor is it *undirectional through time.* Since the Holocene as well has been characterized by repeated if moderate ecological change (see Figure 18–5), every surface soil dating back beyond perhaps 3,000 years has at least been modified by variable rates of soil formation in response to small- or intermediate-scale environmental changes and small- or large-scale cultural disturbance. The majority of soils on surfaces exposed for at least 10,000 years have been significantly modified by major, Pleistocene environmental changes (climate, biota, drainage). And many of the soils on Tertiary surfaces, exposed for 2 million years or more, may reflect to a large degree upon one or more ecosystems unlike those prevalent today. In other words, a deep ("mature") soil profile represents a net accumulation of weathering products that more often than not is the cumulative result of different soil-forming trends and balances. Such a profile normally persists longer than the conditions or sum of conditions responsible. Soil development is a multivariate phenomenon that is frequently polygenetic and possibly characterized by convergent evolution within any one environment.

In effect, soil profiles may then be monogenetic or polygenetic. If they are the latter, they may show one or more subsurface horizons dominantly modeled by an episode of past climate and that will not develop under existing pedogenetic conditions. In this case only the topsoil may be significantly modified by contemporary processes, even though the entire profile would appear to be in a steady state with the existing environment. Such properties define a *relict* soil. Obviously distinct is a *buried* soil. Finally, older soils may also be destroyed by erosion but partially reconstituted elsewhere as soil sediments.

Relict, buried, and derived soils are grouped as *paleosols*. They owe their origins to differing balances of erosion and weathering, to intervening periods of erosion and sedimentation, or to changes of regional ecology with intensification or diminution of the pedogenetic process. Any of these paleosols, when found at the surface, are significant in terms of land use as well as in terms of contemporary landscape ecology. They consequently merit serious study in the same way as do patently "modern" soils, particularly since the available information from many environments does not yet allow a safe distinction between modern and "paleo" soils.

16-5. LANDSCAPE STABILITY AND GEOMORPHIC PERIODICITY

Studies of historical geomorphology demonstrate that most world regions have been subject to repetitive or cyclic change. Large expanses of mid-latitudes were alternatingly glaciated and ice free during the course of the Pleistocene. River systems draining these ice sheets alternately alluviated broad sandy-to-gravelly floodplains and then incised their valleys to form narrower floodplains mantled with flood silts. Analogous terrace systems, but of different origins, accompany most streams in now arid environments. Immobile fields of dunes or sheets of loess cover extensive subhumid landscapes in mid- and lower latitudes; they are now weathered by deep soils and fixed by vegetation; closer examination reveals multiple alternations of eolian activity and soil formation in the past. Finally, some landscapes also include multiple flat upland surfaces whose margins are now being eroded by lower-lying river plains. There is then a built-in periodicity to landscape development, and historical geomorphology is concerned with the recognition, dating, and interpretation of such events.

The most common impulses responsible for this tendency to "stop and go" include climatic change, fluctuations of relative sea level, and orogenic deformation. A general concept that provides useful perspectives for periodicity in ecosystems in general and geomorphology in particular was first developed by Erhart (1956). Based on African experience, Erhart argued for alternating periods of biological equilibrium (*biostasis*) and disequilibrium (*rhexistasis*) in order to explain the soils record through geological time. Biostasis, on the one hand, is characterized by maximum development of vegetation, optimal infiltration of water, deep chemical weathering, and limited erosion. Leaching and eluviation are favored, including mobilization and removal of bases, colloidal silica, and possibly also iron and manganese. Residual components include resistant heavy minerals, kaolinite, quartz, gibbsite (aluminum hydroxide), and ferric oxides. The impact on the pedogenetic record is the development of deep residual soils. Erosion is sufficiently slow and gentle to provide mainly ions and suspended matter for sedimentation, and coarse clastic components are rare. Rhexistasis, on the other hand, leads to disruption of this equilibrium, with disappearance of the vegetation "filter" as a result of climatic change, increased potential energy, or human interference. The result is a period of erosion, leading to net destruction of existing soil mantles and favoring coarse clastic sedimentation.

Erhart's model, like any other model, is oversimplified and not universally applicable. But it can be readily modified to be of considerable value as a principle of landscape periodicity:

1. Periods of landscape stability are characterized by an effective mat of vegetation, a balance between weathering and denudation, and development of relatively deep and continuous soil mantles. Such *morphostatic* conditions are today characteristic of undisturbed tropical forests, mid-latitude woodlands, and grasslands.

2. Periods of landscape instability are characterized by attrition or destruction of the vegetation mat, an active slope balance, with denudation at least initially outstripping weathering, thus promoting widespread denudation and possible truncation or even total removal of existing soils. Such *morphodynamic* conditions are accompanied by frequent lateral transfers of soil and regolith and a trend to coarse-grade sedimentation on lower slopes or valley floors or both. In a qualified way, geomorphic dynamism is at a premium in polar and arid regions and, above all, in glaciated areas.

Periodicities would occur in the geomorphic record as result of ecological shifts between woodland or grassland on the one hand and icecap, polar tundra, or semidesert on the other. At least that is what all analogies of human interference, such as deforestation and cultivation, imply for the Pleistocene sedimentary record.

The various principles discussed in Chapter 16 will be applied in the subsequent sections to analyze the more representative geomorphic landscapes, in terms of both their contemporary environmental systems and their past heritage. This environmental-genetic approach complements the structural-topographical chapters (Chapters 14 to 15, Appendix B).

16-6. SELECTED REFERENCES

Ahnert, Frank. Functional relationships between denudation, relief, and uplift in large mid-latitude drainage basins. *American Journal of Science,* Vol. 268, 1970, pp. 243–263.

Büdel, Julius. Das System der klima-genetischen Geomorphologie. *Erdkunde,* Vol. 23, 1969, pp. 165–182.

Bunting, B. T. *The Geography of Soil.* London, Hutchinson, and Chicago, Aldine, 1965, 213 pp. See Chap. 6.

Butler, B. E. Periodic phenomena in landscapes as a basis for soil studies. *Commonwealth Scientific and Industrial Research Organization* (Melbourne, Australia), *Soil Publication* No. 14, 1959, pp. 1–20.

Butzer, K. W. *Environment and Archeology,* 2nd ed. Chicago, Aldine, and London, Methuen, 1971, 703 pp. See Chaps. 18 to 23 for evidence of Pleistocene environmental change.

———. Pluralism in geomorphology. *Proceedings, Association of American Geographers,* Vol. 5, 1973, pp. 39–43.

———. Pleistocene climates. In R. C. West, ed. *Ecology of the Pleistocene. Geoscience and Man,* Vol. 13, 1976, in press.

Derbyshire, Edward, ed. *Climatic Geomorphology.* London, Macmillan, and New York, Harper & Row, 1973, 296 pp.

Dury, G. H. Relict deep weathering and duricrusting in relation to the paleoenvironments of middle latitudes. *Geographical Journal,* Vol. 137, 1971, pp. 511–522.

Erhart, H. *La Genèse des Sols en tant que Phènomène géologique;* 2nd ed. Paris, Masson, 1967, 177 pp.

Ganssen, Robert, and F. Hädrich. *Atlas zur Bodenkunde.* Mannheim, Bibliographisches Institut, 1965, 85 pp. (with key in five languages).

Holeman, J. N. The sediment yield of major rivers of the world. *Water Resources Research* 4, 1968, pp. 737–747.

Jansen, J. M. L., and R. B. Painter. Predicting sediment yield from climate and topography. *Journal of Hydrology,* Vol. 21, 1974, pp. 371–380.

Meade, R. H. Errors in using modern stream-load data to estimate natural rates of denudation. *Bulletin, Geological Society of America,* Vol. 80, 1969, pp. 1265–1274.

Rathjens, Carl, ed. *Klimatische Geomorphologie.* Darmstadt, Wissenschaftliche Buchgesellschaft, 1971, 485 pp.

Selby, M. J. Rates of denudation. *New Zealand Journal of Geography* 56, 1974, pp. 1–13.

Stoddart, D. R. Climatic geomorphology: review and re-assessment. *Progress in Geography,* Vol. 1, 1969, pp. 159–222.

Trewartha, G. T. *An Introduction to Climate,* 4th ed. New York, McGraw-Hill, 1968, 408 pp.

Tricart, Jean, and André Cailleux. *Introduction to Climatic Geomorphology.* Translated from the French original of 1965 by C. Kiewiet de Jonge. London, Longmans, and New York, St. Martin, 1972, 274 pp.

Trimble, S. W. Denudation studies: can we assume stream steady state? *Science,* Vol. 188, 1975, pp. 1207–1208.

Wilhelmy, Herbert. *Klima-geomorphologie in Stichworten.* Kiel, Hirt, 1975, 374 pp.

Chapter 17

PERIGLACIAL LANDFORMS

Periglacial geomorphology refers to the unglaciated and treeless environments of high latitudes and high elevation. Intensive soil-frost dynamics, particularly over permanently frozen subsoil, create some totally unique landforms while accentuating certain varieties of mass movements, surface runoff, and stream discharge. The result is a complex assemblage of small, medium and large-scale landforms tied to lowland plains, slopes, or upland surfaces. Although many of these features are not restricted to periglacial settings, their associations are distinctive and, when selectively superimposed upon a given terrain, produce characteristic landscapes. Periglacial lowlands on permafrost are swampy, dotted with thaw lakes, crisscrossed by giant polygonal networks, occasionally punctuated by dune fields or groups of ice-cored mounds, and swept by braided rivers. Periglacial uplands are rocky, veneered with a variety of highly mobile sedimentary forms, and broken by steep, straight slopes that terminate in subangular inflections. Backwearing appears to be the dominant trend.

17-1. THE PERIGLACIAL ENVIRONMENT

The Arctic flight of the airship *Graf Zeppelin* in 1928 first revealed the truly bizarre landscape of the polar world. In some areas there were flat plains stretching from horizon to horizon that were dotted with innumerable and inexplicable lakes. In other regions, linear gashes up to a mile or more long intersected to form giant polygonal networks. This bird's-eye view confirmed what were then only incidental surface impressions that unglaciated polar environments were very unusual. A first comprehensive analysis of such landforms was attempted in 1944 (Troll, 1958), and the true scope of *periglacial* geomorphology has only unfolded during the past three decades.

The word "periglacial" was originally devised in 1906 to describe features next to the ice, as the name infers. It has since been extended in a number of, in part contradictory, ways to unglaciated polar and alpine environments in general or to the impact of geomorphically effective soil and rock frost in particular. Despite any ambiguities, "periglacial" does conjure up an impression of extraglacial, polar landscapes. The definition of convenience proposed here follows the argument that vegetation, soils, and geomorphic balance are interrelated: Periglacial geomorphology refers to those landscapes delimited by tundra, alpine and rocky soils beyond the polar treeline or above the alpine timberline. Such a demarcation is difficult to quantify in terms of climate or plant ecology, but it is realistic in singling out a striking geomorphic environment.

The basic environmental parameters that contribute to periglacial processes are preeminently a matter of ground and subsoil climate. These can be outlined for the case of the high-latitude periglacial realm:

1. Winters are long and cold, with protracted or daily frost for 10 months of the year. The snow cover is commonly shallow and frequently incomplete, so that seasonal frost penetrates deeply into the soil (2 to 15 ft, or 0.6 to 4.5 m).

2. During the modest summers subsoil temperatures at a depth of 4 inches (10 cm) seldom rise out of the 40s (5 to 10° C), even when ground surface temperatures are 75 degrees F (24° C) or more. Little or no chemical weathering and, above all, practically no hydrolysis are possible under such low-energy conditions. However, solution of carbonates is active as long as soil and rock temperatures remain above 32 degrees F (0° C).

3. As a result of the short summers, the annual thaw is commonly inadequate to melt the winter frost. The subsoil of about 95 percent of the treeless polar environments remains permanently frozen (*permafrost*) (Figure 17–1). Above the permanently frozen subsoil there is a seasonal *thaw zone*, varying in depth from as little as an inch or two (5

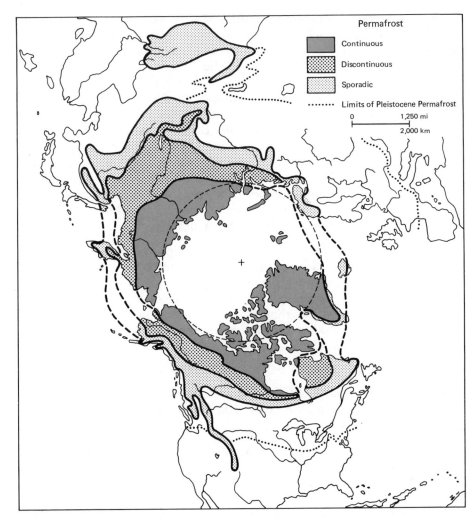

Figure 17–1. Modern Distribution of Permafrost, with Limits of Pleistocene Permafrost. Compiled from various sources.

cm) in boggy soils to as much as 15 feet (4.5 m) or more in well-drained, sandy soils. Below this thaw zone the permafrost layer may have a thickness measured in tens, hundreds, or even thousands of feet.

4. The presence of permafrost or other deeply frozen soils impedes soil drainage during the short summer season. As a consequence of these high water tables, soils tend to be waterlogged almost everywhere, with widespread development of lakes and swamps. Similarly, deep rooting of plants is impeded.

5. Deep frost favors active mechanical weathering in the subsoil, as well as frost shattering of rock outcrops. At the same time, initial freezing of the thaw zone each autumn leads to a variety of frost or cryostatic pressures that produce mixing, churning, or heaving of the regolith. One mechanism of mixing is a result of the differential conductivity and expansion of stones compared with those of wet, fine soil aggregates. Another is the formation of ice crystals underneath stones when frost penetrates the soil; as such crystals grow into ice needles, stones are pushed up toward the surface; thawing ice is later partly replaced by fine soil. Churning or heaving results when pressures build up in the subsoil as the frost penetrates and expands from above; these pressures may squeeze muddy oozes laterally or even vertically, where they can tear open the frozen subsoil. Other forms of disturbance result from solifluction and auxiliary types of mass movements.

Although permafrost is spotty in occurrence among the high mountains of mid- and low latitudes, the above conditions apply in some degree or other to alpine environments in mid-latitudes. However, in the tropics, freeze-thaw alternations occur on a daily rather than a seasonal basis, so that frosts seldom penetrate more than 2 inches (5 cm); soil mixing is correspondingly restricted to the topsoil.

17-2. WEATHERING AND SOIL FORMATION

The air and soil climate of the polar and alpine world strongly inhibits hydrolytic and biochemical weathering but favors mechanical disintegration. The results are that:

1. Minerals are altered slowly; few fine silts and clays form, and soils are mainly sandy.

2. Mechanical weathering provides abundant rock waste but little finer than sand-sized particles. Together with the mixing effects of soil frost, this favors sandy-stony soils, with limited development of soil horizons of distinctive texture. Patterning within the soil profile or at the surface mainly reflects sorting due to frost activities or mass movements, not soil development.

3. Excessive moisture at most times, together with unaerated, waterlogged, or permafrost horizons at shallow depths, limits root penetration, deep profile development, and illuviation in all but well-drained soils.

4. The restricted biotic activity, by those bacteria and certain fungi that tolerate this cold and waterlogged environment, is insufficient to decompose organic matter, which accumulates through time and becomes increasingly acidic. Raw humus and peats are therefore common.

5. Most periglacial environments were repeatedly glaciated during the Pleistocene. Consequently, with such low-intensity soil development, profiles remain shallow and often undistinctive.

The most representative tundra soil on properly drained surfaces is the *arctic brown soil* (see Figure 18–1,A). An average profile might look as follows:

O2 horizon: 0–2″ (0–5 cm) thick. Dark-brown to black dense mat of roots and plant stems. Very acid. Some fine sand and silt.

A horizon: 12–20″ (30–50 cm) thick. Dark-brown sandy soil with abundant organic matter and roots. Acid.

C horizon: variable thickness. Brownish or yellowish stony soil, over compact rock or permafrost. Neutral reaction.

If drainage is exceptionally good and climate sufficiently wet, there may be evidence of weak podsolization. Such profiles would show very thin E and B horizons, both seldom more than 1 inch (2.5 cm) thick, interjected just below the A.

At very high latitudes in the Arctic, beyond 70 or 75° N, vegetation cover is at a minimum and the wind strips off fine soil and loose humus. The resulting lithosols have an incomplete 02 horizon, resting right on the C. Permafrost is found at 1 or 2 feet (30 or 60 cm) beneath the surface.

Arctic brown soils, whether podsolized or not, are relatively rare in the tundra world and seldom account for as much as 10 percent of the surface. Because of the rapidity of mass movements and frost weathering, moderate and steep slopes are usually bare of soil. Similarly, crude rock rubble abounds wherever rocks outcrop.

Last but by no means least, the extensive lowlands are in general poorly drained, giving rise to a variety of waterlogged or bog soils. Where lateral drainage is poor, *tundra gleys* develop with a thick g horizon beneath the A horizon. Typically, these gley horizons are gray in color, sometimes mottled with rusty stains, and often foul smelling. During the thaw season, they turn to an almost liquid mud, prone to frost churning and "squeeze-outs" in the autumn. When drainage is completely impeded, *bog soils* develop with a thick, highly acid 01 horizon of peat moss (8–40 in., 20–100 cm thick) resting on top of permafrost. Because of frost heaving the surface is often hummocky. Such peat hummocks or *palsas* may have a relief of 3 to 22 feet (1 to 7 m) or more and a diameter of 30 to 150 feet (10 to 50 m), with a core of churning mud, peat sludge, or ice.

In the case of alpine environments, poor drainage is far less common, but lithosols or bare rock are widespread. *Alpine turf soils* are closely comparable to arctic brown soils in most aspects.

17-3. PERIGLACIAL PROCESS AND FORM

The gradational agents of the tundra and other treeless polar and al-pine environments are intensive and effective. They do not match with the potential of glacier ice, but collectively they rank second best. For most of the year the ground is frozen solid, and, except for possible wind erosion, little change takes place. But during the first month or two of the thaw season, mass movements and running water become remarkably active. Abundant water is supplied by the melting snow cover and by fresh precipitation. Later, as the thaw season comes to an end, refreezing of the soil from above recreates the frost dynamism that heaves, mixes, and sorts the soil (*cryoturbation*).

Overland Flow and Throughflow. Melting snow and rain provide sur-face runoff that cannot percolate into the soaked or frozen soil and that is hardly diminished by evaporation. The vegetation mat is thin, incomplete, or absent on intermediate slopes, thus accelerating runoff. Finally, thawing soil is highly erodible—it is loose, uncohesive, and saturated with water. If there is an incomplete A horizon, throughflow flushes the stony subsoil to wash out fine materials. As a result, diffuse runoff is effective in rill cutting, sheet wash, and subsoil removal.

Cryoturbation Phenomena. Freeze-and-thaw alternations, particularly the intensive form typical over permafrost, favor lateral segregation of rock and soil. The rock waste pushed outward from pockets of soil and needle ice is not retracted after thaws, whereas finer sand and silt are. The cumulative effect is to create stone rings that begin to concentrate in polygonal networks on level ground. Throughflow can then flush out

Stone Rings → Stone Garlands → Stone Stripes

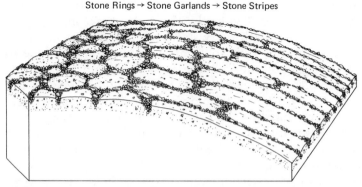

Figure 17–2. Small-Scale Patterned Ground with Stone Rings on Level Ground, Stone Garlands on the Gentle Convexity, and Stone Stripes on the Slope. (From C. F. Stewart Sharpe, *Landslides and Related Phenomena*, New York, Columbia University Press, 1938. With permission of the author.)

the remaining fine matrix. The resulting stone rings form *patterned ground* that may eventually extend vertically down to the base of the annual freeze-and-thaw layer (Figures 17–2 and 17–3). Similar selective segregation can affect the vegetation mat as uneven frost heaving creates tears that are selectively eroded by deflation and diffuse runoff. This type of patterned ground has high centers of turf surrounded by low polygons of stony soil.

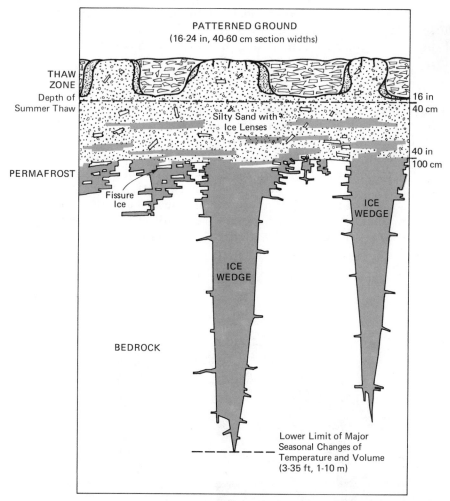

Figure 17–3. Patterned Ground, Permafrost Contacts, and Ice Wedges. This generalized Spitsbergen example, based on Büdel (1961), includes a thick zone of silty soil that was seasonally thawed at a time of warmer climate in the recent past. These particular ice wedges are therefore "fossil," since they do not extend to the top of the permafrost zone (see Figure 17–4).

More conspicuous ground patterns form on wet ground in the heart of the permafrost belt. Intense winter cold (at least − 2° F, or − 15° C) leads to contraction in the upper permafrost horizon, creating minute fissures. These subsequently fill with water, extracted from the adjacent soil or percolating from above, that subsequently refreezes. Each year the crack reopens, fills, and freezes, creating a wedge of ice (Figures 17–3 and 17–4). These can grow to a width of 2 feet (60 cm) or more and depths of 3 to 15 feet (1 to 5 m) and more. Such ice wedges are almost never found where mean air temperatures are above 32° F (0° C), and they actively form only in the continuous permafrost belt. Isotopic dating indicates that most existing ice wedges began to form during Pleistocene times. Also, inactive or filled-in ice wedges outside of the continuous permafrost zone imply former mean annual temperatures below 18 to 22° F (− 6 to − 8° C). Active ice wedges have conspicuous topographic expression, with raised rims creating a relief of 3 feet (1 m) or more. Since active ice wedges commonly crisscross the surface, they intersect to form *ice-wedge polygons* (see Figure 17–6) with typical diameters of 65 to 350 feet (20 to 100 m).

The uppermost permafrost horizon is rich in moisture, in the form of subhorizontal ice lenses. In wet, lowland plains as much of 75 percent of the volume of the topmost 10 feet (3.5 m) of the permafrost is represented by such segregated ice. Consequently, if the permafrost table is somehow lowered, liquid water is released and the ground surface sags. Such melting can happen as a result of human disturbance (deforestation, vehicle tracks, house construction), periodic surface flooding, an unusually warm summer, or climatic change. The impact

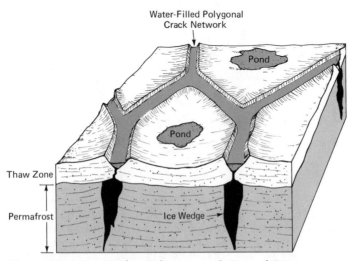

Figure 17–4. Ice-Wedge Polygons with Raised Rims.

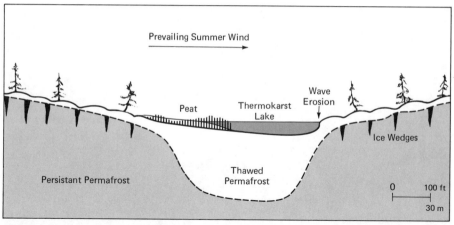

Figure 17–5. A Thermokarst "Thaw" Lake.

is creation of a *thermokarst topography*, including large circular depressions, sinkholes, or lakes. Once established, such lakes enlarge by wave erosion and intensified, local permafrost melting to depths of 200 feet (60 m) or more. Swarms of thermokarst "thaw" lakes (Figure 17–5) are characteristic of many flood and coastal plains (see Figure 17–8). Another feature of wet ground is the forcing up of liquid water into and through the permafrost. This water freezes as it pushes up the topsoil, forming mounds with a core of ice, liquid mud, and water. The principle is that of an ice laccolith, and the resulting feature is called a *pingo*. Relief can be as much as 300 feet (90 m), although much smaller varieties are far more common. Pingos can form in a few years, although others, in Siberia, have formed landmarks for three centuries or more. Pingos can deteriorate and collapse (Figure 17–6) if the water supply is cut off or when permafrost melts out. This leaves an open-centered, circular rim with a diameter of up to several 100 yards or meters.

A last cryoturbate feature of landform status is provided by *string bogs*, a widespread surface type in the forest-tundra mosaic. Elongated, wavy lines of turf run approximately parallel to the contours on gentle inclines, damming back water. Origins are uncertain and clearly complex, and string bogs are common outside of the permafrost belt.

Nivation. Snowpatch processes are locally important in snowline proximity. Seasonal or perennial snowpatches and firn create erosional hollows: Frost weathering is accentuated around their margins, while mass movements continue to erode their floors. Dirty snow avalanches may additionally build up nivation ridges that simulate small

Figure 17–6. Collapsed Pingo Among Ice-Wedge Polygons on Old Lake Floor of the Mackenzie Delta. (Courtesy of T. L. Péwé.)

end-moraine arcs (see Figure 5–8). In addition to forming *nivational niches* of variable size, snowpatch erosion below rock ledges can combine with accelerated mass movements to form stepped, *altiplanation terraces* in mountainous terrain (Figure 17–7).

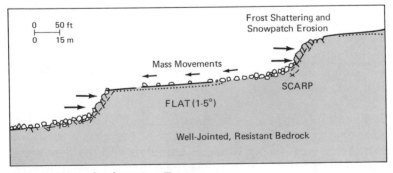

Figure 17–7. Altiplanation Terrace.

Mass Movements. Plastic flow of supersaturated soil is characteristic of polar slopes with gradients of 2 to 20 degrees. With little or no percolation possible in water-soaked regolith, particularly above permafrost, the thawed soil mantle becomes a muddy ooze. Large-scale solifluction (see section 5-5), affecting entire hillsides at appreciable rates, is the most typical form of denudation and generally acts to obliterate rills on gentle to intermediate slopes. Patterned ground is gradually elongated to form stone garlands where solifluction affects slopes as gentle as 2 degrees or less (Figure 17–2). If gradients exceed 6 degrees or so, the garlands break off and parallel lines of stone run downhill, perpendicular to the contours. Such *stone stripes* become increasingly disorganized on inclines of 15 degrees or so and are generally destroyed by active sheet and rill erosion when slopes approach 25 degrees. Where patterned ground is absent, solifluction with microslumping of the vegetation mat may produce terracettes (section 5-5). Solifluction can locally involve large masses of soil and regolith in repeated movements that develop into lobate tongues. Such *solifluction lobes* may approach earthflows in their development. In fact, rapid earthflows and local debris slides are not uncommon on suitable substrates with permafrost and moderate inclinations.

On steeper slopes talus development and rockfalls are active, sometimes developing into block fields (see Figure 3–2) or rock glaciers or continuing downhill as ridges of boulders that slide on muddy substrates, with slope inclinations of as little as 2 to 5 degrees. The overall effect of well-developed mass movements on permafrost is to create a host of peculiar slope phenomena.

Fluvial Features. Streams are provided with abundant sediment by overland flow and mass movements. Flow is markedly seasonal and correspondingly torrential. Additional sands and gravel are added to the load by bed erosion and undercutting of thawing, frost-riven bedrock. Braiding is common, in response to a preponderance of bedload sediments. As a result of high competence, turbulence, and an abrasive load, streams are hyperactive and efficient at both dissection and lateral planation in bedrock. The shallow, sandy or pebbly floodplains resemble those of outwash streams.

Some stream valleys of intermediate order are systematically steeper on one side than on the other, forming *asymmetric* ridges or valleys. Such asymmetry can have many possible origins, although differential exposure to snow-bringing winds or radiation is of possible regional importance in the periglacial world. In general, shade slopes tend to be snowbound and frozen longer than sunny slopes, favoring undercutting and steepening of the thawed, sunny side.

Common to some periglacial areas are *dry valleys* at the head of first-order drainage lines cut into bedrock. These are attributed to a combination of rill cutting, sheet wash, and mass movements, with solifluction repeatedly obscuring any incipient channel. Such dry valley heads may also converge with or simulate nivational niches.

Eolian Phenomena. Wind, or eolian, erosion involves both deflation of fines and sandblasting of surface rock or pebbles. The effect is to concentrate a residual *lag* or pavement of crude rock or pebbles on the surface, many of which have polished or faceted edges due to wind abrasion (*ventifacts*). Deflation attacks all rocky surfaces but derives most of its material from sandy floodplains during late summer and autumn, as runoff wanes but before a new snow cover insulates the

Table 17–1. CHARACTERISTIC PERIGLACIAL LANDFORMS

LOWLAND PLAINS

Soil-frost features (commonly or exclusively linked to permafrost)
    Stone and vegetation rings (small scale)
    Ice-wedge polygons (medium to large scale)
    Thermokarst lakes and ponds (medium to large scale)
    String bogs (medium to large scale)
    Peat hummocks (palsa) (small to medium scale)
    Pingos (medium to large scale)

Eolian features (not unique to periglacial areas)
    Dune fields (large scale)
    Deflation scars (small to medium scale)
    Lag surfaces (small scale)

Fluvial features (not unique to periglacial areas)
    Braided channels (medium to large scale)

SLOPES

(Features accented by or exclusively linked to frost dynamics)
    Stone rings and garlands (small scale)
    Soil or vegetation terracettes (small scale)
    Stone or vegetation stripes (medium scale)
    Solifluction sheets (medium scale)
    Solifluction lobes (medium scale)
    Earthflows and debris slides (medium scale)
    Block streams (medium scale)
    Block fields (medium to large scale)
    Rock glaciers (large scale)
    Angular, frost-riven facets (small to medium scale)
    Nivation ridges (medium scale)
    Nivation scars and niches (medium to large scale)
    Asymmetric ridges (medium to large scale)
    Dry valley heads (medium to large scale)
    Altiplanation terraces (medium to large scale)

UPLAND SURFACES

Most small- to medium-scale soil-frost features of lowland plains

Medium- to large-scale features can be recognized on aerial photography.

ground. If the deflated material is sandy, dunes may form downwind; if silts predominate, the dust is carried far in suspension, deposited miles away as thin increments of loess particles.

17-4. PERIGLACIAL LANDSCAPES

The composite landscapes that emerge from periglacial sculpture are highlighted by a selection of the appropriate landforms discussed previously and listed in Table 17–1. These features are not all unique to periglacial environments, but the assemblages are.

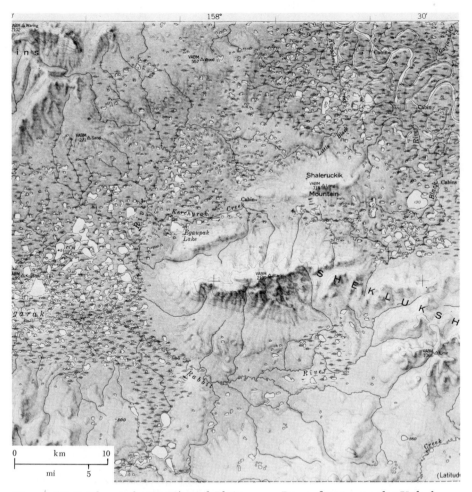

Figure 17–8. Thermokarst Alluvial Plains over Permafrost near the Kobuk River, West-Central Alaska (66°45′ N, 158° W). (Excerpted from 1:125,000 topographic map "Shungnak" of the U.S. Geological Survey, 1958. With permission)

Two basic land-form situations can be singled out: (1) flat to undulating plains created by marine or fluvial deposition, or glacial scour; and (2) low to high mountain terrain originally modified or sculptured by glaciation. In other words, the gross setting (see Figures 17–8 and 17–10) is generally inherited and dates from the Pleistocene or the early Holocene. But in the course of several millennia, periglacial processes have superimposed a colorful assemblage of small- to large-scale landforms to fashion distinctive lowland or upland landscapes.

The lowland variant is marked by incredibly poor drainage and extensive swamp, clusters of thermokarst lakes, great spreads of ice-wedge polyons, occasional groups of pingos, local dune fields, and broad, dynamic, braided streams. Interspersed residual hills or mountains exhibit a host of solifluction forms on their lower slopes, possibly surmounted by accumulations of crude rubble. The overall picture is illustrated by Figures 17–8 to 17–10, exemplifying two landscapes found in no other geomorphic environment.

The upland counterpart consists of slopes, cliffs, terrace facets, and valleys veneered by an unusual variety of highly mobile, sedimentary

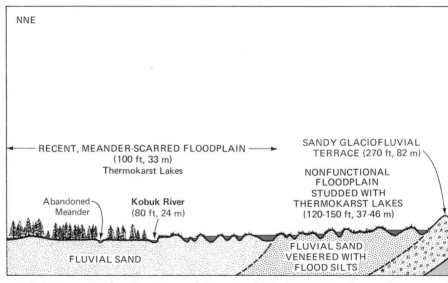

Figure 17–9. Topographic Section and Photogeologic Interpretation to Figure 17–8. Based on 1 : 63, 360 topographic map (Shungnak D-4, 1959) and aerial photography. Thermokarst is locally present both on the active Kobuk floodplain and on remnants of an old outwash terrace; it is best developed on a nonfunctional floodplain averaging 35 feet (10 m) above modern flood. This extensive alluvial surface may relate to more extensive glaciation upstream about 3,000 years ago. Extensive dune fields, deflated from the Kobuk, are found northeast of the modern floodplain.

forms. There is an abundance of bare rock or crude detritus on all convex slopes. Many soils are arranged in circular, elliptical or linear patterns, with lobate or terracette forms stepping up many lower slopes. Slopes tend to be rectilinear and steep, marked by subangular upper and lower inflections. The composite profile of Figure 17–11 is representative of one type of periglacial upland. As slope and relief increase and soil-based depositional forms become scarce, the upland periglacial landscape grades almost imperceptibly into the montane countrysides found in most high mountain ranges at an latitude. These alpine landscapes are clearly more familiar, but they are nonetheless periglacial as well.

In general, given sufficient relief, periglacial valleys deepen and widen as a result of stream undercutting and rapid slope transfer. There is no excessive build-up of soil, regolith, or slope debris, since mass movements and running water are effective on all slope segments. The net effect is that slope angles remain steep and inflections prominent. This is clearly shown by the impact of 10 millennia of periglacial sculpture in former glacial troughs, where *backwearing* is

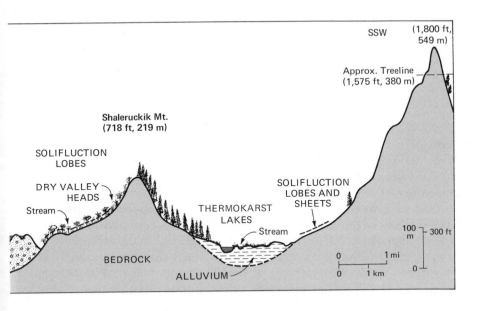

Figure 17–10. Landform Analysis of a Periglacial Landscape, Canadian Arctic Islands (78°45′ N, 103°30′ W). Ice-wedge polygons are shown by several intricate geometric patterns; gentle to steep slopes by various intensities of directional lines; low to high cliffs by thin to heavily marked, barbed lines. Note the extension of silt- to sand-veneered flats (dotted) at several elevations, as well as the broad, braided streams. (Excerpted from D. A. St. Onge, *Géomorphologie des Environs d'Isachsen, Ile Ellef Ringnes*. Ottawa, Geographical Branch, Ministry of Mines and Technical Surveys, 1964. With permission.)

both conspicuous and relatively rapid (see Figure 17–11).

Periglacial landscapes today are restricted to high latitudes and high mountain environments, which are extensive but remain thinly settled. Many of the dynamic processes that operate on hillslopes or plains pose serious obstacles to every human artifice, from house and road building to complex urbanization with all its ramifications. This applies to permafrost in particular. But periglacial geomorphology is far more than an engineering problem in the great northland. It is deeply impressed on the scenery of mid-latitudes, where Pleistocene cold created a legacy of periglacial landforms that often persist to this day.

17-5. SELECTED REFERENCES

Büdel, Julius. Die Abtragungsvorgänge auf Spitzbergen im Umkreis der Barentsinsel. *Abhandlungen, Deutscher Geographentag*, Cologne 1961. Wiesbaden, F. Steiner, 1961, pp. 337–375.

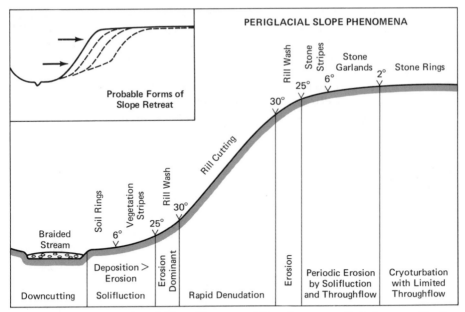

Figure 17–11. Generalized Slope Forms and Development, Spitsbergen. Modified after Büdel (1961). The original valley was a glacial trough.

———. Der Eisrinden-Effekt als Motor der Tiefenerosion in der exzessiven Talbildungszone. *Würzburger Geographische Arbeiten*, Vol. 25, 1969, pp. 1–41.

Butzer, K. W. *Environment and Archeology*. Chicago, Aldine, and London, Methuen, 1971, 703 pp. See Chap. 7.

Davies, J. L. *Landforms of Cold Climates*. Canberra, Australian National University Press, and Cambridge, Mass., M.I.T. Press, 1969, 200 pp.

Embleton, Clifford, and C. A. M. King. *Periglacial Geomorphology*. London, Edward Arnold, and New York, St. Martin, 1975, 240 pp.

Hamelin, L. E., and F. A. Cook. *Illustrated Glossary of Periglacial Phenomena*. Québec, Presses de l'Université Laval, 1967, 237 pp.

Péwé, T. L. *Permafrost and Its Effects on Life in the North*. Corvallis, Ore., Oregon State University Press, 1966, 40 pp.

———, ed. *The Periglacial Environment: Past and Present*. Montreal, McGill-Queen's University Press, 1969, 487 pp.

———, and J. R. Mackay, eds. *Permafrost: North American Contribution, Second International Conference*. Washington, D.C., National Academy of Sciences, 1973, 783 pp.

Price, L. W. *The periglacial environment: permafrost and man*. Washington, D.C., *American Association of Geographers, Commission on College Geography, Resource Paper No. 14*, 1972, pp. 1–88.

Rapp, Anders. Talus slopes and mountain walls at Tempelfjorden, Spitsbergen. *Norsk Polarinstitutt Skrifter*, Vol. 119, 1960, pp. 1–96.

Sharpe, C. F. S. *Landslides and Related Phenomena*, 2nd ed. New York, Pageant Books, 1960, 137 pp.

Suslov, S. P. *Physical Geography of Asiatic Russia.* Translated from the Russian by J. E. Williams and N. D. Gershevsky. San Francisco, Freeman, 1961.

Tricart, Jean. *Geomorphology of Cold Environments.* Translated from the French by E. Watson. London, Macmillan, and New York, St. Martin, 1970, 320 pp.

Troll, Carl. Structure soils, solifluction, and frost climates of the earth. Translated from the German original of 1944 by H. E. Wright. Wilmette, Ill., U.S. Army Snow, Ice and Permafrost Research Establishment, Translation 43, 1958, 121 pp.

———. Der subnivale oder periglaziale Zyklus der Denudation. *Erdkunde,* Vol. 2, 1948, pp. 1–21.

Washburn, A. L. *Periglacial Processes and Environments.* London, Edward Arnold, 1972, 328 pp.

Chapter 18

HUMID LANDSCAPES OF MID-LATITUDES

Humid mid-latitudes can be defined as naturally forested environments with perennial stream flow, situated between approximately latitudes 25 degrees and 65 degrees. Both mechanical and chemical weathering are effective, and continuous, relatively deep soil mantles are the rule. Podsols and bog soils are dominant in the cold, boreal or coniferous woodlands where cryoturbation is active. In the temperate, mixed or deciduous woodlands there are podsolic soils, interspersed with brown forest, calcimorphic, or hydromorphic soils. In the mixed, subtropical woodlands red-yellow podsolic soils are most representative, and the dominance of relatively deep and clayey soils infers a lower threshold for slope instability. In general, denudation under intact woodland vegetation is slow and

primarily a matter of (1) removal of ions and clays by throughflow and groundwater and (2) creep and other, slow mass movements. This ecosystem is morphostatic without human interference, and downwearing would appear to be the dominant trend in slope development. Deforestation, grazing, and cultivation greatly accelerate denudation, by rainsplash, sheet wash, and gullying. Such land use over many millennia in Europe and the Far East, and since the eighteenth century in North America, have greatly modified the soil landscape and the pattern of geomorphic processes, particularly in the subtropical woodlands. As a result, denudation now is relatively rapid and may favor backwearing of slopes. The passage of diverse paleoclimates in Tertiary, Pleistocene, and Holocene times requires careful appraisal of the landform assemblages in humid mid-latitudes for inherited traits or hybrid forms. Glacial and periglacial imprints are distinctive and are only very gradually modified under morphostatic conditions. Nonetheless, it remains difficult to evaluate the nature of polygenetic landscape evolution in the Pleistocene periglacial zone. Equally interesting but more uncertain is the survival of planation surfaces and inselbergs in Europe and the American Piedmont Plateau that may show influences of sculpture in a semitropical, Tertiary environment. Although the details are not satisfactorily understood, the landform assemblage of humid mid-latitudes cannot be adequately explained by contemporary processes alone.

**Still glides the stream, and shall forever glide;
The form remains, the function never dies**

WILLIAM WORDSWORTH, *Afterthought*

18-1. PERSPECTIVES
ON HUMID MID-LATITUDES

Most standard geomorphologic concepts evolved in temperate mid-latitudes. The work of the pioneer geologists of the eighteenth and early nineteenth centuries reflected the environments of Britain and continental Europe, and Davis' cycle of erosion was based on a perceived model for the eastern United States. The experience of Richthofen and Penck was gained in similar settings of Europe and China, and Powell and Gilbert in the arid American West worked primarily along perennial rivers, fed by discharge from more humid mountain ranges. Specific investigations of desert, tropical, and periglacial settings began about 1900 and were never adopted into the conceptual framework of general geomorphologic theses. Even general texts of recent vintage are written with the bias of humid mid-latitudes and continue to consider other environments anomalous.

As a definition of convenience, humid mid-latitudes can be delimited by simple criteria: (1) Latitudinal location between 25 and 65 degrees, where cool winters alternate with warm summers, and (2) humid climate, as characterized by a natural vegetation cover of woodland and by perennial stream flow. On the world soils map (Figure 16–1) these criteria would demarcate about one-third of the continental area of North America and Eurasia, including those ecounits labeled as boreal, temperate and subtropical woodlands. In actual fact, these areas support about 75 percent of the world population and are intensively modified by human activity.

In terms of contemporary processes, the humid mid-latitude environments are broadly similar. Both chemical and mechanical weathering are effective. Soil processes are sufficiently intensive to produce continuous and distinctive soil mantles that mainly exhibit some degree of podsolization. A wide range of mass movements and fluvial processes operate throughout the year, except when the hydrosphere is frozen. Ironically, these same environments also exhibit a complex legacy of glacial, periglacial, or tropical landforms dating from the Pleistocene or the Tertiary. Perhaps in no other setting is the disjunction of modern processes and inherited forms as apparent. Consequently, humid mid-latitudes are far from ideal for synthetic models of composite landscape evolution. Yet even now most of the available observational data linking process and small- to intermediate-scale form comes from this one ecological system. It is therefore essential that ongoing research that is directed toward a better synthesis should (1) recognize differences of scale, (2) evaluate the quantitative role of human disturbance, and (3) develop objective criteria to distinguish inherited and contemporary landform components.

18-2. WEATHERING PATTERNS

Humid mid-latitudes span a range of climatic provinces between Hudson Bay and the Gulf of Mexico in North America, or between Lapland and the Riviera in Europe. Yet in all respects the impacts of mechanical and chemical weathering are "intermediate."

Frost is commonplace, even if sporadic, and frost weathering is certainly prominent on exposed rock in the boreal and temperate woodlands. Soil frost also plays a role in accelerating mass movements in those regions cold enough to support a winter snow cover, and cryoturbation is active in the permafrost belt that reaches well into the boreal woodlands.

Chemical weathering is possible whenever water is present in liquid form, so that the duration of the frost-free season is important. Solution of carbonates can operate at all temperatures above 32° F (0° C). However, the key agents of hydrolysis and oxidation are seldom effective at low temperatures, particularly when monthly averages are below 50 or 60° F (10 or 16° C). The period of major chemical weathering is therefore restricted to less than 3 months in the boreal woodlands but increases to a range of between 4 and 10 months in warmer or more maritime situations. On the whole, depth and intensity of chemical weathering are moderate.

Perhaps the most characteristic process of chemical weathering in the mid-latitude woodlands is podsolization (see section 4-5). This dissociation of clays, with eluviation of sesquioxides and clay-humus from the topsoil, is related to a specific biological microenvironment made possible and partly conditioned by climate and parent material. This microenvironment is acidic (pH ≤ 5), with organic acids produced by lichens and fungi or released by roots and decaying organic compounds. Some of these acids combine with the iron, aluminum, magnesium, and silicon ions in the soil and then form mobile, organic-mineral combinations that move down in the profile until pH is once again above 5. At this point they are precipitated in the subsoil (see Figure 4–3).

The immediate factor favoring an acidic soil environment is vegetation that requires few mineral bases from the soil and returns correspondingly few. Most conifers and heath plants release resins and acids as their litter decays. These do not break down into mild humus, since acidity is uncongenial to most of the more beneficial soil microorganisms. Parent materials of high permeability favor percolation and accelerate podsolization; similarly, parent material deficient in mineral nutrients is prone to podsolization. The ultimate influence is climatic, since the overall distribution of coniferous vegetation is more or less climatically controlled, even if only in a very generalized way.

Podsolization is a biochemical process that becomes a zonal characteristic in well-drained coniferous forests of cooler climates. In warmer mid-latitudes and the tropics, advanced podsolization is restricted to parent materials exceptionally poor in mineral bases or unusually permeable. Thus, also, in sandy soils the planting of commercial conifers has promoted or accelerated podsolization, both in western Europe and the United States. In these warmer settings, deeper and more intensive hydrolytic weathering is favored as summers become long and hot. Ultimately, in the humid, subtropical climates of the southeastern United States and southern China, the coincidence of the wet season with the warmer half of the year permits five to seven months of high-intensity chemical weathering. At this point, podsolization as a biochemical process becomes subordinate to other chemical processes that also are effective well down into the regolith.

18-3. THE SOIL LANDSCAPE

Soils are an important component of the geomorphic landscape in humid mid-latitudes. The more representative profiles and other relevant soil properties can be briefly outlined with reference to Figure 16–1.

Podsols and Bog Soils of the Boreal Woodlands. Characteristic of fully podsolized soils is the white to ash-gray E horizon, from which the Russian designation *podsol* (ashen soil) is derived. An average profile might look as follows (Figure 18–1,B):

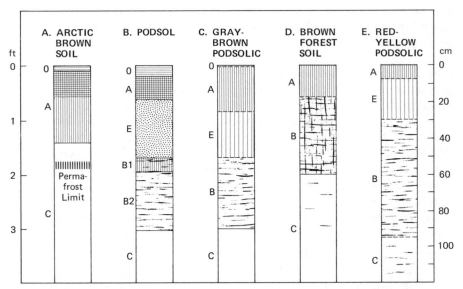

Figure 18–1. Representative Soil Profiles of Periglacial and Humid Mid-Latitude Environments.

O horizon: 1 to 4 inches (2 to 10 cm) thick. Mat of highly acid, undecomposed humus.

A horizon: 4 to 6 inches (10 to 15 cm) thick. Gray or pale-brown
sandy soil with abundant organic matter and roots. Highly acid,
partly eluviated.

E horizon: 6 to 20 inches (15 to 50 cm) thick. Light-gray or white,
ashlike sand. Highly acid, fully eluviated.

B1 horizon: 0 to 6 inches (0 to 15 cm) thick. Dark-brown, dense illuvial zone with clay-humus and sesquioxides. Acid.

B2 horizon: 6 to 20 inches (15 to 50 cm) thick. Brown, often yellowish or reddish, dense, clayey illuvial horizon of sesquioxides.
Slightly acid.

Both B subhorizons are fine textured and have a dense structure, with
little aeration. This impedes downward percolation of soil waters, so
that temporary waterlogging is not unusual and may be indicated by
rust-colored mottles. When very clayey podsols are cultivated, they are
prone to alternating dehydration (mid-summer) and waterlogging
(spring and autumn); this favors compaction of the iron-rich B2 horizon into a hardpan, while the incohesive A and E horizons remain
quite erodible.

Drainage is a major problem in large parts of the boreal woodlands, particularly on ice-scoured plains and in the West Siberian lowlands. *Bog soils,* with 3 to 20 feet (1 to 6 m) accumulations of acid
peat moss, are widespread and regionally dominant in parts of the
forest-tundra transition belt. Swamps, grown over with bog moss and
acid-tolerant shrubs and trees (tamarack, black spruce), are in fact a
hallmark of the subarctic. Gley soils, with deep, organic O horizons are
equally prominent.

Podsolic Soils of the Temperate Woodlands. Podsolization in the temperate, mixed deciduous woodlands of mid-latitudes is less advanced
and extreme, at least with "average" parent materials. Such intermediate soils include gray-brown podsolics and gray wooded soils. They
are characterized by organic (O1, O2) and E horizons less developed
and less acid than those of true podsols. In fact, the E horizons of most
podsolic soils retain appreciable quantities of clay-humus and sesquioxides. Correspondingly, the B horizon is less dense and compact
than in a true podsol.

A representative *podsolic* soil has an intermediate type of profile
(Figure 18–1,C):

O horizon: distinct but shallow.

A horizon: 10 inches (25 cm) thick. Brown-gray, loamy soil with
abundant, soil-integrated organic matter, moderate acidity,
abundant microorganisms, and some base nutrients.

E horizon: 10 inches (25 cm) thick. Brown-gray, loamy soil with less humus and fewer nutrients. Acid.

B horizon: 16 inches (40 cm) thick. Yellowish or reddish-brown clayey soil with abundant nutrients. Slightly acid.

Typically, the horizons are deeper but less distinctive than in a true podsol, with gradual transitions from one horizon to the next.

In parts of western Europe, podsolic soils form a mosaic with rather fertile soils showing little or no evidence of eluviation and illuviation. Known as *brown forest soils* (or braunerdes), they lack an A2 horizon and are incompletely leached. There may also be shallow zones of lime enrichment at the base of the B horizon. A typical profile is as follows (Figure 18–1,D):

A horizon: 2 to 12 inches (5 to 30 cm) thick. Dark gray-brown, loamy soil, rich in mild humus, microorganisms, and base nutrients. Neutral.

B horizon: 10 to 24 inches (25 to 60 cm) thick. Brown loamy soil, rich in microorganisms and base nutrients, some mild humus. Neutral.

All brown forest soils have developed under deciduous woodland, on parent materials with bases or lime; they are restricted to areas with low-intensity rainfall (commonly only 20 to 30 in, or 500 to 750 mm) distributed evenly through the year.

Tracts of bog soils and gleys are also present in the temperate woodland belt, but they are less common and less extensive, being restricted to glacial plains. In environments rich in lime, bog soils are replaced by well-decomposed, neutral *muck soils* in sedge and reed marshes. Also widespread in the temperate forest, side by side with posdolic soils, are waterlogged *planosols*. These are podsolics with compacted, clay B horizons that seriously impede soil drainage.

Not all the intrazonal soils of humid mid-latitudes are hydromorphic. Calcimorphic soils are also common, on limestones and dolomites. Particularly where liable to temporary drought, microorganisms are limited in variety, while leaching is ineffective. The result are *rendzina* soils with thick organic horizons, often poorly decomposed, that rest directly on bedrock or on a shallow, lime-rich B horizon.

Podsolic Soils of the Subtropical Woodlands. The intensive chemical weathering of the humid subtropics favors development of striking reddish or yellowish soils with deep profiles. These are commonly designated as *red-yellow podsolics*,[1] although podsolization is only inci-

[1] Other designations are *red* and *yellow latosolic-podsolic* soils, *krasnozem*, *rotlehm* (reddish), and *braunlehm* (yellowish).

dental to intensive alteration and oxidation of the subsoil.

A typical red-yellow podsolic profile from Alabama or the Carolinas may look as follows (Figure 18–1,E):

A horizon: 2 to 4 inches (5 to 10 cm) thick. Dark gray-brown, crumb-structured loamy soil, with modest nutrient and organic levels. Highly acid.

E horizon: 6 to 12 inches (15 to 30 cm) thick. Yellowish-brown, loamy soil with limited nutrients and humus. Highly acid.

B horizon: 16 to 36 inches (40 to 90 cm) thick. Red, yellowish-red, or yellow clayey soil. Compact structure, abundant sesquioxides, modest nutrient level. Moderately acid.

Both the A/B and B/C horizon contacts are gradual and indistinct. The color of the B horizon is generally mottled red and yellow, the dominating color depending on subsoil drainage, sesquioxide level, and the available mineral nutrients of the parent material. Yellowish variants are less well drained, more acid, and found primarily on rocks with few bases. Reddish variants are better drained, have higher sesquioxide ratios, and are developed on parent materials with more abundant carbonates or ferromagnesian minerals. In composition the B horizon consists largely of clay minerals and oxides (formed in place by decomposition of weak minerals), as well as residual quartz. Only a small part of the clays are illuviated from the A horizon. The dominant clay material is kaolinite, which has few available valences for metallic ions and consequently retains only a modest number of base nutrients—even when relative clay content is high.

Just as any compact, clayey subsoil, the red-yellow podsolics are often subject to impeded drainage after rains, and periodic groundwater oxidation and reduction may be evident in the morphology of the B horizon. The moderately compact soil structure allows sufficient aeration and water infiltration but combined with a high clay content favors accelerated soil erosion. When soaked, these soils are unusually prone to sheet wash, mass movements, and gullying. The last is particularly dangerous—when unchecked—since the parent material is commonly soft, altered by decomposition well below the B horizon. As a result, the red-yellow podsolics of the American South and the South China Uplands have been intensively and extensively eroded in the wake of agriculture. Consequently, recent alluvium and foot-slope deposits consist primarily of reworked soil and regolith.

Of restricted distribution in the humid subtropics are exceptionally deep, reddish-brown soils, with B horizons some 40 to 200 inches (100 to 500 cm) thick. These *reddish-brown lateritic* soils have profiles that are only slightly acid. They may have up to 80 percent clays and 9 to 12 percent free sesquioxides, with iron-rich concretions below the groundwater table. Such soils are analogous to tropical latosol profiles (see section 20-3) and link the red-yellow podsolics to the general family of tropical soils. In the United States such deep profiles are found on ancient surfaces in Georgia and the Carolinas and are at least in part relict, reflecting semitropical conditions during Tertiary times.

18-4. GEOMORPHIC PROCESSES UNDER FOREST

Where not disrupted by human land use, the average landscape of the humid mid-latitudes is mantled with soil and forest on gentle to intermediate slopes. Except where slopes are undercut or where rough terrain exceeds angles of 15 to 25 degrees, soil accumulates on crest- and foot-slopes. This soil mantle is bound and protected by a vegetative mat, and rainsplash is impossible, sheet wash minimal. Instead, denudation typically proceeds by (1) slow removal of ions and clays in saturation throughflow and groundwater, and (2) creep and other slow mass movements that shift topsoil aggregates downslope. Locally, there may be periodic "slope failure" where soil and regolith are deep, clayey, and wet on impermeable parent material or in seepage zones. Such sudden displacements are most commonly due to rapid earthflows or debris slides. In general, foot-slopes exhibit little or no sediment accumulation, in part as a result of limited erosion upslope, in part because net accumulation is retarded by sheet wash and rill wash in this clayey zone of reduced infiltration capacity.

These broad generalizations are supported, for example, by detailed study of a small drainage basin in northwestern Luxembourg by Imeson and Jungerius (1974). Slopes were gentle to intermediate, rarely exceeding 12 degrees; the soil was a slightly podsolized, brown forest type; precipitation amounted to 34.5 inches (880 mm), with modest submaxima in spring and fall and periodic rains of high intensity. Monitoring of forested slopes showed no net erosion, in part as a consequence of the rapid, seasonal accumulation of leaf litter and gradual incorporation of organic products into the soil. Profile sampling further showed a total lack of correlation between slope angle and soil profile thickness, implying long-term steady-state conditions. Only locally were there shallow accumulations of colluvium, attributed to temporary woodcutting in past centuries. The surprising degree of slope stability is attributed to high soil porosity and aggregate stability induced by the high organic content of clay-humus plasma.

This apparent picture of steady state applies only to intermediate environments and relatively gentle slopes. A similar study of a hilly, forested watershed (average slope greater than 15 degrees) in the central Appalachians showed thin, in part discontinuous, soils and regolith or debris mantles ranging widely from 0 to 20 feet (0 to 6 m) in thickness (Hack and Goodlett, 1960). Here there was a high rate of activity on slopes, by creep and rain wash, with periodic debris slides after major rainstorms.

A more dynamic form of slope equilibrium is also indicated for the boreal and subtropical variants of the humid mid-latitude model. In the case of the boreal woodlands, permafrost is relatively widespread, and soil frost is generally effective. Talus transfer and other mass movements consequently inhibit vegetation and soil development on slopes steeper than 25 degrees. At the same time, creep is rapid on gentler slopes, frequently grading into slow earthflows or solifluction. A proportion of this detrital material reaches the stream network during the spring thaw, when soils are least cohesive and supersaturated with water, or following heavy rainstorms. Stream activity is also spectacular when thick, winter snow covers and ice-bound rivers thaw rapidly, creating strong flood surges and conspicuous floodplain modeling. In the case of the subtropical woodlands, the incidence of slope failure is increased by the prevalence of deep, clayey soils, implying a lower mean angle of slope stability. Furthermore, the vegetation is to some degree dormant at the end of the cooler, drier part of the year, while soil organic content is lower than in cooler latitudes. The soil is therefore less protected by a plant cover in early spring and generally more erodible. Heavy rainstorms, particularly those associated with hurricanes, are therefore effective and often destructive (Williams and Guy, 1973).

Despite these regional and altitudinal variations, humid mid-latitudes provide morphostatic environments, characterized by a fair degree of geomorphic equilibrium. Rates of denudation are low to moderate, and erosion is generally balanced by soil development. It may be recalled that microexperimental studies (Schumm, 1966) showed a reduction of slope convexity by creep in permeable materials. This suggests that downwearing should be the dominant trend in slope development. Over very long periods of time rectilinear mid-slopes should then be consumed into smooth, concavo-convex profiles. However, the rates involved appear to be so slow that Pleistocene inherited forms retain great longevity, while even some Tertiary features have survived into the present.

18-5. THE IMPACT OF DEFORESTATION AND CULTIVATION

The idealized model outlined above helps explain the generalities and the anomalies of humid landforms today. But contemporary processes are seldom in a state of "normal" dynamic equilibrium, except perhaps in thinly settled parts of the boreal forest. Elsewhere, human settlement has played a major geomorphic role over centuries or millennia, and recent advances in technology have created both extensive concrete surfaces and great strip-mine scars that almost destroy the existing landscape.

Once man enters the picture, nothing remains quite the same. Deforestation and pasturing lead to a marked increase in the proportion of surface runoff, with intensified denudation and a greater supply of increasingly coarse sediment added to the stream network. Then the moment that the grassy vegetation and organic soil horizons are plowed up, sheet wash and rill cutting become practicable. If soil structure deteriorates and organic matter is greatly reduced, soil density increases, infiltration and permeability decrease, and clays begin to seal the surface. The result is that overland flow becomes a reality, increasing in importance as the porous A horizon is stripped and less permeable subsoils exposed. As sheet wash and rill erosion are accelerated by a factor of 10 or more, the equilibrium of weathering and erosion is upset, and soils are frequently removed faster than they can develop. Soils are stripped from crest-slopes while mid-slopes and foot-slopes are rilled or gullied. The various patterns of accelerated soil erosion that ensue (see section 4-8) fall well outside the range of variability of what happens under undisturbed woodland. They conform to the sheet-wash model of observed evolution of microslopes and, in the long run, will probably favor backwearing.

In North America, accelerated soil erosion is a comparatively recent phenomenon, since Indian farming was sporadic in the eastern woodlands. Early signs of accelerated soil erosion were apparent in the hilly farmlands of New England by the early eighteenth century, and remedial measures were already the subject of debate in the mid-nineteenth century. As the woodcutters slashed westward and the settlers flooded across the Appalachians and shifted from farm to farm in their wake, ecosystems were thoroughly disrupted. This has been tangibly documented for southwestern Wisconsin by Knox (1972), who resurveyed stream widths first recorded in the Federal Land Survey of 1832–1833. Existing forest cover was reduced 80 to 90 percent by the earliest settlers, increasing the frequency and magnitude of flood flows and disrupting the balance between slope, sediment, and discharge. Soils were truncated and the eroded silts and clays systematically distributed over downstream floodplains. This increased sediment yield created alluvial fans at tributary confluences and significantly increased bed-load fractions and channel width. In an older paper, Gottschalk (1945) describes the impact of soil erosion since colonial times on sedimentation in Chesapeake Bay, where sediment has partly filled in the estuarine river mouths.

Historical documentation of accelerated soil erosion has been provided for another region, the southern Piedmont, by Trimble (1974). Settlement of the bottomlands began there between 1700 and 1830, but erosion first became apparent when sloping lands were cleared and was locally accentuated by the impact of plantation agriculture after the 1840s and 1850s. In contrast to the older cotton plantation areas, major erosion was delayed until the 1880s, when forest acreage was significantly decreased and row crops, especially cotton, became dominant. By the 1930s, when the efforts of the U.S. Soil Conservation Service began to take effect, the impact of man on the soil mantle, hydrography, and sedimentation had exceeded that of any natural climatically induced ruptures of equilibrium experienced in the southeastern United States during all of Pleistocene time.

In temperate Europe accelerated soil erosion began much earlier. Loessic slope soils were first swept into the river valleys about 750 B.C., when Celtic tribes began to disperse in Germany, expanding areas of cultivation to sloping soils and introducing the iron-tipped plowshare (Butzer, 1974). Older, Bronze Age valley settlements were buried by as much as 3 to 6 feet (1 to 2 m) of soil wash, and accelerated floodplain sedimentation locally destroyed riverine forests. Celtic and Roman cultivators continued to concentrate their plots on upland plains and some intermediate-slope hillsides. By the time of the Germanic migrations the uplands of Britain and Gaul had lost much of their topsoil to erosion, and the Germanic tribes colonized the mainly unused lowlands, where they were able to till the heavy-textured soils with a heavier plow first introduced into northwestern Europe in Roman times. Thus in southern England cultivation shifted to the lowlands, with the old, eroded upland fields partly abandoned to grazing. Population pressure built up in central Europe during the thirteenth century, leading to extensive deforestation and cultivation on the rougher uplands. By the time the more marginal of these settlements were abandoned, about 1350–1450 A.D., most of the topsoil had been destroyed and a second wave of loessic soil sediments inundated footslopes and floodplains. The truncated B horizons of abandoned fields can still be recognized under the forests of today, and some German valleys are mantled by 3 to 25 feet (1 to 8 m) of corresponding sediment.

The third episode of soil erosion in temperate Europe was between about 1760 and 1880, a time of renewed rural population pressure, consolidation of fields, shifts to monocultures, resurveying along geometric grids, upland deforestation, and destructive herding in once protected woodlands. This last major wave of destruction was eventually halted by conservation measures such as strip cropping and reforestation.

The tangible impact of soil erosion has not been the same everywhere. In the boreal woodlands, cultivation is sufficiently marginal that soil erosion has by and large remained localized. In the temperate woodlands, erosion has significantly depleted soil resources but, until recent decades, has not been readily apparent in the geomorphic landscape. But a trained eye can spot the truncated, yellowish B horizon exposed on the average convex undulation of relief, or the unusually thick organic soils accumulating in shallow depressions and swept into floodplains. Gullies are present, but their impact is not general and is seldom catastrophic. However in the American South and in the South China Upland, where rainfalls are very intensive, rainsplash and unchecked sheet and rill erosion have denuded many uplands to angular, jagged profiles, while rolling country may be crisscrossed with gully systems to form badland topography. Small alluvial fans are common wherever stream systems debouch on floodplains and other lowlands, and here the red or yellow soil products are deposited after rains.

Accelerated soil erosion will remain a chronic problem in all but the boreal forests of the subarctic. Indeed, man has become the single most potent agent of erosion in the premium lands of mid-latitudes.

18-6. PALEOCLIMATES OF HUMID MID-LATITUDES

For about 90 percent of the past 600 million years (see Table 10–1) there were no glaciers on Earth. Instead, all the evidence indicates that subtropical conditions normally prevailed in mid-latitudes, while polar regions enjoyed a temperate climate. It can therefore be concluded that continental and highland glaciation is an unusual phenomenon and that ice ages are abnormal events.

Yet the geological record shows that relatively brief ice ages have recurred periodically. One such ice age can be vaguely recognized at the beginning of the Cambrian period, a little after 600 million years ago. Glacial deposits of that period have been found on both the Northern Hemisphere (Greenland, Scandinavia) and the southern continents (South Africa, Brazil, Australia). This is often called the Infra-Cambrian Ice Age. The next major spasm of glaciations began at the end of the Carboniferous, about 290 million years ago. At least five major glacial episodes are recognized, covering a time span of 10 to 20 million years. Evidence of continental glaciation is widespread in what are now southern Africa, eastern South America, Australia, India, and Antarctica. However, continental configurations were quite different at the time, and it is now widely accepted that all of these land masses were more or less contiguous and arranged around the South Pole (see Figure 13–7).

Since early Permian times, world climate has been on the warm side and there is no conclusive evidence of glaciers anywhere prior to the Miocene (Table 10–2). During this 260-million-year time span rapid evolution took place in the animal and plant world, the Mesozoic age of reptiles passing into the Cenozoic age of mammals. The higher-latitude continents became covered by woodlands, with temperate forests in northern Greenland and Antarctica and tropical rainforests in southern England. Estimates based on plant and animal indicators suggest that, during the early Tertiary, mean annual temperatures in the western United States were 36° F (20° C), in western Europe 22° F (12° C) higher than they are today (Figure 18–2). It is probable that seasonal contrasts were small, even in high latitudes, with mild or warm winters (Butzer, 1976).

A noticeable cooling trend is apparent during the Pliocene (Figure 18–2), when glaciers were widespread in the polar world and in many highlands; but there is no firm evidence for continental glaciers in Europe or North America. Repeated oscillations of increasingly colder climate gradually eliminated the tropical or subtropical floras of mid-latitudes during the early Pleistocene, until some critical threshold of cold was passed and ice sheets began to form in Scandinavia and Canada (Table 10–2). For the remainder of the Pleistocene, Earth climate oscillated violently and fairly rapidly between two extremes, glacials on the one hand, interglacials on the other. During each glacial, ice sheets formed afresh in Europe and North America, while highland glaciers expanded or developed in most high mountain ranges. A number of geological and biological indicators suggest that mean world temperatures were at least 9° F (5° C) colder than today, about 27 to 32° F (15 to 18° C) on the higher-latitude continents and 11 to 14° F (6 to 10° C) on the tropical land masses. During the interglacials, the European and North American icecaps disappeared, but ice continued to cover Greenland and Antarctica, very much as it does today. Temperatures averaged close to their modern values.

The number of Pleistocene glacials is uncertain. There have been at least seven or eight glacials during the past 700,000 years, including the Mindel in Europe, the Nebraskan and Kansan in North America.

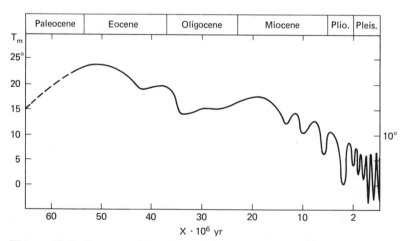

Figure 18–2. Cenozoic Temperature (T_m) Trends in Mid-Latitudes. Note scale distortion for Plio(cene) and Pleis(tocene). (After Butzer, 1976.)

Most details about these glacials is controversial and much in dispute. The oldest tangible glacial that left an extensive record in mid-latitudes terminated about 130,000 years ago. This event can be best called the Penultimate Glacial; many European workers designate it as the Riss, North American workers as the Illinoian Glacial. The succeeding period between 125,000 and 75,000 years was mainly warm and can be described as the Last Interglacial (Figure 18–3). It was followed by the Last Glacial, which lasted about 65,000 years. Events in this time range are fairly well understood and dated by means of the radiocarbon method, based on the decay of the isotope C^{14} of atmospheric carbon. This Last Glacial, known as the Würm in Europe, as the Wisconsin in North America (Figure 18–4), was responsible for the deposition of the Newer Drift. Fossil logs from till beds; charcoal, shell, or bone from the sites of prehistoric man; and shell from beach deposits have often been suitable for radiocarbon dating and provide a chronological framework. It is now known, for example, that there was a temporary warm-up about 40,000 to 25,000 years ago and that the ice subsequently readvanced to its maximum stand by about 20,000 years ago. Thereafter, the ice sheets began to dissipate, step by step, creating a number of recessional moraines. A last major readvance all across the Northern Hemisphere began 11,000 years ago, coincident with a brief oscillation of cold and exceptionally severe climate. A millennium later, the icecaps went into final retreat, and most geologists date the beginning of the Holocene or Recent period at about 10,000 years.

The Scandinavian icecap completely disappeared about 8,500

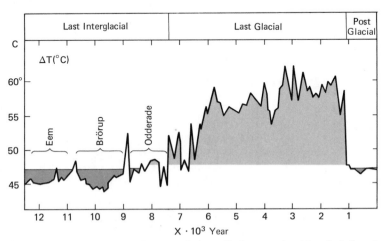

Figure 18–3. Temperature Gradients (ΔT) over the North Atlantic Basin Since 125,000 B.P., Assuming Limited Air Temperature Deviations at the Equator. (After Butzer, 1976.)

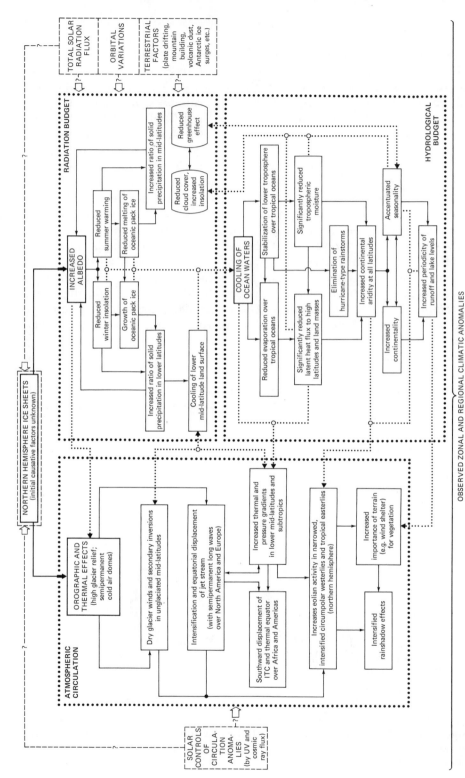

Figure 18–4. An Explanatory Model for Glacial Climates. (After Butzer, 1976.)

OBSERVED ZONAL AND REGIONAL CLIMATIC ANOMALIES

years ago, the residual Canadian ice sheets some 2,500 years later. But already by about 6000 B.C. summer temperatures were conspicuously warmer than today in Alaska and mid-latitude Europe—probably by at least 3° F (about 2° C), judging by the distribution of a number of temperature-sensitive plants (Figure 18–5). This trend was less apparent in other parts of the world, and about 3000 B.C. the climate began to become a little cooler once again. Since then highland glaciers have periodically advanced and retreated a little. In particular the sixteenth century A.D. ushered in what is often called the Little Ice Age. Early during the 1500s or 1600s a number of large Alpine glaciers began to advance, a phenomenon paralleled in western North America according to recent radiocarbon and lichenometric determinations. A slight warming trend during the eighteenth century was reversed between about 1820 and 1850, after which highland glaciers the world over again advanced. Beginning in the 1890s and particularly since 1920 a general retreat of the glaciers has become evident both on the Northern and Southern hemispheres. In fact, glaciers in the Alps, Scandinavia, the Rocky Mountains, Alaska, in the Andes of Chile and Argentina, as well as in New Zealand have retreated by several hundred yards (meters) or even several miles over the decades since 1920.

A survey of climatic variation during the recent geological past is of far more than academic interest. An appreciation of the time element in geomorphology is essential. Some land surfaces have been exposed since the mid-Tertiary, that is, for 25 million years or more. The landforms created over such time spans must be seen in perspective when compared, for example, with the Newer Drift, which is less than 25,000 years old. In addition, as climate has changed, so has the rela-

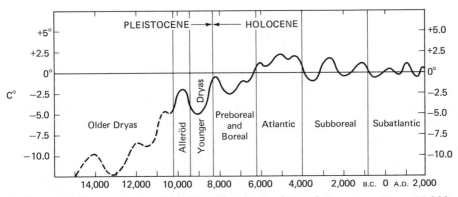

Figure 18–5. Temperature Trends in Europe and North America Since 15,000 B.C. The substage designations refer to the European sequence of pollen zones. Holocene dates have been approximately corrected for atmospheric radiocarbon flux.

tive importance of various geomorphologic processes. Consequently, change through time, as both a direct function of time and a result of environmental changes with time, is a vital geomorphologic parameter. The significance of more recent climatic fluctuations lies in the opportunity they provide to observe the balance of forces in nature and in the speed and degree to which responses are made. Of all geomorphic phenomena, glaciers have probably reacted most dramatically. But hydrologic equilibrium must have readjusted concurrently through much of mid-latitudes. This provides a valuable perspective on the implicit need for repeated readjustments of dynamic equilibrium among most components of the geomorphologic balance operating in many, if not most, environments.

Last but not least, the Pleistocene is by no means remote in terms of human prehistory. The existence of early Pleistocene hominids of direct relevance to the line of human ancestry has recently been shown from deltaic beds in southwestern Ethiopia. On the basis of the gradual decay of a potassium isotope found in volcanic deposits, these strata can be dated at 2 to 3 million years. Similar dating techniques show that a little over 2 million years ago true men, adept at making stone tools, butchered animals on the shores of a stream draining into Lake Rudolf, in northern Kenya. These human ancestors were anatomically quite similar to modern man. In fact, biologists believe that most human traits and emotional responses developed during 2 million years of Pleistocene prehistory when man roamed the Old World continents as a hunter and gatherer, increasingly efficient and adapted to a diversity of environments.

18-7. INHERITED AND POLYGENETIC LANDFORMS

Returning to the landscape assemblage of humid mid-latitudes, inherited landforms provide the stage for the Holocene processes and balances outlined earlier. Just how prominent inherited landforms are will depend on whether past geomorphic systems were distinctive and whether Holocene conditions are morphostatic or morphodynamic. There can be little argument that past glacial and periglacial regimes will have left a distinctive record in the northern half of this environmental complex and that subsequent conditions have been morphostatic. Less certain is the comparative role of paleoclimatic impulses in the evolution of humid subtropical landscapes.

1. *Glacial* landforms were created in abundance during one or more of the Pleistocene glaciations. They include the whole spectrum from ice-scoured plains to plains of glacial deposition with ground moraines, drumlins, eskers, end moraines, outwash fans and terraces, spillways, and proglacial lake beds (Figure 18–6). These landform assemblages, which have already been discussed in Chapter 10, remain

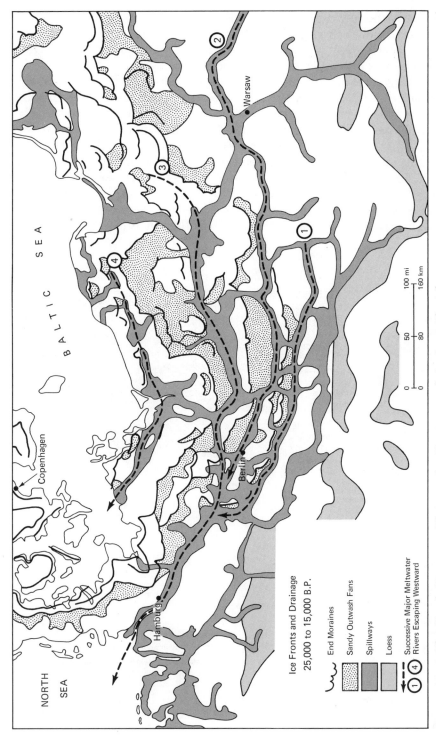

Figure 18–6. The North European Plain at the Maximum of the Last Glacial. (Compiled from various sources.)

prominent today despite partial dissection, attrition by mass movements, and development of moderately deep soils. Especially on ice-
scoured plains, unsystematic drainage and widespread bog or lake
country reflect the heritage of Pleistocene glaciation.

2. *Periglacial* landforms, in the widest sense of the concept, are
often preserved in those areas once affected by Pleistocene permafrost
or by other forms of effective cryoturbation. In the United States (Figure 18–7) permafrost was restricted to a narrow belt adjacent to the
Pleistocene glaciers, but periglacial sculpture was also effective in unglaciated parts of the Cascades and Northern Rockies, and the central
Appalachians. No systematic inventory of these features has yet been
attempted, but they include ice-wedge casts, collapsed pingos, crude
blockfields, altiplanation terraces, solifluction mantles, coarse-grade
alluvial terraces, and extensive loess mantles. In Eurasia, Pleistocene

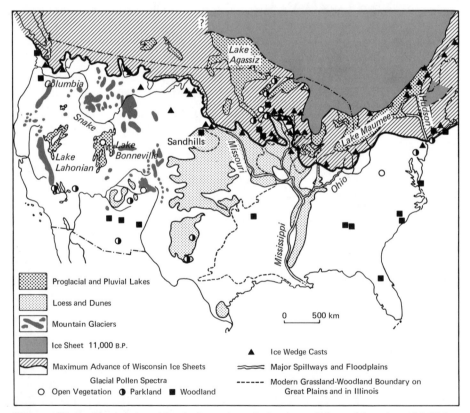

Figure 18–7. Glacial-Age North America. Only those lakes older than 11,000
B.P. are shown. (After Butzer, *Environment and Archeology.* Copyright © 1971
by Aldine Publishing Co., Chicago.)

permafrost extended as far south as latitude 46° N across most of the length of the continent, with further periglacial sculpture and local glaciation throughout the Alpine-Himalayan Cordillera. Regional inventories of Pleistocene periglacial features exist for many countries and, in addition to those enumerated for the United States, include dune fields, dry valley heads, and evidence for systematic ridge asymmetry in some areas.

Whereas the impact of valley aggradation and loess covers is incontrovertible, the evolution of polygenetic slope assemblages (see Peltier, 1949) is less clear. Slope detritus of various grades and arrangements proves the existence of episodes with accelerated slope dynamism under cold conditions (see Rapp, 1967), often in areas with resistant rock where subsequent change has been negligible. However, it provides little indication as to whether basic slope forms were significantly remodeled by periglacial backwearing or, perhaps, altiplanation of older, in part Tertiary, topographies. A discriminating study of the former ice margin in the upper Susquehanna drainage (King and Coates, 1973) illustrates just how intricate the problem is. For the time being, therefore, it is no more than a reasonable working hypothesis that the majority of hillslopes in resistant bedrock found in the former periglacial belt of mid-latitudes represent hybrid forms that reflect both Pleistocene periglacial backwearing and Tertiary to Holocene downwearing under more temperate conditions.

3. Possible *tropical* influences (see sections 20-5 and 20-6) have been invoked by several European geomorphologists to explain some of the ancient, Tertiary erosional surfaces that cut across uplands such as the old fold mountains of western and central Europe (see Bakker and Levelt, 1965; Rathjens, 1971). These were once called "peneplains" and ascribed to erosion and planation, almost down to sea level; renewed dissection followed uplift during the Alpine orogenies. However, there are good records of humid, tropical vegetation in Europe during early to mid-Tertiary times, and relics of tropical soils have been found on such planation surfaces. Furthermore, the residual hills found on such planation surfaces have many similarities with those of the African savannas.

An analogous problem of interpretation can be raised for the Appalachian Piedmont Plateau in the southeastern United States. Here the issue is not obscured by Pleistocene periglacial sculpture, since temperate-humid processes continued to operate during periods of glacial climate, when the regional vegetation consisted of spruce and pine forests (Whitehead, 1973). Morphologically, some parts of the piedmont can be readily replicated in parts of Africa, and Stone Mountain, Georgia, is the prototype of a savanna inselberg (see section 20-5), down to the details of its vertical macrogrooves (Figure 20–9). The contemporary soil mantle has unmistakable affinities with tropical soils, and the occurrences of "reddish-brown lateritic soils" (see Giddens et al., 1960), with their characteristic clay mineralogy of kaolinite and gibbsite, suggest relict distribution. Finally, examination of sections shows polygenetic soil profiles and reworked paleosols in multiple generations of foot-slope colluvium (Parizek and Woodruff, 1957), with "stone lines" like those of tropical Africa (see Figure 20–9). Although the case is not proved beyond reasonable doubt, it is quite probable that Tertiary climates did leave a legacy of tropical deep weathering and related savanna planation in parts of humid midlatitudes (Dury, 1971).

All in all, the landform complex of humid mid-latitudes is a mosaic and overlay of contemporary and inherited processes and forms. We are far from understanding the details satisfactorily, but any attempt to explain the visible landform assemblage solely by invoking contemporary processes would be inadequate and naïve.

18-8. SELECTED REFERENCES

Bakker, J. P., and T. W. M. Levelt. An inquiry into the probability of a polyclimatic development of peneplains and pediments in Europe during the Senonian and Tertiary period. *Publications, Service Géologique de Luxembourg*, No. 14, 1965, pp. 27–75.

Bunting, B. T. *The Geography of Soil*. London, Hutchinson, and Chicago, Aldine, 1965, 213 pp.

Butzer, K. W. *Environment and Archeology: An Ecological Approach to Prehistory*, 2nd ed. Chicago, Aldine, and London, Methuen, 1971, 703 pp.

———. Accelerated soil erosion: a problem of man-land relationships. In I. Manners and M. W. Mikesell, eds. *Perspective on Environment*. Washington, D.C., Association of American Geographers, 1974, pp. 57–78.

———. Pleistocene climates. In R. C. West, ed. *Ecology of the Pleistocene. Geoscience and Man*, Vol. 13, Baton Rouge, La., 1976, in press.

Carson, M. A., and M. J. Kirkby. *Hillslope Form and Process*. London and New York, Cambridge University Press, 1972, 475 pp. See Chap. 11.

Denton, G. H., and W. Karlén. Holocene climatic variations—their pattern and possible cause. *Quaternary Research*, Vol. 3, 1973, pp. 155–205.

Dury, G. H. Relict deep weathering and duricrusting in relation to the paleoenvironments of middle latitudes. *Geographical Journal*, Vol. 137, 1971, pp. 511–522.

Giddens, J., H. F. Perkins, and R. L. Carter. Soils of Georgia. *Soil Science*, Vol. 89, 1960, pp. 229–238.

Gottschalk, L. C. Effects of soil erosion on navigation in upper Chesapeake Bay. *Geographical Review*, Vol. 35, 1945, pp. 219–238.

Hack, J. T., and J. C. Goodlett. Geomorphology and forest ecology of a mountain region in the central Appalachians. *Professional Papers, U.S. Geological Survey*, No. 347, 1960, pp. 1–66.

Imeson, A. C., and P. D. Jungerius. Landscape stability in the Luxembourg Ardennes as exemplified by hydrological and micropedological investigations of a catena in an experimental watershed. *Catena*, Vol. 1, 1974, pp. 273–295.

King, C. A. M., and D. R. Coates. Glacio-periglacial landforms within the Susquehanna Great Bend area of New York and Pennsylvania. *Quaternary Research*, Vol. 3, 1973, pp. 600–620.

Knox, J. C. Valley alluviation in southwestern Wisconsin. *Annals, Association of American Geographers*, Vol. 62, 1972, pp. 401–410.

Parizek, E. J., and J. R. Woodruff. Description and origin of stone layers in soils of the southeastern United States. *Journal of Geology*, Vol. 65, 1957, pp. 24–33.

Peltier, L. C. Pleistocene terraces of the Susquehanna River, Pennsylvania. *Bulletin, Pennsylvania Geological Survey* (4th Series), No. G23, 1949, pp. 1–158.

Rapp, Anders. Pleistocene activity and Holocene stabiltiy of hillslopes with examples from Scandinavia and Pennsylvania. In P. Macar, ed. *L'Evolution des Versants*. Liège, Université de Liège et Academie royale de Belgique, 1967, pp. 229–244.

Rathjens, Carl, ed. *Klimatische Geomorphologie.* Darmstadt, Wissenschaftliche Buchgesellschaft, 1971, 485 pp. Includes German classics by Otto Jessen and Julius Büdel on the effects of former periglacial and tropical conditions in mid-latitudes.

Sartz, R. S. Effect of land use on the hydrology of small watersheds in southwestern Wisconsin. *International Association of Scientific Hydrology, Publication 96*, 1970, pp. 286–295.

Schorger, A. W. *Prairie, Marsh and Grove: The Natural History of a Midwestern County.* Madison, Wis., University of Wisconsin Press, 1973, 376 pp.

Schumm, S. A. The development and evolution of hillslopes. *Journal of Geological Education*, Vol. 14, 1966, pp. 98–104.

Trimble, S. W. *Man-induced Soil Erosion on the Southern Piedmont, 1700 to 1900.* Ankeny, Iowa, Soil Conservation Society of America, 1974, 185 pp.

Whitehead, D. R. Late Wisconsin vegetational changes in unglaciated eastern North America. *Quaternary Research*, Vol. 3, 1973, pp. 621–631.

Williams, G. P. and H. P. Guy. Erosional and depositional aspects of Hurricane Camille in Virginia, 1969. *U.S. Geological Survey, Professional Paper 804*, 1973, 80 pp.

Chapter 19

DESERT LANDSCAPES

Arid and semiarid geomorphic environments can be defined by sporadic and seasonal stream flow respectively. The core deserts can further be singled out as hyperarid on the basis of receiving no rainfall whatever during some years. Mechanical outweighs chemical weathering, and in the desert the soil mantle is a matter of coarse lags (due primarily to winnowing out of fines by deflation and sheet wash) over subsurface concentrations of lime and salts, sometimes indurated to caliche. Slow but cumulative eolian processes and periodic but effective activity by running water combine to fashion a distinctive landscape. Component parts include deflation hollows; rocky or pebbly lags, often blackened by desert varnish; sandblasted rock surfaces; various dunes or sand sheets; steeply walled valleys alternating with alluvial fans and rock-cut pediment surfaces; interior basins where alluvial aprons grade into mixed alluvial and lake beds, mantled by dune fields, and terminating in evaporation pans. The composite landscape is marked by steep rectilinear

slopes, with angular inflections, that maintain constant mid-slope angles in all stages of dissection and that separate extensive, flat uplands from undulating or irregular upland surfaces. Drainage patterns are often poorly integrated, and texture is coarse to very coarse, even in dissected areas of moderate relief. Paleoforms are common and include fossil lake beds and shorelines, alluvial terraces, and evidence of former chemical weathering (clayey paleosols, cavernous rock hollows). These features blend into the landform assemblage, and past climatic changes may be essential to explain the origin of the great sand seas or of valleys and pediment generations in the low-latitude desert cores. Rates of landform development are rapid in the semiarid transitional environments but very slow in the hyperarid core deserts, where wetter intervals periodically accelerated the general trend to backwearing. The mid- and low-latitude grasslands and the mediterranean-type, open woodlands provide transitional environments where arid paleoforms such as eolian mantles, deflation hollows, and angular slope constellations are prominent and where cultural disturbance leads to accelerated erosion.

> . . . **desert air, and the nakedness of space, pure as a theorem, stretching away into the sky drenched in all its own silence and majesty.** . . .
>
> LAWRENCE DURRELL, *Balthazar*

19-1. A VIEW OF DRY LANDS

Polarizing world scenery, it is not difficult to sense the peculiar quality of desert landscapes. In more humid environs the land surface is carpeted with vegetation or picturesque farms, to the degree that rock seems incidental to form. In the periglacial realm, perhaps equally austere, the geometry appears less angular because of the abundance of water, the organic mat, and the winter mantle of snow. Mountain country is certainly angular and rocky, but the impression of order is absent among forms that are largely vertical and confining. It is in the desert that geomorphology emerges as a systematic science of rock forms, with hills conveniently sectioned to expose geological history and arranged to demonstrate pattern and regularity. Yet no two deserts are alike. The structural and lithological ingredients vary too widely, and time, measured now in millions of years, produces landscapes that are both similar and different.

Perhaps the unifying theme is provided by the contrast of flat plains and steep slopes and by the accentuation of rock details. Only the dune fields are rounded, and even these are jagged in profile. Dry lands are, then, dominated by mechanical processes. Chemical weathering is subordinate, and soils become incidental landscape components as aridity increases. Wind is a potent force at times, but water, seemingly scarce, periodically plays a dynamic role that more often than not matches rates of weathering. Change is slow, giving the impression of immobility, until powerful forces are suddenly activated over wide areas. This stop-and-go mechanism appears to have produced landscapes with distinctive attribute assemblages over long periods of time. It is evident that most desert forms are of great age and that climatic changes have recurred again and again. Yet the diagnostic attributes appear homogeneous from one horizon to the next. Landform components of widely different ages in similar settings blend readily. Therefore, it becomes somewhat academic to distinguish inherited and contemporary features.

Dry lands vary in their degree of aridity, and, correspondingly, they exhibit a greater or lesser degree of distinctiveness. It is standard to subdivide the arid zone according to several climatic indices that somehow compare available moisture and net evaporation. Such measures of water deficiency are of practical value, if for no other reason than that 30 percent of the continental land masses can be described as dry, although the semiarid variants generally have high agricultural productivity. Subdivisions are also essential to a geomorphologic appraisal in order to distinguish desert landscapes from the transitional realms defined by grasslands and semiarid woodlands. To this end it is simplest to use a geomorphic criterion, stream flow. Aridity is thus conveniently defined by sporadic or ephemeral stream flow, with classic desert topography best developed in hyperarid settings where there may be no rain whatever in any one year. The transitional geomorphic environment of the mid-latitude steppes, the grassy thorn savannas, and open mediterranean-type woodlands, is demarcated by intermittent stream flow on a predictable, seasonal basis covering several months of the year. Exotic rivers, fed from external areas of heavier rainfall, or streams in favored hydrogeological situations complicate this picture. But as a general organizational device, the distinction of sporadic and seasonal discharge is useful.

The subsequent discussion focuses first on arid and hyperarid environments (Figure 19–1), where desert landforms are best developed. The transitional landscapes of semiarid type are then considered individually.

19-2. WEATHERING IN ARID AND SEMIARID CLIMATES

As aridity increases, mechanical weathering becomes prominent, if not preeminent. Chemical weathering is important in semiarid environments but subordinate in desert regions. Instead, frost weathering (in higher latitudes only), salt weathering, as well as pressure unloading produce regolith on rocky surfaces and hillslopes. As the variety and number of microorganisms in the soil are reduced, the limited organic debris decomposes slowly.

Soils of arid and semiarid environments are commonly prone to carbonate enrichment rather than leaching and eluviation. This process of *calcification* produces the class of soils known as *pedocals*. It involves high or complete base saturation, in combination with net accumulation of carbonates or water-soluble salts. Several factors favor calcification:

1. Under conditions of a generally dry climate or one with a long dry season, the subsoil is seasonally or permanently dry, while rainwater percolation is limited. Intensive drought periods draw up capillary moisture from the subsoil or from a water table charged with lime. As a result, some of the solubles washed into the subsoil by percolating rainwaters are carried upward until precipitated in a zone of lime or salt accumulation.

2. Nutrient-demanding vegetation types, particularly deep and densely rooted grasses, also promote calcification. A plant cover that supplies abundant organic matter, rich in mineral nutrients, favors high base saturation of the clay-humus molecules. Leaching is therefore impeded, since leached bases are rapidly replaced, thanks to the healthy nutrient cycle. Grasses each year provide more organic matter than trees, much of it added directly to the soil through slow decomposition and replacement of the intricate root network. Burrowing rodents also aid in rapid humus dispersal from the O2 into the A horizon. In cooler, semiarid environments the result is intensive and deep humification.

3. Nutrient-rich parent material is another factor. Bedrock that contains carbonates (e.g., limestone) or that weathers to a residual product rich in mineral bases (e.g., basalt) also favors a healthy nutrient cycle with high base saturation and limited leaching.

The morphological features that result from calcification can be outlined as follows (see Figure 4–4):

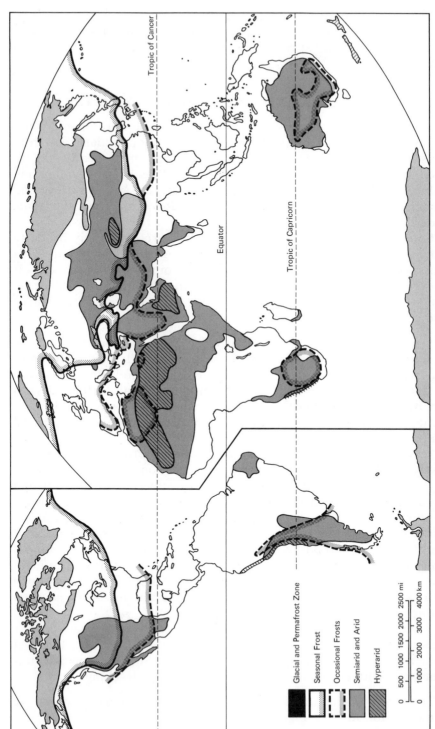

Figure 19–1. Aridity and Frost as Potential Geomorphic Parameters. (Modified after Butzer, 1971.)

First, a horizon of calcium carbonate or salt enrichment in the subsoil is commonly found just above or in the top part of the C horizon. This may take the form of concretions, nodules, white veneers on soil aggregates, or soft, powdery accumulations. Well-defined, whitish subsoil horizons of this type are labeled carbonate (Ca) or salt (Sa) horizons. Since salts are more soluble than lime, they wash out of well-drained soils in all but arid environments. They include gypsum (hydrated calcium sulfate) and the more soluble chlorides (sodium, calcium, magnesium) or bicarbonates (sodium, potassium).

Second there is little or no disassociation of clay minerals, with little or no eluviation. Illuvial B horizons are usually absent, and humification and calcification outweigh biochemical weathering in the soil, so that oxidized B horizons are also poorly developed or absent. In effect, many calcified soils lack typical B horizons and instead have A-C or A-Ca-C profiles.

Third, in temperate latitudes, humification is prominent under subhumid or semiarid conditions. The organic content of mid-latitude grassland soils is at a maximum, in terms of both quantity and quality: A horizons may be 6 feet (1.8 m) deep with as much as five times more organic matter (per unit volume) than the A horizon of an average podsolic woodland soil.

19-3. DESERT SOILS

Humification is, of course, minimal in arid regions, since the vegetation cover is too sparse to provide much organic matter. Deflation may also destroy what little humus veneer there is. Chemical weathering is inhibited by the scarcity of soil moisture at all seasons. However, frost weathering is active in mid-latitude deserts and, together with other forms of mechanical comminution in low-latitudes, provides a stony regolith. Resulting soils are thin, rudimentary, and often little more than an accumulation of lime or salts.

In the mid-latitude semideserts, such as those of the American West or Central Asia, *gray* or *red semidesert soils* (also known as serozems) are representative. These have shallow (2-to-8-in, 5-to-20-cm) A horizons, very low in organic matter, nitrogen compounds, or clay-humus. Subsoils normally show soft Ca or Sa horizons, 16 to 24 inches (40 to 60 cm) thick. Sometimes the Ca horizon is exceptionally thick or indurated or both. Such *calcrete* may resemble a travertine in structure but appears to form over many millennia as soil water seepage or surface runoff distributes lime from a source area. Deep, red, clayey soils in semidesert environments are polygenetic or relict and reflect moister Pleistocene climates. Demonstrably recent soils in Arizona or New Mexico, developed on surfaces dated at 2 to 5 millennia, are little more than weak examples of the gray semidesert variety.

In truly arid or hyperarid settings, soil profiles are frequently limited to a lag or pavement of surface rock or pebbles that protects subsurface concentration of soft gypsum or lime. Such lags are commonly produced as wind and sheet wash winnow out fine materials. Soil dynamism can also contribute to lag formation. When a desert topsoil is fine grained and rich in expandable, montmorillonitic clays, alternating wetting and drying have a similar effect as soil frost, pushing stones upward to reinforce the lag effect and, in some instances, favoring the growth of small polygonal networks along desiccation cracks (Figure 19–2). Lags and any rocky outcrops or talus are almost invariably blackened by a skin of ferromanganese compounds known as *patina* or desert varnish. Salts precipitated on or just below rock surfaces or in fissures are locally important in salt weathering, particularly on shade slopes or under overhangs. Cavernous hollows or tafoni may develop in this way. Alteration of sand dunes, fields, or veneers is generally minimal. However, many dune fields even in the Sahara are inactive today and are partially stabilized by reddish soil profiles, attributed mainly to moister climates or to oxide enrichment from aerosolic dust.

Salt enrichment, particularly of sodium bicarbonate and chloride,

Figure 19–2. Polygonal Crack Networks Developed in This Montmorillonitic Flood Silt, as a Result of Alternating Expansion and Contraction. Later the desiccation cracks were filled with a calcrete of lime and gypsum, and fossilized. Pleistocene Nile floodplain near Kom Ombo, Upper Egypt. (The hammer in the middle is 12 in., 30 cm, long.) (Karl W. Butzer)

is favored by evaporation of irrigation waters or of salt-charged waters in the shallow pans or playas or closed basins. The salt is originally derived from simple weathering products, from groundwaters emerging from salty marine sediments, or from atmospheric aerosols. Instead of being leached and carried away in solution as in more humid climates, desert waters are insufficient to flush out soil or groundwater salts. Saline profiles are referred to as *white alkali soils,* or solonchaks.

19-4. THE ACTIVITY OF WATER AND WIND IN DESERTS

The sparse shrub vegetation and the lack of sod in semidesert environments provide ground conditions resembling those of an exhausted and compacted, plowed field. Rainfall is sporadic but intensive, and infiltration small. Most of the water sweeps the surface in the form of sheet wash, rill wash, or sheet floods, little impeded by vegetation. The scattered, shrub vegetation also concentrates water in countless rills, but precipitation is localized and of limited duration, so that local floods dissipate rapidly and do not always provide defined channel flow in major streams. This unintegrated and relatively brief type of fluvial activity favors denudation by sheet and rill wash rather than by stream erosion.

Limited chemical weathering provides no complete soil mantle on hillslopes, and fines are commonly deflated as fast as they form. At the same time, rill erosion prevents a build-up of talus and other mechanically weathered debris on foot-slopes. The result is that foot-slopes are cut and expanded in all directions at the expense of the uplands, while hillslopes are undercut and tend to maintain steep angles. This form of intensive denudation favors backwearing and the cutting of rock-floored lowland plains known as *pediments* (Figures 19–3 and 19–4, also Figure 15–1). Thin alluvial veneers and local alluvial fans, at the exit of major upland streams, mantle these pediments. Deflation of fine materials also provides lag surfaces here, with microdunes tied to obstacles such as bushes or rock outcrops.

In the great desert interiors of Africa and Arabia fluvial processes hardly operate with the contemporary hyperarid climate. Even though fluvial forms such as alluvial spreads, pediments, and stream-cut valleys are prominent in the landscape, wind erosion and deposition hold the upper hand at the moment (Figure 19–5). But wind is a slow and an ineffective agent of denudation, so that it takes a long time to obscure inherited fluvial landforms that reflect on moister paleoclimates.

Strictly speaking, wind can operate as a gradational agent in any environment. However, significant wind erosion is limited to dry, finely divided materials and where vegetation is scant or absent. Ideal

Figure 19–3. Valley Pediment Cut into Alluvial Fill and Now Dissected by Gully, near Cornelia, South Africa. (Karl W. Butzer)

Figure 19–4. Broad Pediment Plain in the Red Sea Hills of Egypt, with Rilled, Alluvial Veneer. Complex igneous and metamorphic rocks form a jagged skyline. (Courtesy of C. L. Hansen.)

Figure 19–5. Sandblasted Sandstone on Pediment Surface, Nubia. (Courtesy of C. L. Hansen.)

situations are provided by deserts, beaches, exposed river beds, and parched, plowed fields.

Deflation can (1) winnow silt and fine sand, reducing the surface level and concentrating rock and coarse sand in a protective surface lag; (2) scoop blow-outs, the largest of which may be miles across and up to several hundred feet (100 m) deep (Figure 19–6); (3) carve out sharp, sinuous ridges (*yardangs*) and parallel depressions in dry silts, with a local relief of up to 30 feet (9m) (Figure 19–7); and (4) *sandblast* rock surfaces, sometimes undercutting rock outcrops within 1 to 2 feet (30 to 60 cm) of the ground (see Figure 19–5). Although blow-outs and yardangs are restricted in distribution, lag surfaces mantle most stony deserts. Lags of crude, mechanically weathered regolith are character-istic of the *hamada* or rocky desert, with pavements of rounded gravel and coarse sand produced by deflation of alluvial deposits to form *serir* or pebbly deserts.

At its most potent, operating over long periods of time, deflation appears to be responsible for the removal of sediment or chemically weathered materials, so preventing closed depressions from filling up with detritus. In part, deflation has necessarily played a critical role in creating great erosional basins, such as the major oasis depressions of the Libyan Desert.

Wind or *eolian transport* involves:

Figure 19–6. Large Blow-out Dunes (*Right Center*) in the Kurkur Oasis, Libyan Desert. Note mesaform hills on skyline and limited talus of foot-slope (*foreground*). (Karl W. Butzer)

Figure 19–7. Yardangs Excavated in Pleistocene Nile Silts, Kom Ombo, Upper Egypt. (Karl W. Butzer)

1. Suspension, to transport silt-sized particles over great distances and to great heights in the atmosphere
2. Saltation, as sand grains hop, land, and rebound, imparting fresh impetus to other sand particles—a motion confined to short distances and maximum heights of 5 to 8 feet (1.5 to 2.5 m)
3. Surface creep, when coarse sands and small pebbles inch forward with the momentum gained from jumping sand particles.

Sandstorms are actually restricted to the lowest layers of the atmosphere and confined in area. *Dust storms,* on the other hand, are far more effective. Dense, high clouds of dust of a single storm have been known to carry more than 100 million tons of material over distances of up to 2,000 miles (3,200 km), and medieval "blood rains" over England resulted from Saharan dust washing out of British skies. During the dry 1930s some of the storms generated in the Dust Bowl of western Kansas and Oklahoma dumped as much as 15 to 30 tons of dust per square mile (35 to 75 tons/sq km) over the distant state of Wisconsin. Across the centuries and millennia, such dust can potentially build up into thin sheets of loess.

Eolian *deposition* can take the form of *sheets* of sand or loess or of distinctive *dunes.* Dunes may occur singly or in fields, and may be "free" or "tied," in the lee of surface obstacles. Finally, dunes can be "live" and active, or they can be "fixed" or immobile under vegetation or a soil mantle.

The major types of dunes (Figure 19–8) include:

1. Sand drifts or tied dunes, of variable size, blown up in the lee of bushes, rocks, or isolated hills
2. Parabolic *blow-out dunes,* with a sand ridge located downwind of a deflation scar, the arms pointing upwind (Figure 19–6)

Table 19–1. GENERALIZED WIND ACTIVITY IN DIFFERENT ENVIRONMENTS

VEGETATION COVER	EOLIAN EROSION	EOLIAN DEPOSITS
Desert	Active: deflation lags (hamada, serir), yardangs, blow-outs	Dunes (barkhan, seif, transverse), sand seas, sand sheets
Grassland and tundra	Localized: blow-outs deflation lag, dry channel-bed deflation	Loess and sand sheets, local blow-out dunes
Forest	None	Negligible, with some eolian dust
Littoral vegetation	Localized blow-outs and general deflation	Restricted blow-out dunes or fields of transverse dunes

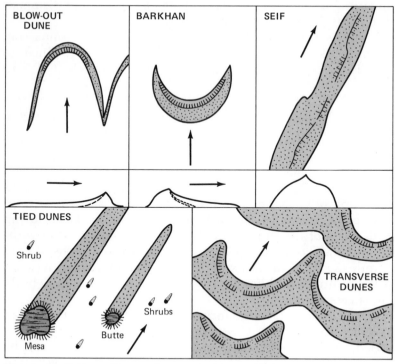

Figure 19–8. Dune Types. Arrows indicate effective wind directions.

3. Ridges of *transverse dunes*, running perpendicular to the effective wind direction and forming where fields of barkhans merge
4. Longitudinal *seifs*, constituting sinuous ridges running parallel with the effective wind direction
5. Crescent-shaped *barkhans*, with horns facing downwind
6. *Sand seas* of complex origin, with several superimposed generations of different dune forms, frequently dominated by transverse or pyramid shapes.

Whereas blow-out dunes form only where there is some vegetation, the other types are normally linked with bare, flat surfaces. In general, dunes only develop when there is a source of sand, such as dry-lake or stream beds in the desert, or sandy beaches on the coast. The great sand seas of the northwestern Sahara, southern Arabia, the Tarim Basin, or the Nebraska Sandhills, may have a local relief of over 300 feet (100 m) and thicknesses of as much as 1,000 feet (300 m). Other dunes are smaller: Some of the rare seifs are over 300 feet (100 m) high and as much as 65 miles (100 km) long; blow-out dunes, barkhans, and transverse dunes (Figures 19–9 and 19–10) seldom exceed 100 feet (30 m) in relief, 1,300 feet (400 m) in plan.

Although eolian processes can theoretically operate almost anywhere on a local scale, eolian landforms are almost confined to arid and coastal settings. However, whereas deflation significantly affects the majority of desert surfaces, sandy desert or *erg* accounts for only 25 percent of the Sahara and little more than about a quarter of the world deserts.

19-5. DESERT LANDFORMS

The landform assemblage of the desert is truly distinctive. To illustrate the characteristic features we can take a hypothetical cross-section of desert topography (Figure 19–11) and describe what is most pertinent. This model is based on the assumption of horizontal sedimentary rocks.

The *uplands and interfluves* consist of a gently undulating, rock-strewn hammada developed on horizontal caprocks. Surface drainage lines are few, very shallow, and poorly defined, connecting a number of broad, flat-floored depressions. Many of the depressions, the sides of which slope almost imperceptibly at 1 to 2 degrees, lack outlets and were originally excavated by eolian processes. Peripheral drainage

Figure 19–9. Transverse Dunes, Namib Desert, Southwest Africa. (Karl W. Butzer)

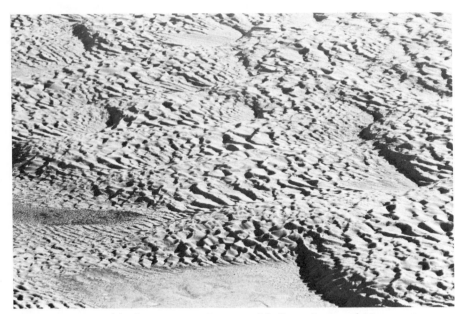

Figure 19–10. Field of Transverse Dunes, with Superimposed Macropatterns, Yuma Desert, Arizona. Effective wind direction is from the right. (Karl W. Butzer)

lines are deeply incised, with steep gradients and V-shaped valleys (*wadis*); even here drainage density is low.

The *hillslopes* linking the lowlands and uplands are steep, main-

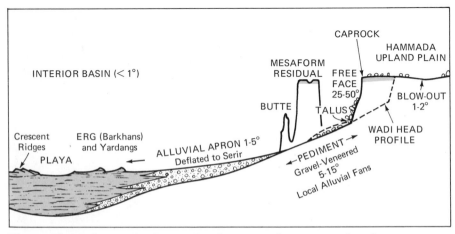

Figure 19–11. Composite Desert Landscape, with Horizontal Sedimentary Rocks. Note strong vertical exaggeration.

taining rectilinear mid-slopes with angles of 25 to 50 degrees. Undercut wadi walls may be vertical, while foot- and crest-slope inflections are angular. Mid-slopes are veneered with thin and incomplete mantles of talus (Figure 19–12).

Lowland *pediment plains* are cut across bedrock, steepening to as much as 5 to 15 degrees near their upper terminus, where pediments are veneered with alluvium and some slope wash. Further downslope, inclinations decrease to as little as 1 to 5 degrees as alluvial aprons thicken. Shallow alluvial fans are found where wadis debouch from the uplands, and piedmont alluvial plains may develop below fault-block scarps. The alluvial veneer is only superficially dissected by rills and shallow draws, the surface often deflated to a serir and studded with sand drifts or small tied dunes. Pediments are commonly interrupted by residual hills near the fretted, upland margins. Such residuals may form steep pinnacles (*buttes*), flat-topped hills (*mesas*), bell-shaped "domes," or simply heaps of blocks and boulders—depending on their size and lithology.

Interior basins form the focal point of pediment and piedmont alluvial plains. Normally of tectonic origin, they remain nonoutlet through deflation and a negative evaporation-precipitation balance. Slope inclinations are less than 1 degree, developed on alluvium, lake beds,

Figure 19–12. Weathered Sandstone Free Face, with Limited Talus Grading into Colluvium That Overlies Dissected Valley Fill. Hamada in foreground. Nubia. (Karl W. Butzer)

or evaporation pans. Eolian remodeling of these deep fills produces deflation surfaces, yardangs, blow-outs, and free or tied dunes. Permanent or seasonal salt pans (*playas*) are typical of the lowest points, with arcuate ridges of sediment deflated from the pan floor accumulating downwind. Only rarely are playa basins linked to through-rivers.

This landscape model can be readily adjusted to warped sedimentary rocks—with cuestaform plains and escarpments conspicuous—or to complex igneous and metamorphic rocks—where uplands are highly irregular and where pediment and playa systems commonly coincide with downfaulted basins or areas of weaker rock. Structure and lithology are consequently prominent and precondition the landform composite. But both the details and the landform assemblage produce a distinctive landscape. In fact, several of the functional landforms enumerated here are characteristic or even exclusive to the desert. The difference between arid and humid landforms is as fundamental as the contrast of fluvial and glacial landscapes. With sufficiently detailed work this can be defined more precisely and possibly expressed in quantitative terms. The most distinctive aspects of arid landscapes can be given as follows:

1. Extensive lowland and, in part, upland areas of remarkably smooth slopes, with limited relief and dissection, and disorganized or interrupted drainage
2. Coarse drainage texture, even in areas of appreciable relief and marked dissection
3. Steep, rectilinear slopes in areas of accentuated relief, offset from flat uplands and lowlands by sharp breaks of gradient
4. Persistence of fairly constant mid-slope angles in all stages of dissection—from a fretted upland escarpment to a residual butte or mesa (Figure 19–13)
5. Bimodal distribution of generalized slope classes in gentle (less than 5 degrees) and steep (greater than 25 degrees) categories, with only restricted surface areas of intermediate slope
6. Replacement of a "normal" soil mantle by lag surfaces, coarse alluvial spreads, crude slope wash, or talus veneers.

Although some of these landforms reflect eolian modeling or remodeling, they are mainly a result of the fluvial processes peculiar to arid environments.

19-6. DESERT PALEOFORMS

Pleistocene climates repeatedly affected landform development in desert regions. In part, cold glacial-age climate reduced evaporation or increased the frequency of the now rare cloudbursts, thereby revitalizing drainage lines and allowing the lakes of closed basins to expland.

Figure 19–13. Constant Slope Angle Is Maintained by These Rectilinear Free Faces Cut into Limestone, Kurkur Oasis, Libyan Desert. (Karl W. Butzer)

In part, too, warm climatic intervals saw an expansion of tropical monsoonal rains well into the desert belts, favoring the development of loamy soils, a higher water table, and the sustenance of ponds, marshes, and lakes in depressions, valley floors, and intradunal swales.

These paleoforms include:

1. *Fossil Lake Beds and Related Shorelines.* In mid-latitudes such features relate to colder climates with reduced evaporation and little or no increase in precipitation. The great expanse of "pluvial" lakes Bonneville and Lahontan in the American West (see Figure 18–7) or of the Aral and Caspian seas (see section B-3) are examples of this type, where a detailed record of lake-level fluctuations shows a close correlation between expanded lake volume and colder climatic trends. In lower latitudes, such as the Sahara, pluvial lakes more commonly relate to a poleward expansion of the monsoonal rain belt and are of early to mid-Holocene age. Substantial increases of rainfall would be required to maintain the huge lakes such as the one that covered much of the Chad Basin between 7500 and 3000 B.C.

2. *Alluvial Terraces.* Most larger stream systems and even many wadis, no matter how inactive they may seem today, are accompanied by sets of dissected, alluvial fills (Figure 19–14). Some of these consist of crude, poorly sorted angular rubble, others of large, rounded

Figure 19–14. Pleistocene Alluvial Terraces near the Red Sea Coast of Egypt. These fills were deposited at a time of falling sea level in response to accelerated discharge and sediment transport from the upper watershed. (Karl W. Butzer)

cobbles, and still others of silts or sands. General interpretations are difficult to give, but in hyperarid areas the majority of these features relate to more frequent and more prolonged rainstorms, that is, increased morphodynamism (Butzer and Hansen, 1968). Much of the material reworked in such alluvial spreads was prepared either by slow weathering during the preceding period of more inert runoff or by a coeval acceleration of mechanical weathering, particularly frost.

3. *Spring and Marsh Deposits.* Higher water tables periodically supported lime-rich ponds or marshes in small depressions or fed seasonal or perennial springs in limestone country, with the result that massive tufas and travertines built up along foothills. Such features, now blasted by wind erosion, ring many contemporary or former oases.

4. *Paleosols.* Relatively striking and clayey, red soils are not uncommon on top of the alluvial terraces in desert regions (Figure 19–15). In individual cases, strong arguments can be offered that these are not only more "mature" than modern desert soils but that they could never form under present conditions of erosion and minimal chemical weathering. They would require increased soil stabilization by vegetation as well as intensified biochemical alteration. More effective chemical weathering may also be responsible for sculpturing ta-

Figure 19–15. Alluvial Terrace in a Nubian Wadi. Older fill (late Pleistocene) is cemented by a white honeycomb of calcrete; younger fill (early Holocene) is weathered to a red paleosol, now buried under a colluvium of crude gravel. (Karl W. Butzer)

foni as well as shaping those inselbergs that resemble savanna counterparts.

The role of past climates is not limited to isolated, often disjunct features. There is every reason to believe that paleoclimates affected the evolution of composite landforms such as the great ergs and many pedimented landscapes.

In the case of the sand seas, it is believed that the sand was eroded from old marine sediments and washed into depressions by running water, under relatively moist conditions. It was subsequently sculptured into great dune fields, on several occasions, by wind action under hyperarid conditions. Repeatedly, too, red paleosols developed, increasingly stabilizing the dunes, so that many such sand seas are in good part inactive today (Tricart and Cailleux, 1969). In some cases the paleosols can be linked to now defunct, intradunal lakes.

In the case of pediments the problem is that slope retreat is negligibly slow in dry areas where frost is absent, and even in colder deserts pediments are commonly choked by relatively thick sheets of alluvium that indicate little or no rock beveling underway. This should not imply that pedimentation is inoperative today but that its prevalence and effectiveness in the past demand an explanation. Some authors feel that pediments are polygenetic in their development (Birot, 1968), with chemical weathering (more vegetation and moisture) first preparing a residual mantle, which is subsequently removed by a phase of accelerated runoff (greater cloudburst frequency); frost weathering of rock faces (colder climate) would aid in slope retreat. Such a sequential development could include long periods of limited dynamism of any kind. Pediments in Nubia repeatedly developed as lateral, erosional plains graded to alluvial terraces of the Nile in early to mid-Pleistocene times. However, the great coalescent pediment plains that reach far into the deserts are similarly linked to the development of the Nile Valley in Miocene or Pliocene times, when all evidence suggests a subarid to semiarid climate (Butzer and Hansen, 1968).

In the heart of the great deserts many geomorphic features clearly are inherited, while others are polygenetic. Nonetheless, landforms everywhere seem to be fairly consistent, and relatively moist paleoclimates appear to have changed the rate and efficacy rather than the overall direction of landform development. If it is correct that pediment cutting is most effective under subarid or semiarid conditions, particularly when intensive frost weathering aids in slope retreat, then it could be argued that past periods of cooler or moister climate served to endow low-latitude deserts with features similar to those currently forming in the semideserts of the western United States and central Asia. Thus the desert landscapes would appear to be a composite related to dominance of backwearing processes through most of the geological past. Inherited landforms are evident everywhere, but these blend with features forming under contemporary conditions. As a result, the alternation of wetter or drier climates has never changed the fundamental character of arid zone landscapes. They remain more or less homogeneous, even with superficial remodeling by wind, on the one hand, and accelerated fluvial processes or chemical weathering, on the other.

19-7. GEOMORPHIC EQUILIBRIUM IN MID-LATITUDE GRASSLANDS

The arid zone is diverse in many ways, not the least in terms of geomorphic processes. Geomorphic change is slow and almost imperceptible at the hyperarid and humid ends of the spectrum, while in between, near the arid-semiarid boundary, erosion rates are the highest of any major environment. Based on contemporary data from drainage basins in the United States, Schumm (1965) has argued that mean annual sediment yield and, indirectly, the mass of soil eroded, is greatest in areas with 12 inches (300 mm) of precipitation (Figure 19–16). At this critical value, more than 800 tons of sediment are eroded per square mile (275 metric tons/sq km)—which is equivalent to stripping an average thickness of 12 inches (30 cm) of soil from all surfaces every 1,000 years. However, in very dry and in humid parts of

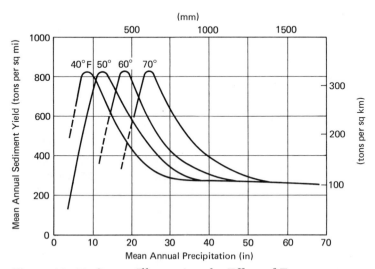

Figure 19–16. Curves Illustrating the Effect of Temperature on the Relation Between Mean Annual Sediment Yield and Precipitation. (Adapted from Schumm, 1965.)

the United States, the sediment yield is less than one-half this value, that is, under 400 tons per square mile. This means that rates of erosion increase steadily across the Great Plains, from Chicago to Denver, but decrease again between Denver and Phoenix, Arizona.

In the undisturbed state, the tall grasses, organic sod, and dense rooting network of the prairie vegetation inhibit erosion. Cultivation exposes the underlying soils to heavy rains that often follow long periods of dry weather. Especially in areas of unconsolidated sediments—on loess and till plains—accelerated soil erosion becomes prominent, with extensive gullying. This double effect has distorted the curve relating precipitation to sediment yield. In fact, it is more likely that lush, undisturbed grasslands would normally provide a geomorphic equilibrium comparable to that of the mid-latitude woodlands. On the other hand, the short grasses of the drier half of the same macroenvironment (Figure 19–17) provide a less effective cover, with much bare soil exposed between tufts of grass. This applies even where land use is restricted to moderate grazing. Lag can often be seen among the grass tufts, and it seems that such soils do indeed erode rapidly.

The role of grassland vegetation in stabilizing soil is consequently important, both on the North American prairie and the Ukrainian steppe. The gradation of humus content (from a maximum of 5 to 20 percent in chernozems), calcification, and degree of leaching closely

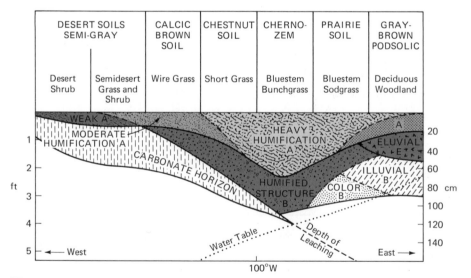

Figure 19–17. West to East Transition of Soils and Vegetation Across the North-Central United States.

follows moisture availability and vegetation type (Figure 19–17). Today these landscapes are dominated by well-developed gully networks that effect upland dissection, while the major river floodplains are choking in silty alluvium. Spells of dry years favor active deflation and blow-outs on the interfluves, particularly when cultivation and grazing remove the grass cover, for instance, the Dust Bowl of the 1930s. In fact, the mid-latitude grasslands happen to coincide in large part with Pleistocene depositional plains (loess, sand, till) that provide both fresh parent material for high base-status humified soils and soft sediment for ready erosion. As a result, this landscape is relatively stable under native vegetation but particularly unstable in the wake of land use. In effect, then, the mid-latitude grasslands constitute a geomorphic landscape that is transitional between the humid and arid spheres, with both the legacy of Pleistocene forms and high initial soil stability linking it to humid mid-latitudes.

19-8. SOILS AND LANDFORMS OF LOW-LATITUDE GRASSLANDS

Temperatures are high all year in the semiarid environments of the subtropics and the tropics, so that organic matter is rapidly decayed and oxidized, and humification restricted. The wet season is marked by intensive rains that leach all but clayey soils, so that Ca horizons are poorly developed on most old erosional surfaces or fixed dune fields. At the same time, there is a certain amount of hydrolysis and oxidation, producing striking B horizons with some clay minerals. Consequently, the typical grassland soils between latitudes 35° N and S range from *calcic* and *noncalcic brown* to *reddish-brown soils* that are more erodible than their mid-latitude counterparts.

Almost equally important are dark clay soils found on many clayey parent materials or where silt-covered lowlands are prone to seasonal flooding and temporary waterlogging. These intrazonal *vertisols*[1] of the tropics and the subtropics are due to the strong, vertical mixing dynamism that results from alternating swelling and cracking. The A horizon, although poor in humus, is dark gray and uniform to a depth of 3 to 6 feet (90 to 180 cm)—the depth of the cracking horizon. Typical vertisols tend to be situated on upland or lowland flats, so that they frequently have direct geomorphic significance as representative landscape components.

Lower-latitude semiarid environments comprise a variety of Tertiary-to-Pleistocene erosional plains of angular form and coarse drainage in western Texas, Patagonia, parts of Australia, and in South

---

[1] Other names are *black cotton soil, regur, tirs, margalitic,* and *grumusol.*

Africa. These may include eolian veneers or blow-out depressions that strengthen the arid flavor of such landscapes. Extensive Pleistocene sand sheets or dune fields take on regional significance in the thorn savanna along the southern margins of the Sahara or in the Kalahari. In all cases the contemporary environment is transitonal between the humid and arid zones, a fact often obscured by the legacy of inherited forms. But the moment the vegetation cover is disturbed by human activity, accelerated processes resume and continue to accentuate the arid flavor of the geomorphic landscape.

19-9. MEDITERRANEAN LANDSCAPES

In the summer-dry, subtropical woodlands of the Mediterranean Basin, winters are cool enough to inhibit chemical processes such as hydrolysis but insufficiently cold to promote significant frost weathering. In fact, chemical weathering is largely limited to the moist but relatively cool transitional seasons—summers are too dry, winters too cold. Soil development is correspondingly modest and commonly inadequate to produce distinctive soil profiles other than semiarid variants of temperate woodland soils. Instead, relict or polygenetic soils are widespread.

There are abundant examples of deep, reddish or yellowish soils. On igneous or metamorphic rocks these are red-yellow podsolics;[2] on limestone they constitute red *terra rossas* or yellowish *terra fuscas*.[3] These limestone variants are neutral in reaction and, like the other red Mediterranean soils, have more base nutrients. Decomposition of compact limestones and possibly also of some schists may produce residues not unlike the materials found in the B horizon of reddish Mediterranean soils. However, such weathering has been too slow to produce distinctive soil profiles recognizable as red-yellow podsolics, terra rossas, or terra fuscas during the past 10,000 to 50,000 years. In this sense, all such soils in the Mediterranean region are relict. They formed during moist, warm Pleistocene interglacials. In practical terms, they do not develop afresh once they are eroded. Since the red-yellow podsolics of the Mediterranean Basin are highly erodible, their occurrence is now sporadic and broad alluvial plains are composed of reddish soil wash. Similar relicts can be found in parts of California, central Chile, and southwestern Australia.

The most widespread soils of the mediterranean-type environments are brown and humified. On many rock types they resemble brown forest soils, although they are sandy by comparison (*noncalcic*

---

[2] Also designated as *rotlehm* and *braunlehm*, German for "red" and "brown loam" respectively.
[3] Italian for "red earth" and "yellow earth" respectively.

brown soils).[4] On limestone there are calcimorphic rendzinas, with dark humus horizons over zones of lime accumulation. These various brown mediterranean soils differ in texture, clay-humus content, and fertility according to the parent material, the degree of soil aridity, and the abundance of soil microorganisms. Clayey relict soils and derived soil wash are now largely confined to lowlands and stream valleys of the Mediterranean Basin (Figure 19–18), and geological evidence suggests that rough terrain in compact rocks has been devoid of deep soil mantles since the mid-Pleistocene. In fact much or most of the erosion of red-yellow podsolics and terra rossas can be dated to the successive glacial periods, when sheet washing and colluvial processes were particularly active (Butzer, 1974, 1975).

Active soil erosion in the Mediterranean Basin was long delayed by careful land management. Only in post-Roman times did stripping and gullying of terraced slopes and older valley fills produce glaring examples of accelerated soil erosion (see Figure 7–20). In California and central Chile, where general landform evolution appears to have many similarities with that of the Mediterranean world, man-induced soil erosion is a matter of the past century or two. For all the mediterranean environments, the relict nature of the deep soil mantles makes soil conservation vital, since fresh alluvial or colluvial beds often are sandy or stony and poor in organic matter.

The overall impact of limited hydrolysis, repeated morphodynamic episodes during the Pleistocene, and historical soil erosion is to provide a decidedly semiarid geomorphic environment. This is reflected in both the landscape assemblage and in contemporary pro-

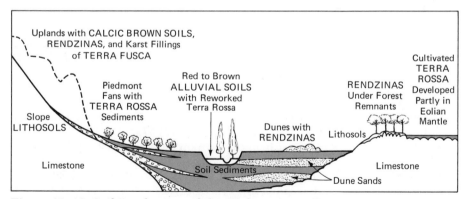

Figure 19–18. Soil Landscapes of the Mediterranean Basin.

[4] Another name is *meridional braunerde*, German for "southern brown earth."

cesses. In areas of both limited and great relief, slopes tend to be steep and rectilinear, inflections and crestlines angular, with extensive development of erosional plains. The latter are in part inherited from marine abrasional surfaces of Pliocene or Pleistocene age, but they have been and occasionally continue to be enlarged by pedimentation. Pleistocene eolian deposits of sandy to silty texture mantle many such surfaces, both in proximity of the coast or major rivers. Piedmont accumulations of coarse gravel and reworked red soils also recall arid landscapes. Contrary to some impressions, karst activity is subdued today, and even in the subhumid Yugoslavian karst, most macroforms such as polja were created in late Tertiary or early Pleistocene times (Büdel, 1973). Finally, all the evidence suggests that mediterranean geomorphic landscapes have long been dominated by sheet erosion or dissection or both, so as to favor backwearing.

19-10. SELECTED REFERENCES

Birot, Pierre. *The Cycle of Erosion in Different Climates*. Berkeley and Los Angeles, University of California Press, 1968, 144 pp.

Büdel, Julius. Reliefgenerationen der Poljebildung im dinarischen Raum. *Geographische Zeitschrift*, Beiheft 34, 1973, pp. 134–142.

Butzer, K. W. Desert landforms at the Kurkur Oasis, Egypt. *Annals, Association of American Geographers*, Vol. 55, 1965, pp. 578–591.

———. *Environment and Archeology*, 2nd ed. Chicago, Aldine, and London, Methuen, 1971, 703 pp. See Chaps. 19 and 20.

———. Accelerated soil erosion: a problem of man-land relationships. In I. Manners and M. W. Mikesell, eds., *Perspective on Environment*. Washington, D.C., Association of American Geographers, 1974, pp. 57–78.

———. Pleistocene littoral-sedimentary cycles of the Mediterranean Basin. In K. W. Butzer and G. L. Isaac, eds., *After the Australopithecines*. The Hague, Mouton, and Chicago, Aldine, 1975, pp. 25–71.

———, and C. L. Hansen. *Desert and River in Nubia*. Madison, Wis., University of Wisconsin Press, 1968, 564 pp.

———, and D. M. Helgren. Late Cenozoic evolution of the Cape Coast between Knysna and Cape St. Francis, South Africa. *Quaternary Research*, Vol. 2, 1972, pp. 143–169.

———, G. L. Isaac, et al. Radiocarbon dating of East African lake levels. *Science*, Vol. 175, 1971, pp. 1069–1076.

Cooke, R. U., and A. Warren. *Geomorphology in Deserts*. Los Angeles, University of California Press, and London, Batsford, 1973, 373 pp.

Engel, C. G., and R. P. Sharp. Chemical data on desert varnish. *Bulletin, Geological Society of America*, Vol. 69, 1958, pp. 487–518.

Ganssen, Robert, and F. Hädrich. *Atlas zur Bodenkunde*. Mannheim, Bibliographisches Institut, 1965, 85 pp.

Glennie, K. W. *Desert Sedimentary Environments*. Amsterdam, Elsevier, 1970, 222 pp.

Goudie, Andrew. *Duricrusts in Tropical and Subtropical Landscapes*. Oxford, Clarendon, 1972, 192 pp.

Hastenrath, S. L. The barchans of the Arequipa region, southern Peru. *Zeitschrift für Geomorphologie*, Vol. 11, 1967, pp. 300–331.

Jennings, J. N., and J. A. Mabbutt. *Landform Studies from Australia and New Zealand*. London, Cambridge University Press, and Canberra, Australian National University Press, 1967, 434 pp.

Kaiser, K. H. Prozesse und Formen der ariden Verwitterung am Beispiel des Tibesti-Gebirges. *Berliner Geographische Abhandlungen*, Vol. 16, 1972, pp. 49–82.

Kirkby, M. J., and M. A. Carson. *Hillslope Form and Process*. Cambridge and New York, Cambridge University Press, 1972, 475 pp. See Chap. 13.

Mabbutt, J. A. Mantle-controlled planation of pediments. *American Journal of Science*, Vol. 264, 1966, pp. 78–91.

Oberlander, T. M. Landscape inheritance and the pediment problem in the Mojave Desert of southern California. *American Journal of Science* 274, 1974, pp. 849–875.

Peel, R. F. The landscape in aridity. *Transactions, Institute of British Geographers*, Vol. 38, 1966, pp. 1–24.

———. Insolation weathering: some measurements of diurnal temperature changes in exposed rocks in the Tibesti region, central Sahara. *Zeitschrift für Geomorphologie*, Suppl. Vol. 21, 1974, pp. 19–28.

Rathjens, Carl, ed. *Klimatische Geomorphologie*. Darmstadt, Wissenschaftliche Buchgesellschaft, 1971, 485 pp. Contains German classics by Horst Mensching, Wolfgang Meckelein, Hans Mortensen, and Leo Waibel.

Schumm, S. A. Quaternary paleohydrology. In H. E. Wright and D. G. Frey, eds., *The Quaternary of the United States*. Princeton, N.J., Princeton University Press, 1965, pp. 783–794.

Smith, G. I. Late Quaternary geologic and climatic history of Searles Lake, southeastern California. In R. B. Morrison and H. E. Wright, eds. *Means of Correlation of Quaternary Successions*. Salt Lake City, Utah, University of Utah Press, 1968, pp. 293–310.

Tricart, Jean, and André Cailleux. *Le Modelé des Régions sèches*. Paris, SEDES, 1969, 472 pp.

Chapter 20

TROPICAL LANDFORMS

The subhumid to humid intertropical zone includes savannas and open woodlands, with a substantial wet season, and a rainforest, with minor dry seasons only. A basic trait of all of these environments is intensive chemical weathering and a deep, continuous soil mantle on gentle to intermediate slopes. Soils on well-drained surfaces of Pleistocene age are red and yellow loams similar in many ways to red-yellow podsolics. Deep latosols are restricted to Tertiary surfaces where they integrade with primary laterites, defined by an indurated horizon followed at depth by mottled and colored zones. Secondary laterites form in hillslope colluvial situations, through ferricretion of reworked lateritic rubble. Other ironpans that do not qualify as true laterites can form in lowlands through lateral seepage of iron-rich waters. The totality of exposed ironpans are examples of duricrusts that have great

geomorphic significance in stabilizing horizontal or mesaform landscapes.

Savanna slopes retreat by a special form of tropical pedimentation, through chemical undermining, throughflow eluviation, and sheet wash. Low-order valleys in the wetter savanna environments are also distinctive, frequently lacking a continuous stream channel. Such dambos are extremely shallow and combine marginal planation (by chemical weathering, throughflow, and sheet wash) with valley axis aggradation. A combination of tropical pedimentation and dambo enlargement appears to be fundamental to the slow evolution of characteristic savanna plains during the course of the Tertiary. Such plains are studded with block masses (tors) or domed hills (inselbergs) that were exhumed from the uneven weathering base under deep residual mantles. The overall impression of great areal planation appears to result from convergence of once compartmentalized drainage basins over time spans exceeding 50 to 75 million years. Savanna plains in sedimentary strata or flood basalts tend to be mesaform, while limestone solution, given very long periods of time, develops accentuated karst relief of cone and cockpit type, in which the open cockpits represent planation surfaces and large cones or karst towers represent inselbergs. Rainforest environments favor valley incision in deep regolith, aided by subterranean mass movements and leading to multiple, small planation surfaces along the margins of high-order valleys through repeated tributary convergence. In montane rainforests, intensive and prolonged downpours lead to large-scale mass movements, including frequent debris slides. Slope profiles tend to be concave, therefore, with sawtooth ridges developing on the watersheds. Deforestation of slopes and cultivation of soils everywhere greatly accelerates denudation rates.

> . . . in the smell of dust and of acacias
> under a hot sun. . . .
>
> WILFRED THESIGER, *Arabian Sands*

20-1. ELUSIVE TROPICAL LANDSCAPES

The tropics mean many things to many people. To the ancient Greeks this was the torrid zone, between latitudes 22° N and 22° S. To nineteenth-century travelers these were lands without winter, where rainy seasons alternated with dry seasons. To climatologists these are environments where daily ranges of temperature exceed all seasonal fluctuations to the thermometer. To some geomorphologists the word "tropics" implies vast, open savanna plains dotted with isolated hills, and to others it conjures up an impression of wildly dissected, jungle-clad hills. To a degree, each of these definitions is true. But the disparity of views is symptomatic of the fact that the essence of the tropics is difficult to pinpoint, that they are diverse, and that it remains impossible to reconcile all impressions of what is characteristic or distinctive.

Tropical geomorphology, as applicable to the perennially or seasonally humid innertropical zone, is also not a unity. Perhaps the lowest common denominator is that intensive chemical weathering is the rule rather than the exception. Beyond that, disagreement as to process and form, or inherited versus contemporary features, is vast. Tropical landscape assemblages remain poorly understood. Even the origin of the extreme variants of tropical soils is shrouded in controversy. And depending on their field experiences, geomorphologists will emphasize some aspects over others. Taking a synthetic perspective, then, requires considerable caution. It is inevitable that many problems will remain oversimplified, but it would be inexcusable if a rigid, simplistic formula were superimposed on the disparate data base.

20-2. TROPICAL WEATHERING

Intensive chemical weathering is the hallmark of tropical climates, with the only limitations being imposed by the length of the dry season. In the ever-humid equatorial regions or rainforest such weathering is maximal, with perennial soil moisture and subsurface temperatures remaining in the 70s (21 to 26° C) throughout the year. In the wet-and-dry tropical climates or savannas, deep chemical weathering is possible during the rainy part of the year, a span of 6 to 10 months. The intensity of weathering is reflected by deep soil mantles, with few primary minerals other than quartz. Even below the soil profile, partial decomposition is often apparent to considerable depths in granites and metamorphic rocks. It is not unusual for soil profiles in some parts of the humid tropics to average 30 feet (9 m) in thickness, with partially rotted bedrock (*saprolite*) down to 100 feet (30 m) (see Figure 3–7). Soil formation is so rapid that complete soil mantles, capable of supporting a closed forest, can be found on slopes of as much as 50 to 60 degrees. Furthermore, exposed granites or metamorphics often show vertical fluting or grooving, comparable to microkarst in limestone, and presumably due to a combination of chemical weathering and mechanical wear. Cavernous rotting, such as tafoni, is another microform due directly to weathering. Finally, a poorly understood mobilization of iron and manganese compounds is responsible for the most extreme forms of varnish or patina found on the semiarid margins of the tropics.

Except for high mountains, frost is absent (Figure 19–1). Other forms of mechanical weathering include thermal tension (aided by chemical alteration) on exposed bedrock, salt hydration in sheltered locations, and above all pressure unloading that detaches boulders or slabs (see Figure 3–4). Nonetheless, slope detritus below bedrock outcrops is limited, partly because mechanical breakdown is slower than in frost climates, partly because rock debris is rapidly decomposed. Gravel is also rare in permanent streams of the humid tropics, except where rock outcrops in the channel bed: Pebbles rot about as quickly as they are supplied to the bed load.

The podsolization of cool, humid woodlands involves a form of biochemical weathering that disassociates clay minerals into their component clays, sesquioxides, and colloidal silica. These sesquioxides become highly mobile in acidic soil environments and are eluviated into the subsoil or groundwater. On the other hand, in warm humid environments *latozation* leads to clay mineral disassociation with selective shedding and eluviation of the colloidal silica (see Figure 4–5). Several prerequisites appear to be essential for latozation:

1. Very intensive hydrolysis and oxidation
2. Semipermanent soil moisture and high subsoil temperatures
3. A neutral or slightly acid soil environment (pH 5 to 7), without organic acids but with some nutrient ions

Unlike podsolization, latozation is slow, and a million years or more may well be necessary to produce the basic characteristics of a true latosol. Two features are diagnostic:

1. Free sesquioxides accumulate in the upper B horizon until iron and aluminum compounds reach concentrations of 10 to 50 percent, forming colorful concentrations or slaglike lumps or layers of ironstone, or *ferricrete*.
2. The colloidal silica is largely washed out of the soil or crysta-

lized into inert quartz, so that the clays lose their plasticity; as a result clayey soils feel earthy or friable, instead of sticky, when wet and even massive clays can become curiously permeable.

20-3. THE SOIL LANDSCAPE

Several representative soil profiles can be described and discussed in order to appreciate more fully the geomorphic role of modern and relict soil processes or forms.

Red Tropical Latosols. The final result of prolonged latozation is development of *red latosols*.[1] These are found on well-drained erosional or alluvial surfaces exposed to soil development since Miocene times or earlier, that is, for at least 5 million years. They are best developed and most widespread in humid, rainforest environments (see Figure 16–1). Typical B horizons vary from 3 to 25 feet (0.9 to 7.5 m) or more in thickness, with some degree of iron segregation somewhere in the subsoil. The clays of the B horizons are mainly kaolinitic and do not swell or shrink in response to wetting and drying; at the same time, these clays have a low exchange capacity and retain few mineral ions, even when the soil is saturated with bases. The removal of most of the colloidal silica makes the soil friable, nonplastic, and highly permeable. Because of this high infiltration capacity, latosols do not erode easily.

The remarkably low fertility of latosols is due to the low exchange capacity and their antiquity, with all weatherable minerals decomposed into inert residues. Low fertility is also caused by bacteria that destroy organic matter about as fast as it is added to the soil, breaking it down to inert acids, proteins, waxes, and resins.

Red and Yellow Tropical Loams. Throughout the humid tropics there are extensive areas of extreme tropical soils. These are found on younger surfaces (Pleistocene alluvium or volcanics) or on parent materials that continually release base nutrients (limestone, basalt) or that weather slowly (shales). These intermediate soils are particularly characteristic of the wet-and-dry savanna climates (see Figure 16–1) where hydrolysis is retarded.

The overall profile appearance of the red and yellow tropical loams compares with the red-yellow podsolics, with which these soils intergrade.[2] There is little or no evidence of podsolization on the one

---

[1] Other names for these are *roterde* (German for "red earth") and *ferrallitic soils*.
[2] Synonyms include *rotlehm* and *braunlehm* in the German literature, the *ferrisols* and *fersiallitic* soils of the Belgian and French pedologists, and the *terra roxa* and *arenolatosols* of Brazilian authors. This group has also been called *plastosols*, that is, plastic soils, in contrast to the friable latosols.

hand, while latozation is incomplete on the other. Eluviation of colloidal silica is insufficient to affect soil plasticity, so that soils are dense, impermeable, and sticky when wet. Different classes of clay minerals exist side by side, since weathering has been less protracted; some of these have better exchange capacities. Sufficient sesquioxides have been released in the B horizon to produce vivid colors, although significant iron concretions are rare.

The yellow variety of these intermediate tropical soils is most common on fresher materials or in permanently moist woodlands. The red variety owes its coloration to anhydrous iron oxides (that lose their water molecules after intensive drought) and is most common in seasonally dry savanna environments.

Both varieties are comparatively rich in base nutrients and moderately fertile. However, use tends to deplete organic matter and ruin the structure, which becomes dense, compact, and poorly aerated. Their plasticity and impermeability when wet make the red and yellow tropical loams among the most erodible of soils.

Laterites. The tropics, whether savanna or rainforest, have their share of gley and bog soils, particularly the seasonally flooded vertisols. But most peculiar among the intrazonal soils are hardpans known as *laterites*. These form zones of concretions, pitted oxide masses, or wavy bands of iron or aluminum oxides. Many authors recognize what might be called a *primary* laterite profile (see Thomas, 1974):

> *Surface soil:* Variable thickness, often colluvial in origin.
>
> *Indurated horizon:* 3 to 30 feet (1 to 10 m) thick. Reddish clay interspersed with ferruginous concretions or ferricrete, honeycomb networks.
>
> *Mottled horizon:* 3 to 30 feet (1 to 10 m) thick. Mottled red and white clay, with blocky to prismatic structure, often rich in quartz grains.
>
> *Pallid horizon:* 3 to 80 feet (1 to 25 m) thick. White kaolinitic clay and quartz sand, with limited ferruginous mottling, grading down into saprolite.

These indurated and mottled horizons should be essential to any definition of a primary laterite, as opposed to *secondary* laterite, consisting of indurated fragments of older laterites. Most secondary laterites accumulated as foot-slope colluvia, with matrix ferricretion through iron-rich seepage waters (Figure 20–1).

Primary laterites initially develop near the top of the water table, where oxides are precipitated after being made soluble by oxygen-seeking anaerobic bacteria in permanently waterlogged subsoil or saprolite.

Originally soft, lateritic horizons will only crystallize into

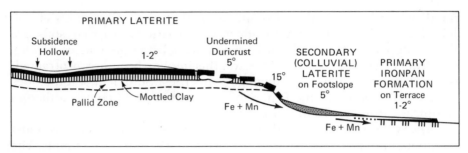

Figure 20–1. Laterites and Relief. (Based on Maignien, 1966, and DeSwardt, 1962.)

hardpans when the water table drops and intensive dehydration ensues. This can happen as a result of natural factors, when the climate becomes drier or when stream dissection lowers the water table. Frequently, too, man is responsible through forest clearance and cultivation, which favor seasonal dehydration. Deliberate burning and accelerated soil erosion can have similar effects. Whatever the cause, induration of laterites is irreversible. When the topsoil is stripped away, such lateritic ironpans are exposed as crusts of rocklike durability (*duricrusts*).

Since there are many different forms of iron enrichment or segregation, some semantic distinctions are in order. First, "primary laterite" should be reserved to latosolic profiles with indurated, mottled, and pallid horizons. Second, "secondary laterite" can be conveniently used for colluvial accumulations of lateritic rubble, newly ferricreted. Third, "ferricretion" refers to any process of oxide cementation, including forms that are not laterites in the original sense. In particular, sesquioxide enrichment and induration can take place in many soils of most latitudes, as a result of pedogenetic eluviation, groundwater enrichment, percolation from surface waters, or precipitation from lateral seepage. The last two forms of nonlateritic ferricretion are common in mildly acidic tropical soil environments (Figure 20–1). And fourth, "ironpan" can be usefully applied to all types of subsurface ferricrete, as distinguished from (ferruginous) "duricrusts" exposed at the surface.

Laterite duricrusts of great geological age (Mesozoic mid-Tertiary) mantle many tropical uplands or survive as caprock on residual hills. They erode slowly and resist weathering, preserving ancient surfaces almost indefinitely (Figures 20–1 and 20–2). When undermined or destroyed, duricrust slabs or iron concretions can collect in poorly drained lowlands, where they may be cemented together into secondary laterites. It appears that even secondary laterites take millions of years to form, and in Uganda they are exclusively of Tertiary age

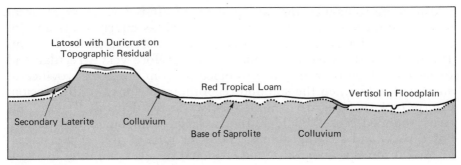

Figure 20–2. Savanna Soil Landscape.

(DeSwardt, 1964). Collectively, duricrusts of various ages, elevations, and origins are significant in the geomorphology of erosional plains and hills in the tropics.

Ironpans or duricrusts that are adequate as building stone or to be mined for iron ore and bauxite constitute a considerable if not insuperable obstacle to cultivation. Even under far less extreme conditions, the presence of incipient laterite horizons in the subsoil also poses the threat of hardpan development once a plot is cleared and cultivated. Nonetheless, true laterites are localized in their distribution and are therefore less restrictive to agriculture than are soils of low fertility or high erodibility.

20-4. GEOMORPHIC SIGNIFICANCE OF TROPICAL SOILS

Weathering processes and soils of the tropics are particularly crucial to an understanding of the geomorphic balance and landforms in general:

1. Deep chemical weathering is a vital prerequisite for several distinctive processes and forms. Deep soil mantles are maintained under natural equilibrium, and the base of the saprolite in igneous and metamorphic rocks is sometimes found at depths of 100 to 500 feet (30 to 150 m). Erosion normally operates in the medium of the soil rather than bedrock, and decomposition outpaces the possible effects of mechanical abrasion by streams, rills, or sheet wash.

2. The thickness, the clayey, impermeable nature, and the limited humus of most tropical soils make them exceptionally prone to erosion, for example, to mass movements on slopes. Only the permeable latosols and various duricrusts are immune.

3. Differential weathering and erosion are more important than in any other environment. Weaker bedrock varieties, rock joints, basic intrusions, and unstable minerals in general (micas, plagioclase

feldspars, ferromagnesians) are selectively altered to clays at great depths within the saprolite. Other rocks such as quartz veins and intrusions, quartzites, and fine-grained granites are far more resistant. The depth of the weathering mantle is therefore very irregular and varies considerably from place to place or from one rock province to another. If and when the soil and regolith are stripped off by erosion, bizarre rock surfaces are exhumed: networks of "depressed" joint patterns, rounded "core" boulders that were once surrounded by decomposed rock, and striking domes, knobs, or hills of greater resistance.

4. Duricrusts can fossilize old erosional surfaces and interfluves almost indefinitely.

Two aspects of how the soil mantle relates to slope forms require

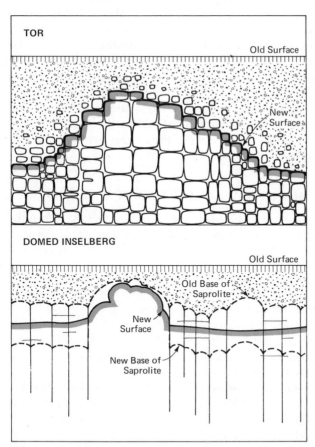

Figure 20–3. Formation of Tors and Domed Inselbergs. Deep weathering is faster in closely jointed rock and continues after the tor or inselberg is exposed. (Based on Thomas, 1974.)

amplification. These involve exhumed features and slope undermining. First, deep weathering with saprolitic rot down to 100 feet (30 m) or more is most common in crystalline rocks like granite and gneiss. The irregular base of the saprolite here provides ready-made landforms the moment the residual mantle is stripped off by erosion. The development of two prominent features of this kind is shown in Figure 20–3. In the case of large jointed blocks a castlelike profile is created referred to as a *tor* (Figures 20–4 and 20–5). When a group of tors decays through ongoing weathering, it may take the form of woolsacks (Figure 20–6). In the case of a much larger, unjointed mass, a dome-shaped hill emerges (Figure 20–7). Such *domed inselbergs* (or bornhardts) may be well over 3,000 feet (1,000 m) in relief. Raindrop impact, surface runoff, and chemical weathering combine to form rill-like macrogrooves (Figure 20–8). The sugar loaf mountains around the bay of Rio de Janeiro are a special form of tall and slender inselberg.

Second, weathering can actively promote slope retreat through basal undermining. Figure 20–9 provides an explanatory model of this process, emphasizing the role of percolating slope runoff and direct rainfall in throughflow. These waters move laterally through the regolith and even the saprolitic bedrock, removing solubles, facilitating hydrolysis, and above all removing clay and silt. This material is deposited farther down the foot-slope by seepage within and on top of any coarse-grained collu-

Figure 20–4. Castellated Tor Group in Granite, Serengeti Plain, Tanzania. Regolith has been stripped from the piedmont. (Karl W. Butzer)

Figure 20–5. Detail of Tor (Figure 20–4), Showing Incipient Vertical Fluting and Extensive Ferromanganese Staining. (Karl W. Butzer)

Figure 20–6. Group of decayed or Woolsack Tors in Granite, near the Erongo, Southwest Africa. (Karl W. Butzer)

Figure 20–7. Domed Granite Inselberg, near Brandberg, Southwest Africa. Relief is approximately 250 feet (75 m). Several areas of massive exfoliation are visible. Late Pleistocene hunter-gatherers settled around seasonal ponds at the base, indicating that the pediment had already been stripped under the still-prevailing arid conditions. (Karl W. Butzer)

Figure 20–8. Vertical Macrogrooving on the Crest of Stone Mountain, Georgia, a Domed Inselberg. The "ribbing" has a relief of 20 inches (50 cm) or more. Vegetation is established on areas of exfoliation. (Karl W. Butzer)

vial wash. This particular form of undermining leads to steepening of inselberg or tor slopes and simultaneously creates a bedrock pediment. The characteristics of this *tropical pedimentation* are chemical weathering, throughflow, and sheet wash. A similar process operates in tropical limestone country to shape karst towers (see Figure 15–5). Many tropical ped-

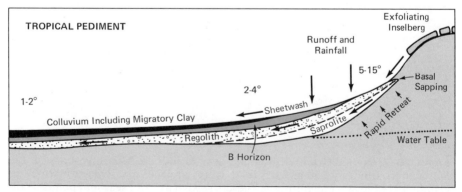

Figure 20–9. Evolution of a Tropical Pediment by Basal Corrosion, Throughflow, and Sheetwash. (Based primarily on Ruxton and Berry, 1961.)

iments in savanna environments are now stripped of their formative cover mantles. The resulting rock floors rise from a gentle flat to a pedestal slope inclined at 5 to 15 degrees or more and are not distinguishable from pediments found in deserts.

20-5. PROCESSES AND LANDFORMS OF TROPICAL SAVANNAS

Basic to the wet-and-dry tropics is a vegetation cover of grass, grass and scattered trees, or open, grassy woodlands. The soil mantle, which although continuous on flats and foot-slopes of as much as 15 degrees, is partially protected through interception, rooting, and a degree of organic matting. However, the vegetation is seasonal, and much bare, desiccated ground is exposed by the beginning of the rainy season. Empirical observations show that the great intensity of tropical rains, combined with seasonal concentration, provide an erosive force 16 times greater than that of average mid-latitude rains (Hudson, 1971). Rainsplash and sheet wash are so effective that the critical gradient threshhold for large-scale erosion from cultivated fields is 2 degrees in Rhodesia, compared with 8 degrees in western Europe (Stocking, 1972). This erosive potential, combined with tropical pedimentation, as outlined previously, has allowed the development of striking landscape assemblages. These can be designated as *savanna plains*. They are characteristic wherever the Gondwana shield is exposed, with limited to moderate relief, in the seasonally humid tropics of Brazil, Africa, India, and Australia.

At first sight, savanna plains appear to belie the concept of moist tropical climates and deep weathering. Broad, pedimentlike plains replace the narrow valleys of humid lands, and butte, mesa, or dome-shaped hills rise above these lowlands (Figure 20–10). Such residuals have angular changes of slope, with steep, rectilinear mid-

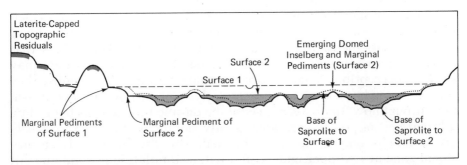

Figure 20–10. Savanna Plains, as Generalized by Büdel (1957). The lower surface (2) was cut by erosion, mainly in soil and saprolite, at a time when deep weathering continued.

slopes—maintaining constant angles—very much as the pediment residuals of the arid zone. But there are differences in detail; some features are unique, and the assemblage itself is often distinctive (compare Table 12–14, Thomas 1974):

1. Mesaform residuals and undissected plateaus are often capped by laterites as much as 30 feet (9 m) or more deep.
2. There may be convex hillocks of laterite rubble, due to undermining of laterite-capped residuals.
3. Talus on hillslopes is rare, although great slabs or rock are evidently being detached by pressure unloading.
4. Many residuals consist of tors, woolsacks, dome-shaped inselbergs, or prominent "sugarloafs" that have clearly been exhumed from beneath a former saprolite.
5. The various exhumed or detached residual hills are often fluted or groved by intensive weathering.
6. Gently convex foot-slopes at the base of inselbergs typically have a mantle of soil and wash, rather than alluvium, resting on bedrock.
7. Foot-slopes go over rapidly onto flat plains underlain by relatively deep soils and saprolite.
8. Streams move within broad and shallow swales that have smooth shoulders, bounding slopes generally less than 5 degrees and a dominantly suspended load, with little gravel in the bed sediment.
9. Valleys that finger into upland masses commonly have steeply concave profiles and abrupt vertical heads, implying that basal weathering complements stream erosion.
10. The numerous interruptions of stream profiles by waterfalls or cataracts reflect barriers of resistant rock.

Büdel (1957) explains the persistence of rock barriers by a lack of sufficient bed load for dissection. The resulting "temporary" base levels of erosion favor areal denudation and cutting of savanna plains.

Savanna planation in Africa that conforms to these basic principles is now restricted to those wet but periodic climates that favor *dambo* development. Dambos are broad, flat, low-order valleys that lack continuous stream channels (Figure 20–11). The margins may slope imperceptibly or abut directly against low bedrock cliffs; they retreat through a combination of hydrolytic undermining, sheet wash, and throughflow. In an average case the dambo boundary is fringed by a broad belt of poorly vegetated, gently sloping ground (0.5 to 2.5 degrees) that is only temporarily flooded. This "wash belt" is an erosional zone at its upper end, with a net accumulation of sand and silt near its lower end, complemented by subsurface ferricretion (Mäckel, 1974). The central "seepage belt" is hydromorphic, with a thick,

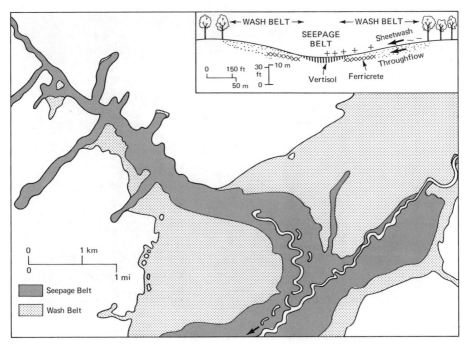

Figure 20–11. Generalized Map of Dambo Topography, Bié Plateau, Central Angola, Developed on the Granitic Shield. (The idealized inset cross-section is based in part on vertical sections illustrated for Zambia by Mäckel, 1974.)

marshy vegetation, and accumulation of dark, organic clays from runoff and seepage waters. Dambos, which were first described in the literature in 1936, are both denudational and depositional. They enlarge by planation, creating lowland plains up to several miles (5 km or more) wide, and at the same time fill in existing valley floors. In intermediate-order valleys the seepage belt grades into a zone of semicontinuous stream channels that locally deteriorates into a maze of oxbow lakes. Dambos in Africa are best developed on plateaus, with 32 to 50 inches (800 to 1,300 mm) of rainfall concentrated in four or five months. They are widespread along the headwaters of the Blue Nile in Ethiopia (Figure 20–12), in upland Angola, and particularly in Zambia. Similar features in the drier Rhodesian high country (Figure 20–13) are not actively forming at the moment but clearly did so in the not too distant past. Dambo formation conforms closely to the model of savanna planation and appears to provide a medium-scale example of the processes involved. Potential counterparts have been described from Brazil and the Guianas and also appear to exist in West Africa.

The savanna plains of subhumid to semiarid environments are different. Foot-slopes are commonly devoid of soil mantles, and mechan-

Figure 20-12. Seepage Dambos in Structural Swales (*Foreground and Right Middle Ground*), Gojjam Plateau, Central Ethiopia. Drainage is to the upper left. (Karl W. Butzer)

Figure 20-13. "Inactive" Dambos Follow These Undissected Valley Floors in the Rhodesian Highlands. (Karl W. Butzer)

ical pedimentation on the semidesert model appears to be operative. This is particularly conspicuous on subhorizontal sedimentary strata that provide permeable subsoils and do not favor deep chemical weathering. Finally, there are typical domed inselbergs, tors, laterite-capped residual hills, as well as patches of red clayey soils well within many deserts. This logically raises the issue of climatic change and polygenetic or inherited forms. At the same time, it has, however, led to considerable controversy as to the origins of savanna plains.

20-6. VIEWS ON SAVANNA PLANATION AND TIME

There is beyond question a continuous gradation between pediment and savanna plains spanning perhaps one-half of the continental land masses and ranging from hyperarid to humid climates or from barren desert to closed woodland. Equally, there is a polarization between dambo planation as in central Africa and pediment cutting as in the American West.

The disparity of views that have been offered for savanna plains and inselbergs is too great to detail but can best be grouped as follows. One view—represented by King (1967), Birot (1968), and Tricart (1972)—downplays the identity of savanna plains, emphasizes the role of duricrusts in preserving planar landscapes, and tends to relate these plains to drier intervals of semiarid pedimentation. Although the role of chemical weathering in cyclic pedimentation or modeling of ex-humed inselbergs is not denied, the interpretation focuses on "pediments and residuals" as semiarid landforms.

The opposing view—represented by Jessen (1936), Büdel (1957, 1965, 1970), Ollier (1959), and Thomas (1974)—admits the existence of distinctive savanna landscapes and emphasizes the role of deep weathering (Figures 20–10 and 20–14). Jessen (1936), in formulating the first model of this type, favored a continual sequence of weathering and erosion, leading to valley deepening and widening, and the gradual reduction of residual inselbergs. Instead, Ollier (1959) argued for a single period of deep weathering (mainly of Mesozoic age), followed later by valley deepening and widening. Büdel (1957) envisages concomitant weathering and denudation, elaborating Jessen's model and giving emphasis to the existence and role of deep regolith and saprolite. He further allowed for tectonic or climatic impulses, by interjecting an interval of accelerated denudation due to uplift or more arid climate. Thomas (1974) follows Büdel to a degree but places greater emphasis on the role of duricrust caps on the regolith while downplaying the role of lateral planation; his end products are correspondingly more irregular and complex.

The preceding sections have provided arguments for why savanna plains are different from desert plains and demonstrated the distinc-

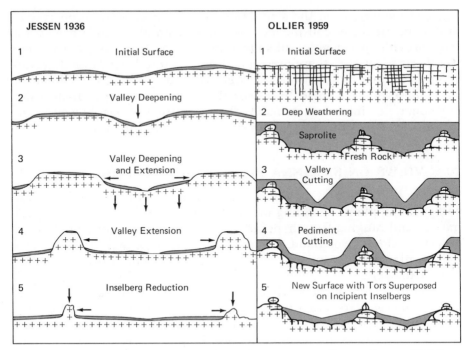

Figure 20–14. Savanna Planation Through Time According to Jessen (1936) and Ollier (1959).

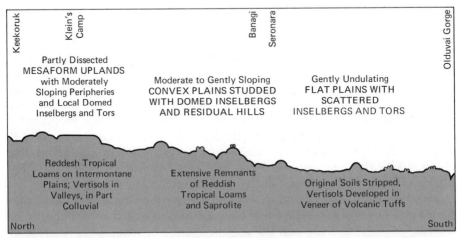

Figure 20–15. Landform Regions of the Serengeti (Kenya-Tanzania). This generalized toposection runs approximately 95 miles (150 km) north-south along the eastern margin of the Serengeti Park. Evolution spans at least the entire Cenozoic (65 million years) and a total relief of about 2,000 feet (600 m) was generated. (Compiled from fieldwork, 1968. Not to scale.)

tive mechanics of tropical pedimentation (under the soil) and valley planation (by dambos). It is nonetheless difficult to apply the "savanna models" to real savanna landscapes. An example to this effect is given by Figure 20–15, a section across a broad region of East Africa. Three composites are recognizable: (1) a high upland area of dissected mesaform hills and low mountains with undulating intermontane plains; (2) a complex peripheral belt of sloping plains, interrupted by selected areas of inselbergs, tors, and mesaform residuals; and (3) a lower relatively flat plainland with scattered residuals. In each case the plains exhibit very shallow V-shaped valleys and very broad rounded interfluves (profile type Ra of section 9-2) (Figure 20–16). At the same time, average longitudinal gradients of the typical interfluve range from less than 0.2 degrees to greater than 15 degrees, with valleys proportionately deeper in areas of greater regional slope. The flat plains readily match with Büdel's model, presuming that regolith has been removed. The other regions do not, since the "plains" are all inclined toward drainage lines of higher order, with the convexity of the interfluves increasing toward inselberg complexes or the mesaform uplands.

In sum, such arrangements argue for progressive denudation accompanied by repeated changes of potential energy, through uplift or stepwise encroachment of drainage networks onto the fringes of the

Figure 20–16. Conspicuously Sloping Plains, with Gently Convex Interfluves, and Complexes of Higher, Residual Topography, Northeastern Serengeti Park. (Karl W. Butzer)

northern upland. The composite of landforms visible today probably emerged between the Cretaceous and late Miocene, over a span of at least 75 million years. Change was obviously slow, and the volcanic veneers in Olduvai Gorge and the adjacent plains argue for little or no significant change since the late Pliocene. The deep reddish soils of the southern, flat plains are now in part incorporated into early to mid-Pleistocene fills at Olduvai Gorge, thus arguing for dry conditions ever since.

Applied to an alternative model (Figure 20–17), the Serengeti case does not require the tenuous argument of broad, regional constraints to dissection, nor does it claim planation across improbably great surfaces. Instead it allows for relatively shallow valleys and their related interfluves converging according to all the laws of drainage-basin development (see Table 9-2). Alternating dambo planation and limited incision can probably account for both the repetitive cross-sectional profiles and the repeated and progressive enlargement of valley floors by tropical pedimentation. The total landscape evolved very slowly by the standards of mid-latitudes, and development was influenced by lithological variation, orogenic impulses, drainage-basin evolution, and climatic fluctuations. The impression of one great savanna plain here is the end product that blurs the complex product of compartmentalized drainage-basin evolution so familiar in other environments.

Such landscapes are plainly polygenetic to the degree that paleoclimatic impulses or inherited landforms need not and cannot be meaningfully isolated, other than that the prerequisites to deep weathering and tropical planation are no longer given in the Serengeti case. In this sense there are inherited savanna landforms in many semiarid or arid environments today (see Figures 20–3 to 20–7 and Figure 20–13).

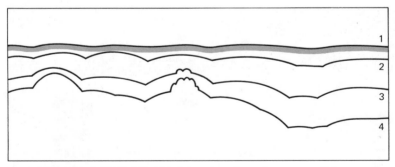

Figure 20–17. A Model for Multivariate Development of Convex Savanna Profiles, in Response to Lithological Variation, Deep Weathering, Denudation, and Potential Energy. (Based on 1968 fieldwork.)

We emphasized already that these nonetheless blend in as an integral, if not essential, part of the landform assemblage. There also are appealing savanna analogs on some Tertiary surfaces in humid midlatitudes (see section 18-7 and Figure 20–8). But in the dambo-forming savannas of today paleoforms are not readily apparent, and a Tertiary mode of morphogenesis continues on the interfluves. This is not to deny climatic change, since the history of valley evolution includes periods of downcutting and some coarse alluvial fills. But the blend of Tertiary landscape evolution under a repeatedly self-adjusting system continues with no discernible shift of long-term directional change. In particular, the disjunctions created elsewhere by glaciation or periglacial remodeling are absent in the heart of the tropics.

20-7. TROPICAL KARST

Not all savanna landscapes are developed in crystalline shields. Sedimentary rocks or flood basalts favor the evolution of mesaform, often laterite-capped landscapes. Limestones develop differently still, particularly because they lack any deep regolith, with soils clayey and shallow. Karst landscapes in the humid tropics offer some striking analogs to savanna plains with inselbergs. Such tropical karst differs from its temperate counterpart in that corrosional forms occupy more space than unconsumed uplands. Polja grow and coalesce to form deep, steep-sided forms called *cockpits*. The higher ground is reduced to detached hills or *cones* (also called pepino hills or hums). Cone and cockpit relief is highly irregular and represents a far more extreme type of corrosional landscape (Figure 20–18).

Cone and cockpit karst is best developed on the Carribbean Islands and South China, with other areas in Southeast Asia. It is by no means the only type of karst, with the "temperate" varieties often found adjacent to the classic "tropical" forms. Along the Puerto Rican coast it is apparent that successively older rings of limestone show an increasing degree of karst development: The mid-Pleistocene coral reefs show a regular, doline-scarred surface, while late Tertiary rocks have been reduced to cone and cockpit karst.

In South China, where cones are usually big (karst *towers* or mogotes), Klimaszewski (1964) demonstrated a similar temporal relationship. Karst towers with a relief of some 650 feet (200 m) are remnants of a high erosional surface that may be of Tertiary age, while the majority of these towers, with an average height of 150 feet or so (50 m) are no younger than mid-Pleistocene. In their morphological details, the majority of the karst towers are steep-sided, fluted, and cavitated, with spired tops and angular undercut bases (see Figure 15–5). Many, however, are surrounded by talus aprons, sometimes resting on short, steep bedrock pediments. This apron debris is linked with an ex-

Figure 20–18. Karst Tower and Polje Plain, Kueilin, China. (Courtesy of M. Klimaszewski.)

tensive gravel terrace, apparently related to cold Pleistocene climate. It can be argued, therefore, that the South China karst landscape is polycyclic (developed in stages) as well as polygenetic (modified in shape by aberrant Pleistocene climates).

20-8. GEOMORPHIC TRENDS UNDER THE TROPICAL RAINFOREST

To what degree does geomorphic equilibrium in the perennially wet tropics resemble that of the wet-and-dry savannas or of humid mid-latitudes? Differences with the forested environments of higher latitudes are clearly apparent, since weathering is very deep and the threshold of slope stability low. Differences with other tropical environments are more difficult to isolate. Savanna models, whether in granites, sandstones, or limestones, are predicated on gently sloping terrain and limited relief. Most forested tropical environments that have been studied instead represent very wet, montane regions at similar latitudes as "savanna plains." Consequently, type examples from East Africa, India, New Guinea, or the Andes all introduce the added variable of relief.

Processess and forms in the perhumid lowland forests of the Amazon Basin have been examined by Bremer (1972, 1973), and they suggest differences of both degree and kind. There are shallow depressions on flat interfluves that have irregular to linear shapes and eluviated soils. These suggest affinities with savanna dambos. Valleys, on the other hand, are deeply entrenched, even in areas of limited relief, as a result of deep weathering. Shoulders are smoothly convex, foot-slopes angular, and bottoms flat. Such valleys have steep heads and maintain irregular floor width, as broad swampy *baixa* alternate with constricted reaches. Drainage density is relatively high, favoring multiple local planation surfaces through tributary convergence along the margins of high-order valley systems. Altogether, soils and related sediments are thick and clayey, favoring subterranean earthflows and allowing rapid physical undermining of valley walls. Sheet wash is the other key agency, made possible by incomplete interception (open canopy woodland, limited ground vegetation). The convex valley shoulders presumably represent the impact of rainsplash and sheet wash by high-intensity rains. Paleoforms are unimportant, since stream terraces, although often capped by iron or aluminum duricrusts, consist of materials identical to modern valley-floor sediments.

It appears that this Amazon rainforest pattern is intensified in mountainous terrain, such as the Serra do Mar (see Appendix B-2). Studies elsewhere, by Starkel (1972) or Temple and Rapp (1972), illustrate that sculpture is greatly accelerated by multiple small debris slides and rapid earthflows on forested slopes in the wake of intense rainstorms. Low shear strength in wet, poorly structured, clayey soils or regolith favors a wide range of other mass movements as well. Devegetation accelerates all processes by a factor of 10. The general impact is to supply abundant suspended sediments to streams, to keep slope soils relatively thin, and to provide concave slope profiles that are interrupted by multiple debris-slide scars or slump masses. A hallmark of such intensively denuded hill or mountain country is the creation of sawtooth ridges along interfluves. Denudation rates in the Indian monsoon mountain belt or on New Guinea, where landsliding is also triggered by earthquake shocks (Ruxton and McDougall, 1967), are consequently among the highest in the world, up to 30 inches (75 cm) per 1,000 years.

Rainforest environments consequently are highly dynamic, even though maintaining remarkably deep and continuous soil mantles. They are correspondingly difficult to fit into either the morphodynamic or the morphostatic model. They also are distinctive enough to warrant inclusion as a subtype within the general constellation of tropical landscapes. As in other environments with red-yellow tropical loams and high rainfall intensities, deforestation greatly accelerates denudation rates, and cultivation on even gentle slopes favors rapid soil erosion.

20-9. SELECTED REFERENCES

Birot, Pierre. *The Cycle of Erosion in Different Climates.* Translated from the

French by C. I. Jackson and K. M. Clayton. Los Angeles, University of California Press, 1968, 144 pp.

Bremer, Hanna. Flussarbeit, Flächen- und Stufenbildung in den feuchten Tropen. *Zeitschrift für Geomorphologie, Supplement* Vol. 14, 1972, pp. 21–38.

_____. Der Formungsmechanismus im tropischen Regenwald Amazoniens. *Zeitschrift für Geomorphologie, Supplement*, Vol. 17, 1973, pp. 195–222.

Büdel, Julius. Die doppelten Einebnungsflächen in den feuchten Tropen. *Zeitschrift für Geomorphologie*, Vol. 1, 1957, pp. 201–288.

_____. Die Relieftypen der Flächenspülzone Südindiens am Ostabfall Dekans gegen Madras. Bonn, *Colloquium Geographicum*, Vol. 8, 1965, pp. 1–100.

_____. Pedimente, Rumpfflächen und Rückland-Steilhänge. *Zeitschrift für Geomorphologie*, Vol. 14, 1970, pp. 1–57.

Bunting, B. T. *The Geography of Soil.* London, Hutchinson, and Chicago, Aldine, 1965, 213 pp. See Chap. 17.

Butzer, K. W. *Environment and Archeology.* 2nd ed. Chicago, Aldine, and London, Methuen, 1971, 703 pp. See Chap. 20.

Cotton, C. A. Theory of savanna planation. *Geography*, Vol. 46, 1961, pp. 89–96.

DeSwardt, A. M. J. Lateritisation and landscape development in parts of equatorial Africa. *Zeitschrift für Geomorphologie*, Vol. 8, 1964, pp. 313–333.

Ganssen, Robert, and F. Hädrich. *Atlas zur Bodenkunde.* Mannheim, Bibliographisches Institute, 1965, 85 pp.

Goudie, Andrew. *Duricrusts in Tropical and Subtropical Landscapes.* Oxford and New York, Oxford University Press, 1972, 192 pp.

Hudson, Norman. *Soil Conservation.* London, Batsford, 1971, 320 pp.

Jessen, Otto. *Reisen und Forschungen in Angola.* Berlin, Reimer, 1936, 397 pp.

King, L. C. *Morphology of the Earth.* 2nd ed. Edinburgh, Oliver and Boyd, and New York, Hafner, 1967, 726 pp.

Klimaszewski, Mieczyslaw. The karst relief of the Kueilin area (South China). *Geographia Polonica*, Vol. 1, 1964, pp. 187–212.

Mäckel, Rüdiger. Dambos: a study in morphodynamic activity on the plateau regions of Zambia. *Catena*, Vol. 1, 1974, pp. 327–365.

Maignien, R. *Review of Research on Laterites.* Paris, UNESCO, Natural Resources Research, Vol. 4, 1966, pp. 11–148.

Ollier, C. D. A two-cycle theory of tropical pedology. *Journal of Soil Science*, Vol. 10, 1959, pp. 137–148.

_____. *Weathering.* Edinburgh, Oliver and Boyd, and New York, Elsevier, 1969, 304 pp.

Ruxton, B. P., and L. Berry. Weathering profiles and geomorphic position on granite in two tropical regions. *Revue de Géomorphologie dynamique*, Vol. 12, 1961, pp. 16–31.

_____, and I. McDougall. Denudation rates in northeast Papua from potassium-argon dating of lavas. *American Journal of Science*, Vol. 265, 1967, pp. 545–561.

Starkel, Leszek. The modelling of monsoon areas of India as related to catastrophic rainfall. *Geographia Polonica*, Vol. 23, 1972, pp. 151–173.

Stocking, M. A. Relief analysis and soil erosion in Rhodesia using multivariate techniques. *Zeitschrift für Geomorphologie*, Vol. 16, 1972, pp. 432–443.

Sweeting, M. M. *Karst Landforms.* New York, Columbia University Press, 1972, 362 pp.

Temple, P. H., and Anders Rapp. Landslides in the Mgeta area, western Uluguru Mtns., Tanzania. *Geografiska Annaler,* Vol. 54-A, 1972, pp. 157–194.

Thomas, M. F. *Tropical Geomorphology: A Study in Weathering and Landform Development in Warm Climates.* London, Macmillan, and New York, Wiley, 1974, 332 pp.

Tricart, Jean. *The Landforms of the Humid Tropics, Forests and Savannas.* Translated from the French by C. Kiewiet de Jonge. London, Longmans, and New York, St. Martin, 1972, 306 pp.

Wilhelmy, Herbert. *Klimamorphologie der Massengesteine.* Braunschweig, Westermann, 1958, 238 pp.

Williams, C. N., and K. T. Joseph. *Climate, Soil and Crop Production in the Humid Tropics.* London and New York, Oxford University Press, 1970, 177 pp.

Chapter 21
PERSPECTIVES

Earth is not just a launching pad for space, or for dreams of interminable expansion. It is the human home in the cosmic scheme of things.

YI-FU TUAN, *"Place: An Experiential Perspective"*

We have come full circle.

In its search for "exact" description and causalities, geomorphology aspires to be a science. The shortcomings that are apparent are shared by other natural sciences. At fault is the complexity of multivariate interrelationships in nature. Yet there also is the element of the human mind with its innate problems of perception and its cultural biases. The pictorial representation of physical space, by the use of shading and geometric perspective, was first accomplished in classical Greek theater for the *skene*, or stage sets. The term survives in *scenery* and should serve to remind us of what physical landscapes consist. They are real in terms of material, but their perception, articulation, and analysis are products of the mind. This is evident even in the quantitative realm of slope measurements or laboratory duplication of processes, as exemplified in the concept of random sampling or in the use of simplifying assumptions.

It is well to remember that geomorphology is also an art, since it is first and foremost a product of visual stimuli that must be translated into a rationalized system of order. Much like the painters of Renoir's generation, geomorphologists filter the real landscape to obtain an impression of its essential components. Landforms, like space, are a continuum, and all abstractions of units are basically arbitrary. Yet units are essential to all organizational hierarchies conceived by the human mind. At any one moment we can perceive on only one scale, big or small, and we can only deal with a limited number of variables in recognizing an entity or postulating a cause-and-effect relationship. It is at this level that the computer has the advantage. Given these limitations, however, the different approaches to the subject matter must be accepted as organizational devices, each of which can be applied to geomorphology with benefit. The more ways of looking at the data, the better the chances of isolating objective reality.

Cultural perceptions not only influence the methods and success with which scientific communities attempt to abstract reality from their data. At the much broader base of mankind they affect the full range of environmental behavior.

> What people do about their ecology depends on what they think about themselves in relation to things around them. Human ecology is deeply conditioned by beliefs about our nature and destiny. . . .[1]

Counteracting the ethos that man was created to have dominion over the Earth and all living creatures will be difficult. It is not enough to invoke sentimental "ecological thinking" or to adopt a stance of outrage. A sober process of education over a generation or two offers the only hope that mankind can learn to appreciate that Earth offers truly unique opportunities within the planetary system, and that these are ours to use rather than abuse. No piece of land can be owned with the privilege of exploitation or destruction. Instead the land passes from generation to generation, and deserves to be tended for the future.

Geomorphologists have a key role to provide in this reeducation. They can evaluate and hopefully measure the physical impact of man as a geomorphic agent. They can hope to influence environmental perception. They can participate in establishing standards for environmental quality and in deciding among alternatives for quality maintenance and improvement. Lastly, they also have an obligation to influence future legislation relevant to the management of the environment.

SELECTED REFERENCES

Clark, Kenneth. *Landscape into Art.* Boston, Beacon, 1961, 148 pp.

Dubos, René. Humanizing the earth. *Science,* Vol. 179, 1973, pp. 769–772.

Lehmann, Herbert. Formen landschaftlicher Raumerfahrung im Spiegel der bildenden Kunst. *Erlanger Geographische Arbeiten,* No. 22, 1968, pp. 1–24.

Tuan, Yi-Fu. *Topophilia: A Study of Environmental Perception, Attitudes, and Values.* Englewood Cliffs, N.J., Prentice-Hall, 1974, 260 pp.

---

[1] Lynn White, The historical roots of our ecologic crisis, *Science,* Vol. 155, 1967, p. 1205.

Appendix A

SOIL CLASSIFICATION ACCORDING TO THE U.S. DEPARTMENT OF AGRICULTURE'S 7TH APPROXIMATION

The U.S. Department of Agriculture currently recognizes 10 major soil classes, designated as *orders*. These are listed here and described in summary form, including subdivisions at the *suborder* level but omitting the lower categories (*great groups, subgroups, families,* and *series*). Revisions since 1964 are incorporated, and to facilitate comparison the classes are rearranged in the approximate sequence followed in Chapters 17 to 20.

1. Inceptisols: soils with weakly expressed horizons found in humid environments on comparatively young surfaces. No evidence of significant eluviation/illuviation or extreme weathering. Include arctic brown soils, alpine turf soils, and brown forest soils, as well as associated pseudogleys, gleys, and wet alluvial soils. Natural vegetation: forest, tundra, and alpine meadow. Subdivisions: wet inceptisols (*aquepts*); inceptisols with thick, dark-colored surface horizons of acidic humus (*umbrepts*); inceptisols with shallow but conspicuous light-colored surface horizons (*orchrepts*); and fresh volcanic ash soils (*andepts*).
2. Spodosols: soils of humid climates with illuvial B horizons. Include podsols, brown podsolics, and gleyed podsols. Forest vegetation. Subdivisions: wet (*aquods*) and well-drained spodosols (*orthods*).
3. Histosols: wet, organic soils of all climates—bog soils. Subdivisions: peat (*fibrists*) and muck soils in which plant remains have been decomposed (*saprists*).
4. Alfisols: soils of humid to semiarid climates with an illuvial B horizon of over 35 percent base saturation. Include gray-brown podsolics, gray wooded soils, noncalcic brown soils, and associated

planosols and gleys. Forest vegetation. Mainly found in temperate latitudes but include many fresh volcanic soils in the humid tropics. Subdivisions: seasonally wet *aqualfs* (pseudogleys), cool alfisols (*boralfs*), alfisols that are usually moist (*udalfs*), and alfisols that are seasonally dry (*ustalfs, xeralfs*).

5. Mollisols: soils of humid to semiarid climates with highly humified, friable A horizons with more than 50 percent base saturation. Include prairie soils, chernozems, chestnut and calcic brown soils, rendzinas, some deep, brown forest soils, and associated hydromorphic and saline soils. Grassland or woodland. Subdivisions: wet mollisols (*aquolls*), both gleyed and saline; cool mollisols (*borolls*) of chernozem type; mollisols that are generally moist (*udolls*), essentially prairie soils and chernozems; dry mollisols with over 80 percent base saturation in subsurface (*ustolls*), essentially chestnut and calcic brown soils; seasonally dry mollisols with less than 80 percent base saturation in subsurface (*xerolls*).

6. Aridisols: soils of arid or semiarid climates with a thin or light-colored A horizon and *either* (a) a Ca or a Sa horizon *or* (b) an illuvial B horizon. Include desert soils, red and gray semidesert soils, some brown (and reddish-brown) soils, white and associated black alkali soils. Natural vegetation: sparse cover of shrubs, grasses, or both. Subdivisions: *argids* with clayey B horizons, and *orthids* with intermediate texture or a Ca horizon.

7. Vertisols: soils of semiarid to seasonally humid climates with large amounts of swelling clays that crack when dry. Natural vegetation: grassland. subdivisions: seasonally wet (*aquerts*) and well-drained (*usterts, torrerts*) vertisols.

8. Ultisols: soils of humid or subhumid, subtropical and tropical environments with illuvial B horizons of less than 35 percent base saturation, in part with subsurface hardpans. Include red-yellow podsolics, terra rosa, red and yellow tropical loams, and associated hydromorphic soils. Forest (and savanna) vegetation. Subdivisions: wet ultisols (*aquults*); ultisols with a highly organic A horizon (*humults*); and ultisols that are usually moist (*udults*) or seasonally dry (*ustults* and *xerults*).

9. Oxisols: soils of humid or subhumid tropical environments with deep clayey B horizons, rich in sesquioxides, dominated by kaolinitic clays of low exchange capacity, and often consolidated to continuous hardpans at or near the surface. Include red latosols and laterites. Forest and savanna vegetation. Subdivisions: wet oxisols (*aquox*), often forming lateritic crusts; oxisols with an organic A horizon (*humox*); and oxisols that are usually moist (*orthox*) or seasonally dry (*ustox*), in part with hardpans (*torrox*).

10. Entisols: poorly developed soils of all climates with inconspicuous horizons. Include azonal types such as lithosols, regosols, and cer-

tain alluvial and waterlogged soils. Subdivisions: waterlogged *aquents:* loamy alluvial soils, in part saline (*fluvents*); sandy *arents;* and finer-grained *orthents* on unconsolidated parent material or compact rock.

REFERENCES

U.S. Department of Agriculture, Soil Survey Staff. *Soil Classification: A Comprehensive System* (*7th Approximation*). Washington, D.C., GPO, 1960, 244 pp.

U.S. Department of Agriculture, Soil Survey Staff. *Supplement to Soil Classification System* (*7th Approximation*). Washington, D.C., Soil Conservation Service, Limited Circulation, March 1967, 207 pp., and September 1968, 22 pp.

U.S. Department of Agriculture, Soil Geography Unit. *Soil Map of the World* (*1:50 million*). Hyattsville, Md., Soil Conservation Service, Limited Circulation, December 1971.

Appendix B
GEOMORPHOLOGY OF THE CONTINENTS

The major land-form units of the continents are discussed in Chapters 14 and 15 and are illustrated by Figures 15–7 to 15–12, with the relative importance of the various structural-lithological surface classes shown by Table 15–1. Although a description of regional geomorphology lies beyond the scope of a topical text, the lack of any convenient source of such information suggests that a summary outline should accompany these maps. This appendix is arranged, therefore, around the maps and organized as a reference for selective consultation, rather than as an analytical discussion. It can best be utilized in conjunction with an appropriate atlas.

B-1. GEOMORPHOLOGY OF NORTH AMERICA

The continental land mass of North America is conveniently divided into three gross units, each with further subdivisions. These land-form provinces, with reference to Figure 15–7, can be listed and characterized as follows:

A. *The Western Cordillera*, part of the Circum-Pacific cordilleran belt, consist of several systems of fold or fault-block ranges, with intermontane basins or plateaus. Maximum width is 1,000 mi (1,600 km) in the United States, narrowing to 500 mi (800 km) or less in Canada, Alaska, and Mexico.
   1. *Pacific Mountain System:*
      (a) Coast Ranges. A series of folded sedimentaries extends along the California coast northward to Vancouver Island. Crestline elevations of these low mountains average 2,000 to 5,000 feet (600 to 1,500 km).
      (b) Central Valley and the Puget-Willamette Valley. Synclinal depressions with flat or dissected, alluvial floors extend through much of California, Oregon, and Washington and continue northward as the interior waterway through or along the fiord coast of British Columbia.
      (c) A high mountain belt extends north from Baja, California, to the Sierra Nevada, the Cascade Range, and the Coast Mountains of British Columbia, terminating in the Alaska Range. This complex belt consists partly of upfaulted intrusive

blocks, partly of folded sedimentaries, and partly of volcanic mountains. Crestline elevations attain 5,000 to 10,000 feet (1,500 to 3,000 m) in Alaska. The peaks of the Coast Mountains and the Alaska Range have mountain and valley glaciers today, and most of western Canada and Alaska, as well as the highest parts of the Cascades and the Sierra Nevada, were glaciated during the Pleistocene.

2. *Intermontane Basins and Plateaus:*
   (a) The Great Basin is a complex of alluvial basins studded with fault-block mountains; it extends from the Mexican Mesa Central through the Basin and Range country of southern Arizona and southeastern California into Nevada and western Utah. The deposits of Pleistocene Lake Bonneville and of the Colorado River delta provide conspicuous areas of younger sediments.
   (b) The Colorado Plateau forms an irregular tableland of intensively dissected sedimentary strata, with prominent cuesta-form landscapes and entrenched meander valleys.
   (c) The Columbia Plateau and Snake River Plains represent partly dissected plains, partly hills or tablelands, cut by stream erosion into early to mid-Tertiary flood basalts that are over 4,000 feet (1,200 m) thick. Wind-borne loess and alluvium mantle parts of this province.
   (d) Fraser and Yukon Plateaus. This intermontane upland of low-mountain topography extends through interior British Columbia and Alaska.

3. *Rocky Mountain System:* Extending from Middle America to the Brooks Range of Alaska, this high-mountain system has an average crestline elevation of 5,000 to 10,000 feet (1,500 to 3,000 m). The western and southern Sierra Madre of Mexico are volcanic and attain elevations of over 15,000 feet (4,500 m). The eastern Sierra Madre and the Southern Rockies consist of up-faulted, folded sedimentaries, as does the Brooks Range. The Northern Rockies and their continuation include folded sedimentaries with a core of block-faulted crystalline rock. Pleistocene glaciation was limited in the Southern Rockies but extensive or universal in the Northern Rockies.

B. *The Interior Lowlands* include coastal plains and the broad interior plainlands and hill country that stretch from the Gulf of Mexico to the Artic Ocean. Average width is about 1,500 miles (2,400 km), and typical elevations range between 1,000 and 5,000 feet (300 to 1,500 m).
   1. *The Laurentian Uplands:* These represent an ancient crystalline shield with rolling plains and hill country, scoured by the Pleistocene ice sheets.

2. *Interior Depositional Plains:*
    (a) Till Plains. Broad tracts of glacial till, interspersed with out-wash and lake beds, include the Newer Drift of the last, Wisconsin Glacial. This terrain is rolling, weakly dissected, and poorly drained. The Older Drift of mid-Pleistocene age is flat and moderately to well dissected.
    (b) Lake Plains. Extensive flat plains of proglacial lake beds mantle parts of the Great Lakes area, the Lake Agassiz plain, and the Clay Belt, south of Hudson Bay.
    (c) Eolian Plains. Dissected, upland plains of wind-borne loess are widespread in the Midwest, and rolling dune country is found in the Nebraska Sandhills. The loess, dating back to the Wisconsin Glacial, mantles much of the Central Plains and extends southward down the eastern edge of the Mississippi River.
    (d) Alluvial Plains. Flat or dissected river plains are primarily found along the lower Mississippi.
3. *Interior Erosional Plains and Uplands:*
    (a) The Southern Plains consist of deeply dissected, upland plains that are cut across warped Paleozoic and horizontal late-Tertiary sedimentaries.
    (b) The Alberta-Missouri Plateau comprises tablelands, as well as the dissected, updomed, and warped Paleozoics of the Ozarks—a mass of hill and low mountain country—and the Ouachitas, an area of upwarped and redissected "old" mountains.
    (c) The Appalachian Plateau is a deeply dissected hill country formed in warped Paleozoic sedimentaries, with a number of domes and basins.
4. *Coastal Plains:*
    (a) Gulf-Atlantic Coastal Plain. Barrier islands, lagoons, coastal marshland, delta alluvium, and Pleistocene coastal deposits form a flat coastal perimeter. Coral reefs are common in Florida and the Caribbean area, with some mangrove tidal flats. Inland, dissected cuestaform plains are developed in late-Cretaceous to Tertiary sedimentaries.
    (b) Karst Plains. Upland limestone plains of Cuba, the Yucatan Peninsula, central Florida, and parts of Puerto Rico and Jamaica are pockmarked with sinkholes or have been reduced to ragged, tower karst country.
C. *The Eastern Roughlands,* originally folded during the Acadian and Appalachian orogenies, average some 250 miles (400 km) in width.
    1. *Appalachian Highlands* (see Figure 14–2):
        (a) Ridge and Valley Country. These exhumed, old fold mountains now have a crestline of 2,000 to 4,000 feet (600 to

1,200 m), with hogbacks of resistant rock from breached
or truncated synclines and anticlines. The Great Valley is
eroded into a broad band of soft rocks. Trellis drainage and
hill to low mountain topography.
  (b) Blue Ridge. These upraised, low fault-block mountains of
crystalline rock have a crestline of 2,000 to 5,000 feet (600 to
1,500 m). Dendritic drainage.
  (c) Piedmont Plateau. Here folded crystalline rocks (Precambrian metamorphics and Paleozoic volcanics) have been reduced, by erosion, to an open plain with hills.
2. *Eastern Highlands of New England and Canada:* This old mountain belt of folded Paleozoic sedimentaries and intrusives, with
some roots or blocks of Precambrian crystalline rocks, includes
the Adirondack Mountains. Dendritic drainage, with a topography of rolling hills and some low mountains, scoured by the ice
sheets.
3. *Greenland:* Some 95 percent of Greenland is covered by a great
ice sheet that reaches from well below sea level to an elevation
of over 12,000 feet (3,700 m). High, rugged mountains fringe the
coasts, including folded Paleozoic sedimentaries along the
northern and northeastern peripheries, crystalline rocks of the
Laurasian shield elsewhere.

B-2. GEOMORPHOLOGY OF SOUTH AMERICA

As in the case of North America, the second New World continent can
be categorized in three gross units. The resulting land-form provinces,
many of which find striking analogies in North America, can be described in outline form (see Figure 15–8).

A. *The Western Cordillera* (Andes), a continuation of the Circum-Pacific
cordilleran belt from Central America, links up with Antarctica by
a widely spaced island arc. Although there are two or three parallel
ranges, maximum width of this belt is 450 miles (720 km), average
width only 200 miles (320 km). There are major longitudinal differences:
1. *Caribbean Island Arc:* This forms a link between the Rocky
Mountain system and the eastern chain of the Northern Andes.
The Lesser Antilles consist of volcanic islands, while the Greater
Antilles include fold mountain belts such as those of Puerto
Rico, Santo Domingo, Jamaica, and southeastern Cuba. Peaks
seldom exceed 2,000 to 5,000 feet (600 to 1,500 m) except on the
largest islands. Coral reefs accompany many of the coasts.
2. *Northern Andes:*
  (a) The Western Andes form high mountains of folded sedimen-

taries, rising from broad, marshy or hilly coastal plains. Crestlines approach 5,000 feet (1,500 m).

(b) The Central Andes are block-fault mountains of Paleozoic intrusives and sedimentaries, capped by dormant volcanoes. Crestlines average 10,000 to 15,000 ft (3,000 to 4,500 m). Marginal grabens are occupied by alluvial plains.

(c) The Eastern Andes comprise mixed block-fault and fold mountains, enclosing the downwarped sedimentary basin of Lake Maracaibo. Crestlines average 5,000 to 10,000 feet (1,500–3,000 m).

3. *Ecuadorian Andes:*

(a) The coast ranges and central lowland represent low mountains of folded sedimentaries, with a parallel, synclinal valley of flat alluvial fill.

(b) The Andes here form a narrow, double chain of complex high mountains, surmounted by numerous dormant volcanoes. Crestlines range from 12,000 to 18,000 feet (3,600 to 5,400 m).

4. *Peruvian Andes:*

(a) The coast range includes low fault-block mountains of Mesozoic intrusives, together with small segments of sandy, coastal plains.

(b) The Western Andes are high mountains, mainly volcanic to the south, folded sedimentaries in the north. Crestlines lie between 15,000 and 20,000 feet (4,500 to 6,000 m), with evidence of considerable Pleistocene glaciation.

(c) The Eastern Andes represent high mountains, separated from their western counterpart by a narrow, intermontane valley system. These folded sedimentaries have crestlines that fall off from 20,000 feet (6,000 m) in the south to 10,000 ft (3,000 m) farther north.

5. *Bolivian Andes:*

(a) Atacama Desert. Low, coastal mountains in Mesozoic volcanics that rise to an irregular plateau mantled with Pleistocene alluvium.

(b) Western Andes. High, volcanic mountains that have crestlines between 15,000 and 20,000 feet (4,500 to 6,000 m).

(c) Altiplano. This intermontane basin of plains and hills includes lakes, salt pans, and Pleistocene lake beds.

(d) Cordillera Real. High mountain chains of folded sedimentaries that rise abruptly from the interior lowlands to crestlines of 15,000 to 20,000 feet (4,500 to 6,000 m). Pleistocene glaciation was important.

6. *Central Chile:*

    (a) The coast range is a belt of low, fault-block mountains of Mesozoic intrusives and folded sedimentaries.

    (b) The interior valley, a graben with alluvial fill and Pleistocene till, continues as an interior waterway into western Patagonia.

    (c) The main Andes are high mountains of folded and faulted volcanics, surmounted by dormant volcanoes. Crestlines are at 15,000 to 20,000 feet (4,500 to 6,000 m), with valleys intensively sculptured by Pleistocene glaciation.

7. *Western Patagonia:* The folded and block-faulted high Andes, here consisting of sedimentaries and Mesozoic intrusives, were strongly modeled by Pleistocene glaciation to a rugged fiord coast.

B. *Continental Lowlands*

1. *Llanos of the Orinoco:* Flat and dissected plains of coarse Pleistocene alluvium are flanked by eroded, horizontal sedimentaries of irregular topography.

2. *Amazon Lowlands:* These compose a great tectonic basin of continuing subsidence. The central parts consist of flat or dissected, Pleistocene to Holocene alluvium, including the world's largest floodplain. The margins include dissected Tertiary alluvium and horizontal sedimentaries. The lower valley follows a syncline, exposing cuestas of Paleozoic sedimentaries, while the Amazon mouth consists of a complex estuarine zone of alluviation.

3. *The Paraná-Chaco Plains:*

    (a) The Paraná-Paraguay lowlands are formed by a series of broad floodplains that terminate in the La Plata estuary.

    (b) The pampas are flat plains, mantled with wind-borne loess of Pleistocene age that rests on a basement of eroded, older mountains.

    (c) The Gran Chaco is a complex of flat and dissected plains of sands and silts that accumulated in a subsiding tectonic basin, with exceptionally broad and shallow floodplains.

    (d) The Monte is the name given to the complex Andean foothill zone of dissected plains, hill country, and folded ranges of low mountains.

4. *The Patagonian Plateau:* This tableland of prominent cuestas is a heterogenous surface of crystalline outcrops, Mesozoic sedimentaries, Tertiary lavas, and Pleistocene alluvium and till.

C. *The Eastern Uplands*, underlain by a part of the crystalline shield of ancient Gondwanaland, are mantled by a discontinuous cover of sedimentaries and volcanics. The Continental Lowlands follow the depressed contact zone between this shield and the Andes.

1. *The Guiana Highlands:* Plains with scattered hills or blocks of

low mountains have crestlines of 3,000 to 8,000 feet (900 to 2,400 m). Dissected, subhorizontal Mesozoic sedimentaries form many of the mountains, with south-facing cuesta escarpments; Precambrian metamorphics and intrusives are exposed elsewhere.

2. *The Interior Plateau of Brazil* (*Mato Grosso*): Dissected plains or tablelands are formed by a subcontinuous cover of subhorizontal Paleozoic and Mesozoic sedimentaries resting on a crystalline shield. Structural cuestas are prominent along the eastern and southern periphery.

3. *Northeastern Highlands:* Plains with hills and low mountains are formed by remnants of sedimentary strata that rest on the eroded crystalline shield.

4. *Serra do Mar* (*"Maritime Range"*): An upfaulted block of crystalline shield that rises above sedimentaries and narrow coastal plains beneath the fault escarpments. Crestlines range from 3,000 to 7,000 feet (900 to 2,100 m), with hill to low-mountain topography.

5. *Paraná-Uruguay Uplands:* Deeply dissected Mesozoic basalt plateaus are partly capped by horizontal sedimentaries. They are offset from the Continental Lowland and from the Paleozoic sedimentaries and crystalline shield of Uruguay by steep escarpments.

Antarctica, with about 75 percent of the land mass of South America, follows the same structural blueprint and is in many respects a continuation of the Americas. The Antarctic Peninsula and the adjacent Pacific coastal ranges mark a continuation of the Western Cordilleras, with crestlines of over 10,000 ft (3,000 m), and widespread volcanics. The great indentations occupied by the Filchner and Ross Ice Shelves correspond to basins that link the folded ranges of Western Antarctica to the crystalline shields of Eastern Antarctica (see Figure 13–6). Although almost all of the continent is covered by ice, averaging 1 mile (1.6 km) thick, there are high, rugged mountains exposed along most of the coastal sectors. The ice sheet itself rises to an elevation of over 14,000 ft (3,300 m).

B-3. GEOMORPHOLOGY OF EURASIA

Eurasia forms a single continental mass, artificially subdivided by the Ural Mountains. Its area is considerably greater than that of the combined Americas and accounts for 37 percent of the world's land surface.

In the New World the spine of the continents runs north-south, along the great cordilleran belt, but the Eurasian cordilleran belt runs west-east. The other broad land-form provinces are organized accordingly. However, the old continental mass is less clearly subdivided into broad zones than is the New World, while the two peninsulas of Arabia and India, derived from Gondwana, further complicate the overall constellation. Consequently, Eurasia is best subdivided into six, rather than three, gross units (see Figures 15–9, 15–10, and 13–6 and Table 15–1).

A. *The Western Roughlands* represent a scatter of remnants from old mountain belts from the Caledonian and Hercynian orogenies. Each has been planed off by erosion, uplifted, and intensively dissected, often several times. Some, such as the Irish Uplands and Bohemian Massif of central Europe, are a matter of hills, small plateaus, and low mountains, with crestlines of 1,000 to 3,000 feet (300 to 900 m). Still others form broad units of hill or low-mountain country, with crestlines of over 3,000 feet (900 m). These include the western peripheries of the Iberian Meseta, the Massif Central of France, the ice-scoured Scottish Highlands and Norway Uplands, as well as the Ural Mountains. The Massif Central of France is terminated to the east by the great fault-valley of the Rhone, while the Central Uplands of Germany are fragmented by the upper Rhine graben and the antecedent valley of the middle Rhine River.

B. *The Eurasian Plain*, the world's largest lowland surface, stretches 3,000 miles (4,800 km) from the Atlantic coast well into the heart of Asia and attains a width of almost 2,500 miles (4,000 km).

   1. *Erosional Plains of Western Europe:* A discontinuous cluster of dissected plains that cut across sedimentary strata, to include the cuestaform English Lowlands, the Paris Basin, and parts of the more irregular South German Plateau. In the case of the central and eastern Iberian Meseta and the Basin of Aquitaine there are horizontal sedimentaries.

   2. *The North European Drift Plain:* Till plains, with broad areas of outwash or loess, and ages and forms comparable to the Newer and Older Drift of North America, are interrupted by the Rhine-Maas delta and alluvial plain (see Figure 8–10).

   3. *The Danubian-Ukrainian Loess Plateaus:* These upland plains, often with considerable horizontal relief and folded sedimentary rocks, are mantled by a deep cover of late Pleistocene loess as well as Older Drift.

   4. *The Fennoscandian Shield:* Ice-scoured plains with hill tracts are developed on an ancient crystalline shield, similar in all respects to the Laurentian Uplands.

   5. *Uplands of Eastern Russia:* Dissected plains that cut across horizontal or warped sedimentaries, emerge east of the drift cover as tracts of irregular plains or hills.

   6. *West Siberian Lowland:* This great tectonic basin, filled with alluvium, till, lake, and marine beds, now provides little or no surface relief.

7. *The Caspian-Turanian Lowlands:* Another broad region of subsidence, this region includes erosional plains as well as extensive veneers of sand dunes, alluvium, salt pans, and old lake beds. The Caspian Sea now stands at 92 feet (28 m) *below* sea level; during the cooler Pleistocene glacials it received water from the Aral Sea and was 250 feet (75 m) higher. This greatly expanded inland sea overflowed into the Black Sea.

C. *The Alpine-Himalayan Cordilleran Belt.* The young mountain belt of Eurasia compares in size, elevation, and magnitude with the New World belts. In the west the Pyrenees and Sierra Nevada of Spain continue to the once intensively glaciated Alps (average crestlines over 10,000 ft, or 3,000 m), the Apennines of Italy, and the Carpathians and Balkan mountain system, each interrupted by synclinal alluvial valleys such as the Po and the Hungarian Plain. In western Asia, complex ranges circumscribe the basins of the Anatolian and Iranian Plateaus, culminating in the peaks of the Armenian Plateau and the Caucasus (with crestlines averaging over 9,000 ft, or 2,700 m) and a record of major Pleistocene glaciation. In central Asia the Himalayan sector of the cordilleran belt runs from Afghanistan through the main ranges of the Himalayas—with the world's tallest mountains and crestlines averaging over 18,000 feet (5,500 m)—across western Burma to the Indian Ocean. From here the belt is continued by a series of island arcs and archipelagos.

D. *The Eastern Roughlands.* The uplands of central, northeastern, and southeastern Asia are geologically the most complex of any major world region. They are also the least understood. Most of them were folded during either the Caledonian or the Hercynian orogeny and have been block-faulted or uplifted several times since. Many parts now rival the Alpine cordilleran belts in relief and irregularity.

1. *The Tien Shan and Tarim Basin:* This complex belt of Hercynian fold mountains has been reduced to broad hilly uplands in the west but rises to a gigantic mountain system, the Tien Shan, in the heart of Asia, with once glaciated crestlines of 10,000 to 20,000 feet (3,000 to 6,000 m). The ranges fan out eastward to merge with those of Tibet, enclosing a number of alluvial depressions of tectonic origin, the largest of which is the Tarim Basin.

2. *The Altai-Sayan Mountains and Mongolian Plateau:* These high ranges, with crestlines of 5,000 to 10,000 feet (1,500 to 3,000 m), were folded or block-faulted during the mid-Paleozoic and then reelevated to attain high relief during the late Tertiary. At that time block faulting created a basin-and-range topography across the Mongolian Plateau, comparable to that of the Great Basin. Pleistocene mountain glaciation was extensive.

3. *The Tibetan Ranges and Plateau:* This system of arcuate and subparallel high mountain ranges includes the Kunlun Shan system that forms the "roof of Asia." The Tibetan Plateau, at elevations of over 10,000 feet (3,000 m), is mantled by Cretaceous sedimentaries, block-faulted during late Tertiary uplift of the underlying Hercynian mass.

4. *Southeast Asia:* Here systems of late Paleozoic and Mesozoic ranges radiate from Tibet across southwestern China (Yünnan Plateau) into the Malay Peninsula and Borneo. These hills and low and high mountains include the Annam Cordillera of Vietnam (crestlines of 3,000 to 5,000 ft, or 900 to 1,500 m) and enclose the erosional and depositional plains of the Mekong lowlands.

5. *Eastern China* is a region of more open landscapes, with numerous alluvial valleys and broad plains.

   (a) The South China Uplands represent a dissected crystalline shield, capped by remnants of warped limestones and volcanics, and forming low mountain topography (crestlines 2,000 to 5,000 ft, 600 to 1,500 m). Numerous small alluvial basins do not compare in size with the plains of the great Yangtze valley system.

   (b) The Red Basin is one of block-faulted Mesozoic sedimentaries, developed in the foreland of the Tibetan Ranges.

   (c) The Hwangho Plateau includes tablelands and low mountains of warped sedimentaries, mantled by Pleistocene sands and loess, and now deeply dissected by the Hwang River.

   (d) The Great Plain of China comprises broad expanses of Holocene alluvium and dissected Pleistocene loess, stretching northward from the Yangtze estuary to beyond Peking.

   (e) The Manchurian Plain is an irregular plain of Pleistocene alluvia, reaching from the Yellow Sea to the Amur Valley.

6. *The Baikal-Amur Highlands:* This belt of fault-block mountains consists mainly of intrusive batholiths stretching from Lake Baikal to the Pacific. Low mountains with a crestline elevation of 3,000 to 6,000 feet (900 to 1,800 m) are characteristic.

7. *The Korean Uplands:* A complex of upraised, crystalline shield and Mesozoic intrusives that are locally capped by warped sedimentaries. Crestlines average about 5,000 ft (1,500 m).

8. *Central Siberian Plateau:* This stable block of the Laurasian shield is mantled by warped Paleozoic sedimentaries and Mesozoic volcanics. Topographic expression ranges widely from hills to plains or low mountains.

9. *The Highlands of Northeastern Siberia:* Here belts of Paleozoic or Mesozoic sedimentaries and volcanics were folded during the

late Paleozoic or Mesozoic and uplifted by Tertiary block faulting. These low to high mountains, with crestlines of 3,000 to 5,000 feet (900 to 1,500 m), have been extensively glaciated.

10. *The Kamchatka Peninsula:* A chain of young folded sedimentaries, widely mantled with young volcanics and dormant volcanoes, forms the terminus of the Pacific archipelagos. The low to high mountains, with crestlines at or above 3,000 feet (900 m), were glaciated during the Pleistocene.

E. *The Pacific Archipelagos* represent a subcontinuous chain of young folded mountains, with countless dormant volcanoes, varying between typical island arcs (Andaman Islands, Ryukyus, Kuriles) and submerged cordilleran belts (Indonesian, Philippine, and Japanese archipelagos). These low to high mountains have crestlines of 5,000 to 10,000 feet (1,500 to 3,000 m), as well as important alluvial valleys and coastal plains.

F. *The Shield Peninsulas.* The Arabian and Indian subcontinents represent fragments of Gondwana, now attached to Eurasia by geosynclinal depressions.

1. *The Arabian Peninsula:* This block of shield rocks is demarcated by a rift valley along its Red Sea margins and further rifted by the deep Dead Sea-Jordan graben (Dead Sea level at 1,286 ft, or 392 m, *below* sea level). The greater part of the land mass is mantled by tilted sedimentaries, local basalt caps, and large expanses of dune sands. Hills and low and high mountains extend from the Levant Uplands down the block-faulted western rim of the peninsula; average crestlines range from 2,000 to 9,000 feet (600 to 3,000 m). The interior consists of rolling sand fields and strongly dissected erosional plains. The dissected alluvial basin of Mesopotamia and the Persian Gulf geosyncline form the link to the Eurasian cordillera.

2. *The Indian Peninsula:* This upfaulted and dissected mass of crystalline rocks is partly mantled by early Tertiary flood basalts and is tilted down to the east. Hills and low mountains are characteristic, and the Western Ghats have crestlines of about 2,500 feet (750 m). A great geosyncline, the Indo-Gangetic Plain, joins the subcontinent to the cordilleran belt; the topography is that of flat or dissected alluvial plains, with some rolling dune country to the northwest.

B-4. GEOMORPHOLOGY OF AFRICA

As a continent, Africa is larger than North America and three times the size of Europe or of the contiguous United States. Yet the land-form provinces are remarkably uniform and often monotonously repetitive over the greater part of this expanse. The reason for this sameness is

that about 80 percent of the surface is either formed by the crystalline
Gondwana shield or by a relatively thin veneer of sedimentaries
resting upon it (see Figure 15–10). Five gross units can be identified:

A. *The Atlas Ranges.* The northwestern corner of Africa forms an extension of the Alpine-Himalaya cordilleran belt, consisting of young,
   folded sedimentaries. Only the High Atlas of Morocco, with its
   crestlines averaging over 10,000 ft (3,000 m), qualifies as high
   mountain country. The Mediterranean coastal ranges (Rif and Tell
   Atlas) and the interior Saharan Atlas have average crestlines of
   3,000 to 6,000 feet (900 to 1,800 m) and the topographic expression
   of low mountains. The higher peaks of Morocco were glaciated
   during the Pleistocene. Alluvial valleys and irregular, intermontane
   plateaus with salt lakes are well developed in Algeria, while the
   Moroccan coastal sector forms a dissected plain cut across warped
   sedimentaries.
B. *The Sahara* comprises a great plain, regionally broken by tracts of
   hills or mountains, marking the northern periphery of the African
   land mass. Three component parts can be recognized in various
   arrangements across an area exceeding that of the continental
   United States.
   1. *The Cuestaform Plains:* Tilted sedimentary rocks form upland
      plains, locally dissected to hills or tablelands.
   2. *Depositional Basins:* Vast expanses of flat, wind-eroded gravel
      plains (*serir*), alternate with undulating dune fields, frequently
      immobile (*erg*). They are generally found in downwarped basins,
      once filled by interior lakes, or in front of the cuesta escarp-
      ments.
   3. *Interior Highlands:* These low or high mountains of crystalline
      rocks are surrounded by volcanoes and lava plateaus and fringed
      by broad hilly piedmont areas cut across the shield. The Hoggar
      Massif and Tibesti Mountains have average crestlines of 6,000 to
      9,000 feet (1,800 to 2,700 m).
C. *The African Shield* is a great, undulating surface of dissected
   uplands and level plains, that rises from an average elevation of
   500 feet (150 m) in West Africa to over 5,000 ft (1,500 m) in the
   southern parts of the continent. In fact, the shield generally dips
   from southeast to northwest. Several component parts can be singled out:
   1. *The Highland Swells and Rises:* The backbone of the shield is
      formed by gently upwarped surfaces that border the oceans and
      isolate a number of interior basins. Dissection has commonly
      eroded these uplands into open hills or low mountains, and
      some areas of block faulting stand out (Fouta Djalon Massif,
      the Jos Plateau of Nigeria, the Bié Plateau of Angola, the Ma-

topos of Rhodesia, and the Damaraland Plateau of Southwest Africa). In addition, the Cameroon Highlands are topped by volcanic mountains that continue out as the volcanic islands of the Gulf of Guinea. In most places the shield borders directly on the coast, providing few harbors and creating cataracts (rapids) near the river mouths.

2. *The Interior Basins:* Downwarped sections of the shield are filled with late Tertiary to Pleistocene alluvium or lake beds. Sands, with extensive, rolling plains of fixed dunes, mark the Chad and Kalahari basins. Fine materials and flat or dissected plains characterize the Middle Niger, Central Nile, and Congo basins.

3. *Coastal Plains:* These are few and poorly developed, with exceptions of the flat Niger delta plain (see Figure 11–15) and the dissected sands of the plain of Mozambique.

4. *Madagascar:* The subcontinental island of Madagascar forms a part of the African shield, with its own continental shelf, still drifting slowly out into the Indian Ocean. Upfaulted crystalline rocks, with average crestlines of 4,000 to 6,000 feet (1,200 to 1,800 m), provide low mountain topography. Cuestaform sedimentary plains fringe the western coast.

D. *The Great Rift System.* The eastern parts of the African shield are rifted apart as a result of slow continental spreading. Horst-and-graben systems, with an average relief of 2,000 to 10,000 feet (600 to 3,000 m) (see Figure 14–3), provide some of the most spectacular scenery in Africa. The fault fractures have tapped deep-seated lavas and flood basalts, and volcanic ash and great composite volcanoes are commonplace (see Figure 14–5). Several separate segments can be recognized.

1. *The Ethiopian Plateau and Rift:* This deeply dissected tableland or mountain country of mid-Tertiary flood basalts (at an average elevation of over 6,000 ft, or 1,800 m) is surmounted by a cluster of rugged, old shield volcanoes reaching well above 12,000 feet (3,600 m). Limited Pleistocene glaciation is evident in the summit area. The plateau is torn apart by a deep rift, with a string of small lakes and alluvial basins (see Figure 14–5).

2. *The Kenya Rift.* The deep graben averages 25 miles (40 km) in width and is filled with Pleistocene lake beds and some modern, nonoutlet lakes. The adjacent horst blocks are mantled with thick lavas and, except for some of the Earth's tallest volcanoes (Kilimanjaro, 19,340 ft, or 5,896 m), provide low mountain topography.

3. *The Western Rift:* This includes Lake Tanganyika and is broadly similar to the Kenya Rift, although longer and more compressed.

4. *The Nyasa (Malawi) Rift:* This rift has only limited, volcanic

uplands and an average relief of less than 5,000 feet (1,500 m).
E. *The South African Uplands.* The southern tip of the continent is
formed by sedimentary rocks, in part tilted, in part folded. In a
miniature way, these landform provinces repeat the cuestaform
uplands and Atlas Ranges of northern Africa.

1. *Cape Ranges:* Sedimentaries, folded during the Mesozoic, form
   subparallel ranges with crestlines that average only 3,000 ft
   (900 m) and are interrupted by broad valleys.
2. *The Drakensberg:* This represents part of a great fault escarpment
   along the coastal rim of the interior plateau. It is extensively
   mantled by Mesozoic flood basalts and generally dissected to
   low or high mountains with average crestlines of 4,000 to 10,000
   feet (1,200 to 3,000 m).
3. *The Interior Plateau:* This includes high-lying, dissected, cuestaform plains cut across tilted sedimentaries and studded with
   erosional hills.

B-5. GEOMORPHOLOGY OF AUSTRALASIA

Although in some respects an appendage to Asia, Australia and New
Guinea form a continent in their own right. Together with Tasmania
they occupy a continental block (Figure 15–12) that is separated from
the island archipelago of central Indonesia by deep seas or trenches.
Including the major oceanic islands and New Zealand, this is another
New World continent now commonly known as Oceania. Total area is
only a little greater than that of the continental United States.

A. *New Guinea and the Oceanic Islands.* The eastern end of the Indonesian Archipelago—including Timor and the Moluccas—consists
primarily of young fold mountains, with widespread extrusive vulcanics and a number of dormant, composite volcanoes and calderas. The New Guinea ranges are formed by folded sedimentaries,
with a core of Mesozoic intrusives, and average crestlines range
from 6,000 to 12,000 feet (1,800 to 3,600 m). The highest peaks
(above 15,000 ft, or 4,500 m) were glaciated during the Pleistocene.
The major Oceanic island groups—the Bismarck Archipelago, the
Solomon Islands, New Caledonia and New Hebrides, and Fiji—are
mountainous, and consist primarily of relatively recent volcanics,
much like the Hawaiian Islands. The smaller island groups are
mainly coralline, with some volcanics. Relatively unique in this island belt are the great spreads of dissected Pleistocene alluvium
and flat, swampy lowlands constituting one-third of New Guinea.
These form part of the shallow Sahul Shelf that provided a land
bridge from New Guinea to Australia as late as 10,000 years ago.
B. *The Australian Shield* is a stable crystalline mass of low-lying but

irregular terrain, incompletely mantled with sedimentary rocks. Average elevation lies between 500 and 1,500 ft (150 to 450 m) and only a few "ranges" of low mountains attain 3,000 to 4,500 feet (900 to 1,350 m).

1. *The Western Uplands:* These eroded and irregular upland plains on crystalline rocks are studded with residual hills and small lacustrine basins formed in closed depressions. Fault scraps delimit the sedimentaries and Pleistocene deposits of the western coastal plains. Dissection of the upfaulted shield margins has produced a hilly topography (Darling Range).

2. *The Northern Plains:* Eroded and irregular plains with several hill-studded swells include the crystalline Kimberley and Arnhem Uplands, and the Barkly Tableland on sedimentary strata.

3. *The Central Plains:* A constellation of shallow basins, mantled by sedimentaries or sand fields, includes the Mesozoic Desert Basin, the flat, karstic Nullarbor Plain, and the expansive rolling surfaces of the Great Sandy, the Gibson, and the Great Victoria deserts. The central part is studded with low ranges of crystalline or moderately folded Paleozoic rocks, enclosing a number of lake basins.

4. *The Southern Highland:* Here several fault blocks of the crystalline shield, ranging in surface expression from irregular plains to low mountains, enclose a number of lake basins or deep bays.

C. *The Interior Basin* is a vast complex of shallow basins, including the Great Artesian Basin, that link the Australian shield to the old cordilleran belt or Eastern Highlands.

1. *The Carpentaria Basin* includes dissected coastal plains of sedimentaries and Pleistocene alluvium, with crystalline rocks exposed on the interior interfluves.

2. *The Eyre Basin:* This great interior basin is filled with Pleistocene alluvium, some lake beds, and complexes of active of fixed sand dunes. Sedimentary rocks form the basin margins. Lake Eyre, at its center, is a great salt pan at 52 feet (15 m) *below* sea level, comparable in size and configuration to the Great Salt Lake.

3. *The Darling River Basin:* This consists of dissected Pleistocene alluvium, with a fringe of sedimentary uplands, and drains to the Murray River.

4. *The Murray River Basin:* This broad expanse of stream-dissected, inactive dune fields (Mallee Plain) grades upvalley into the extensive dissected alluvial piedmont of the Riverina.

D. *The Eastern Highlands.* In part synonomous with the Great Dividing Range, the Eastern Highlands (or Cordillera) include tracts of folded Paleozoic sedimentaries or intrusives, as well as warped

Mesozoic sedimentaries. These were repeatedly upfaulted during the Alpine orogeny, with a moderate degree of volcanic activity. For the most part, they form open hills or low mountain landscapes, with a mosaic of alluvial valleys and small, irregular coastal plains.

1. *Queensland:* Here ranges of hills and occasional mountains alternate with open, irregular basins. The edge of the offshore shelf is formed by the Great Barrier Reef, the Earth's finest coral coast.

2. *New South Wales and Victoria:* This sector includes several clusters of rough, low mountains with average crestlines of 3,000 to 6,000 feet (900 to 1,800 m)—with the New England Range, the Blue Mountains, the Australian Alps, and the Grampians. Only the very highest peaks (Snowy Mountains, 7,316 ft, or 2,234 m) were glaciated during the Pleistocene. The Victoria coastal plains are well developed, and a region of young volcanic hills terminates the ranges to the southwest.

3. *Tasmania:* Separated from Victoria by the shallow Bass Strait, Tasmania was most recently cut off from the mainland about 11,000 years ago. The island consists of low mountains in complex rocks, with crestlines averaging 2,000 to 4,000 feet (600 to 1,200 m). Pleistocene glaciation was moderately extensive.

E. *New Zealand.* With a core of Paleozoic intrusives and sedimentaries, New Zealand was primarily folded during the Mesozoic, with major block faulting and uplift during the Alpine orogeny. Vulcanism remains quite active on North Island.

1. *North Island:* This land of hills and low mountains has some large and many small volcanoes, as well as lakes of volcanic origin.

2. *South Island:* The spine of the island is formed by high mountains with crestlines averaging 4,000 to 8,000 feet (1,200 to 2,500 m). These Southern Alps were intensively glaciated, and the Southwestern coast has deep fiords. The southeastern quadrant forms dissected hill country with a loess mantle, while the Canterbury Plain represents a surface of dissected Pleistocene alluvium.

INDEX

77 78 79 9 8 7 6 5 4 3

BUCK
PASSED

NOT OTHERWISE
STAMPED